T0299078

The Strong Nonlinear Limit-Point/Limit-Circle Problem

TRENDS IN ABSTRACT AND APPLIED ANALYSIS

ISSN: 2424-8746

Series Editor: John R. Graef
The University of Tennessee at Chattanooga, USA

This series will provide state of the art results and applications on current topics in the broad area of Mathematical Analysis. Of a more focused nature than what is usually found in standard textbooks, these volumes will provide researchers and graduate students a path to the research frontiers in an easily accessible manner. In addition to being useful for individual study, they will also be appropriate for use in graduate and advanced undergraduate courses and research seminars. The volumes in this series will not only be of interest to mathematicians but also to scientists in other areas. For more information, please go to http://www.worldscientific.com/series/taaa

Published

Vol. 6 *The Strong Nonlinear Limit-Point/Limit-Circle Problem*
by Miroslav Bartušek & John R. Graef

Vol. 5 *Higher Order Boundary Value Problems on Unbounded Domains: Types of Solutions, Functional Problems and Applications*
by Feliz Manuel Minhós & Hugo Carrasco

Vol. 4 *Quantum Calculus: New Concepts, Impulsive IVPs and BVPs, Inequalities*
by Bashir Ahmad, Sotiris Ntouyas & Jessada Tariboon

Vol. 3 *Solutions of Nonlinear Differential Equations: Existence Results via the Variational Approach*
by Lin Li & Shu-Zhi Song

Vol. 2 *Nonlinear Interpolation and Boundary Value Problems*
by Paul W. Eloe & Johnny Henderson

Vol. 1 *Multiple Solutions of Boundary Value Problems: A Variational Approach*
by John R. Graef & Lingju Kong

More information on this series can be found at http://www.worldscientific.com/series/taaa

Trends in Abstract and Applied Analysis
Volume **6**

The Strong Nonlinear Limit-Point/Limit-Circle Problem

Miroslav Bartušek
Masaryk University, Czech Republic

John R Graef
University of Tennessee at Chattanooga, USA

World Scientific

EW JERSEY • LONDON • SINGAPORE • BEIJING • SHANGHAI • HONG KONG • TAIPEI • CHENNAI • TOKYO

Published by

World Scientific Publishing Co. Pte. Ltd.
5 Toh Tuck Link, Singapore 596224
USA office: 27 Warren Street, Suite 401-402, Hackensack, NJ 07601
UK office: 57 Shelton Street, Covent Garden, London WC2H 9HE

Library of Congress Cataloging-in-Publication Data
Names: Bartušek, Miroslav, 1945– author. | Graef, John R., 1942– author.
Title: The strong nonlinear limit-point/limit-circle problem / by
Miroslav Bartusek (Masaryk University, Czech Republic),
John R. Graef (University of Tennessee at Chattanooga, USA).
Other titles: Strong nonlinear limit point, limit circle problem
Description: New Jersey : World Scientific, 2017. | Series: Trends in abstract and applied analysis ;
volume 6 | Includes bibliographical references and index.
Identifiers: LCCN 2017023900 | ISBN 9789813226371 (hc : alk. paper)
Subjects: LCSH: Limit cycles. | Differentiable dynamical systems. | Differential equations, Nonlinear.
Classification: LCC QA377 .B288 2017 | DDC 515/.352--dc23
LC record available at https://lccn.loc.gov/2017023900

British Library Cataloguing-in-Publication Data
A catalogue record for this book is available from the British Library.

Desk Editors: V. Vishnu Mohan/Kwong Lai Fun

Typeset by Stallion Press
Email: enquiries@stallionpress.com

Printed in Singapore

To Ivana and Frances, without whose help and support this work would never have been finished.

Preface

Since the publication of *The Nonlinear Limit-Point/Limit-Circle Problem*[1] (BDG), there have been a number of new developments in the study of this problem. First, the concept of nonlinear limit-point and nonlinear limit-circle solutions has been extended to equations with p-Laplacians. Secondly, the notions of strong nonlinear limit-point and strong nonlinear limit-circle solutions have been introduced and studied in some detail. The formulations of these ideas for equations of order greater than two and for delay differential equations have also been given.

It is our intention in this book to show the developments described above and then to indicate some new directions that the study of the limit-point/limit-circle problem has taken. Some open problems for future research are also indicated. This volume is not intended to be a sequel to (BDG) and in fact the present work can be read independently of (BDG). However, where appropriate, reference will be made to (BDG).

Chapter 1 gives an introduction to the limit-point/limit-circle problem from its origins with Hermann Weyl. Chapter 2 shows the development of the nonlinear limit-point/limit-circle from the results in the book of Bartušek, Došlá, Graef to the study of equations involving the p-Laplacian. In Chapter 3, the strong nonlinear limit-point and strong nonlinear limit-circle properties are examined beginning with Emden–Fowler equations and then progressing to equations of Emden–Fowler type with p-Laplacians. It is in this study that the terminology of "super-half-linear" and "sub-half-linear" equations is introduced. Generalized Thomas–Fermi equations and equations with forcing terms are also discussed. Second-order equations

[1] M. Bartušek, Z. Došlá, and J. R. Graef, Birkhäuser, Boston, 2004.

with damping terms are considered in Chapter 4 and higher order equations are studied in Chapter 5. Second-order delay differential equations are considered in Chapters 6 and 7. Chapter 8 talks about the use of transformations in studying the limit-point/limit-circle problem. The final chapter contains some open problems of varying degrees of difficulty and looks at extensions of the limit-point/limit-circle problem to different settings.

As was the case with the book (BDG), the connection between the nonlinear limit-point/limit-circle problem and other asymptotic properties of solutions such as boundedness, oscillation, and convergence to zero are interwoven throughout the discussion here.

Contents

Chapter 1

The Origins of the Limit-Point/Limit-Circle Problem

1.1 Introduction

In this chapter, we give a brief discussion of the origins of the limit-point/limit-circle problem including the reason for the choice of the terminology. This also prepares the reader for what is to appear in subsequent chapters.

1.2 The Weyl Alternative

The limit-point/limit-circle problem had its beginnings more than 105 years ago with the publication of Hermann Weyl's classic paper *Über gewöhnliche Differentialgleichungen mit Singularitäten und die zugehörige Entwicklung willkürlicher Funktionen*[1] on eigenvalue problems for second-order linear differential equations. Weyl considered equations of the form

$$(a(t)y')' + r(t)y = \lambda y, \quad t \in [0, \infty), \ \lambda \in \mathbb{C}, \tag{1.2.1}$$

and he classified this equation to be of the *limit-circle* type if every solution is square integrable, i.e., belongs to L^2, and to be of the *limit-point* type if at least one solution does not belong to L^2. In the years that followed, there has been a great deal of work done on this problem due to its important relationship to the solution of certain boundary value problems; see, for example, Titchmarsh [160, 161].

[1] H. Weyl, *Math. Ann.* **68** (1910), 220–269.

Weyl's "limit-point/limit-circle" terminology comes from the proof of one of his fundamental results.

Theorem 1.2.1. *If* Im $\lambda \neq 0$, *then* (1.2.1) *always has a solution* $y \in L^2(\mathbb{R}_+)$, *i.e.*,

$$\int_0^\infty |y(t)|^2 dt < \infty.$$

Proof. For λ with $\operatorname{Im}\lambda \neq 0$, let φ and ψ be two linearly independent solutions of (1.2.1) satisfying the initial conditions

$$\begin{aligned} \varphi(0,\lambda) &= 1, & \psi(0,\lambda) &= 0, \\ \varphi'(0,\lambda) &= 0, & \psi'(0,\lambda) &= 1. \end{aligned}$$

Now the functions $\varphi(t,\lambda)$ and $\psi(t,\lambda)$ are analytic in λ on $\mathbb{C}$, and any other solution y is a linear combination of these two solutions, say,

$$y(t,\lambda) = \varphi(t,\lambda) + m(\lambda)\psi(t,\lambda),$$

where $m(\lambda)$ is to be determined. Choosing $b > 0$ and letting c_1 and c_2 be arbitrary but fixed constants, we need to find $m(\lambda)$ so that y satisfies

$$c_1 y(b,\lambda) + c_2 y'(b,\lambda) = 0. \tag{1.2.2}$$

The value of m depends on λ, b, c_1, and c_2, and moreover, has the form of the linear fractional transformation

$$m = \frac{Az + B}{Cz + D}.$$

The image of the real axis in the *z-plane* is a circle C_b in the *m-plane*. The solution y will satisfy (1.2.2) if and only if m is on C_b. Using Green's identity, it can be shown that this is true if and only if

$$\int_0^b |y(s)|^2 ds = \frac{\operatorname{Im} m}{\operatorname{Im}\lambda},$$

and the radius of the circle C_b is

$$r_b = \left(2\operatorname{Im}\lambda \int_0^b |y(s)|^2 ds\right)^{-1}. \tag{1.2.3}$$

Now if $b_1 < b$, then

$$\int_0^{b_1} |y(s)|^2 ds < \int_0^b |y(s)|^2 ds,$$

and so $r_b < r_{b_1}$, i.e., the circle $\mathcal{C}_{b_1}$ contains the circle $\mathcal{C}_b$ in its interior. Thus, as $b \to \infty$, the circles $\mathcal{C}_b$ converge either to a circle $\mathcal{C}_\infty$ or to a point m_∞. If the limiting form is a circle, then $r_\infty > 0$, and so (1.2.3) implies

$$\int_0^\infty |y(s)|^2 ds < \infty,$$

i.e., $y \in L^2$ for any m on $\mathcal{C}_\infty$. If the limit is the point m_∞, then $r_\infty = 0$ and there is only one solution in L^2. □

In [160, 161], Titchmarsh actually discusses the relationship between the limit-point property and the existence of a unique Green's function for second-order linear differential equations. In the limit-circle case, the Green's function depends on a parameter.

Basic to the study of the limit-point/limit-circle problem is the following result of Weyl.

Theorem 1.2.2. *If* (1.2.1) *is limit-circle for some* $\lambda_0 \in \mathbb{C}$, *then* (1.2.1) *is limit-circle for all* $\lambda \in \mathbb{C}$.

Theorem 1.2.2 holds for $\lambda = 0$, that is, if equation (1.2.1) is limit-circle for $\lambda = 0$, then it is limit-circle for all values of λ. Moreover, if (1.2.1) is not limit-circle for $\lambda = 0$, then it is not limit-circle for any value of λ.

In view of Theorem 1.2.1, the problem reduces to whether equation (1.2.1) with $\operatorname{Im} \lambda \neq 0$ has one (the limit-point case) or two (the limit-circle case) solutions in L^2. This is the famous *Weyl Alternative.*

Thus, the limit-point/limit-circle problem becomes that of determining necessary and/or sufficient conditions on the coefficient functions to be able to distinguish between these two cases. However, it should be pointed out that for equation (1.2.1) with $\lambda = 0$, namely, the equation

$$(a(t)y')' + r(t)y = 0, \tag{1.2.4}$$

Theorem 1.2.1 does not guaranteed that there is at least one square integrable solution.

A systematic extension of these ideas to nonlinear equations began with the papers of Graef [98, 99], Spikes [155, 156], and Graef and Spikes [104]. For the nonlinear differential equation

$$y'' + r(t)|y|^\lambda \operatorname{sgn} y = 0, \tag{1.2.5}$$

the limit-point and limit-circle properties take the following form.

Definition 1.2.1. A solution y of equation (1.2.5) defined on $\mathbb{R}_+$ is said to be of the nonlinear limit-circle type if

$$\int_0^\infty |y(t)|^{\lambda+1} dt < \infty, \tag{NLC}$$

and it is said to be of the nonlinear limit-point type otherwise, i.e.,

$$\int_0^\infty |y(t)|^{\lambda+1} dt = \infty. \tag{NLP}$$

Equation (1.2.5) is said to be of the *nonlinear limit-circle* type if every solution y of (1.2.5) defined on $\mathbb{R}_+$ satisfies (NLC) and it is said to be of the *nonlinear limit-point* type if there is at least one solution y defined on $\mathbb{R}_+$ for which (NLP) holds.

The book [25] discusses some of the early results on limit-point and limit-circle results for linear equations, such as those of Wintner [168], and various attempts at extensions to nonlinear equations by Atkinson [3], Burlak [52], Detki [66], Hallam [110], and Suyemoto and Waltman [158], and Wong [169], so we will not repeat that here. The situation for higher order linear equations is different in that the limit-point and limit-circle cases do not form a dichotomy. This has led to what is known as the deficiency index for self-adjoint differential operators. This also will not be discussed here, but a description of this problem can be found in [25] as can a discussion of its relationship to spectral theory. Some relationships between the limit-point/limit-circle properties and other properties of solutions such as boundedness, oscillation, and convergence to zero can also be found in [25].

The literature contains a number of papers in recent years that use the terminology "limit-point" and "limit-circle" in describing their results for nonlinear equations, but in actuality they are discussing square integrable solutions and do not take advantage of the form of the nonlinear term as it is incorporated in the definitions of nonlinear limit-point and nonlinear limit-circle solutions given above.

Finally, we should point out that the original work of Graef and Spikes [98, 99, 104, 155, 156] on the nonlinear limit-point/limit-circle problem was actually for equations of the form

$$y'' + r(t)f(y) = 0.$$

We refer the reader to [98] as well as to the book [25] for a discussion of the relationship between the functions $f(y)$, $F(y) = \int_0^y f(u)du$, and the form of equation (1.2.5). In this regard, we also refer the reader to Sections 3.11, 6.1, and 6.2 below.

Chapter 2

Equations with p-Laplacian

2.1 Introduction

In this chapter, we begin our study of equations involving a p-Laplacian. Equations with p-Laplacians have been widely studied in various contexts in the last 20 years; see, for example, the books of Došlý and Řehák [71] and Mirzov [138] as well as the papers [55–57, 136, 137]. While our primary interest is in the nonlinear limit-point and nonlinear limit-circle properties of solutions, other asymptotic properties of solutions enter the discussion in a rather natural fashion. We will begin with the study of unforced equations in Section 2.2 and discuss forced equations beginning in Section 2.5. The last section in this chapter contains some examples to illustrate the results.

2.2 Second-Order Unforced Equations

We consider the second-order nonlinear differential equation

$$(a(t)|y'|^{p-1}y')' + r(t)|y|^{\lambda}\text{sgn } y = 0, \tag{2.2.1}$$

where $p > 0$, $\lambda > 0$, $\mathbb{R}_+ = [0, \infty)$, $a \in C^1(\mathbb{R}_+)$, $a^{1/p}r \in AC^1_{\text{loc}}(\mathbb{R}_+)$, $a(t) > 0$, and $r(t) > 0$. Special cases of this equation are the Emden–Fowler equation ($p = 1$),

$$(a(t)y')' + r(t)|y|^{\lambda}\text{sgn } y = 0, \tag{E-F}$$

and the half-linear equation ($\lambda = p$),

$$(a(t)|y'|^{p-1}y')' + r(t)|y|^{p}\text{sgn } y = 0. \tag{H-L}$$

We need to define what we mean by a solution of equation (2.2.1) as well as what it means for a solution to be oscillatory.

Definition 2.2.1. A function $y \in C^1[0,T)$ is a solution of (2.2.1) if $a|y'|^{p-1}y' \in C^1[0,T)$ and (2.2.1) holds on $[0,T)$. A solution y of (2.2.1) is continuable if it is defined on $\mathbb{R}_+$. It is said to be noncontinuable if it is defined on $[0,T)$ with $T < \infty$, and it cannot be defined at $t = T$. A continuable solution y is called proper if it is nontrivial in any neighborhood of ∞.

Definition 2.2.2. A proper solution y of (2.2.1) is called oscillatory if there exists a sequence of its zeros tending to ∞; otherwise it is called nonoscillatory.

For equation (2.2.1) the definition of a nonlinear limit-circle and nonlinear limit-point solution takes the following form.

Definition 2.2.3. Equation (2.2.1) is said to be of the nonlinear limit-circle type if every solution y of (2.2.1) defined on $\mathbb{R}_+$ satisfies

$$\int_0^\infty |y(t)|^{\lambda+1} dt < \infty,$$

and it is said to be of the nonlinear limit-point type if there is at least one solution y defined on $\mathbb{R}_+$ for which

$$\int_0^\infty |y(t)|^{\lambda+1} dt = \infty.$$

We will make use of the following constants in our study:

$$\alpha = \frac{p+1}{(\lambda+2)p+1}, \quad \beta = \frac{(\lambda+1)p}{(\lambda+2)p+1}, \quad \gamma = \frac{p+1}{p(\lambda+1)}, \quad \delta = \frac{p+1}{p},$$

$$\beta_1 = \frac{(\lambda+2)p+1}{(\lambda+1)(p+1)}, \quad \beta_2 = \frac{(\lambda+1)(p+1)}{\lambda - p} \quad \text{for } \lambda \neq p, \quad \text{and}$$

$$\beta_3 = \frac{\lambda - p}{(\lambda+1)(p+1)}.$$

Notice that $\alpha = 1 - \beta$, $\beta\gamma = \alpha$, and $\beta_3 = 1/\beta_2$. We define the functions R, $g : \mathbb{R}_+ \to \mathbb{R}$ by

$$R(t) = a^{1/p}(t)r(t) \quad \text{and} \quad g(t) = -\frac{R'(t)a^{1/p}(t)}{R^{\alpha+1}(t)}.$$

For any continuous function $h : \mathbb{R}_+ \to \mathbb{R}$, we let $h_+(t) = \max\{h(t), 0\}$ and $h_-(t) = \max\{-h(t), 0\}$ so that $h(t) = h(t)_+ - h(t)_-$.

For any solution $y : \mathbb{R}_+ \to \mathbb{R}$ of (2.2.1), we let

$$y^{[1]}(t) = a(t)|y'(t)|^{p-1}y'(t)$$

and define the functions F, $V : \mathbb{R}_+ \to \mathbb{R}$ by

$$\begin{aligned} F(t) &= R^{\beta}(t)\left[\frac{a(t)}{r(t)}|y'(t)|^{p+1} + \gamma|y(t)|^{\lambda+1}\right] \\ &= R^{\beta}(t)\left(\frac{|y^{[1]}(t)|^{\delta}}{R(t)} + \gamma|y(t)|^{\lambda+1}\right) = R^{\beta}(t)V(t). \end{aligned} \tag{2.2.2}$$

For $\lambda > p$ and $g(t)$ not identically a constant on $\mathbb{R}_+$, we let

$$G(t) = F(t)\left(\int_0^t |g'(\sigma)|d\sigma\right)^{-\beta_2} \tag{2.2.3}$$

for $t \geq t_G$ where $t_G \in \mathbb{R}_+$ is such that

$$\int_0^{t_G} |g'(\sigma)|d\sigma > 0.$$

We will often make use of the following assumption:

$$\int_0^{\infty} |g'(\sigma)|d\sigma < \infty. \tag{2.2.4}$$

Notice that this implies that the function g is bounded.

Where it is convenient, we will refer to equation (2.2.1) as being of the *super-half-linear* type if $\lambda \geq p$ and of the *sub-half-linear* type if $\lambda \leq p$.

2.3 Preliminary Lemmas

To prove our main results, we first present some lemmas to be used in the process. Our first lemma shows that solutions of (2.2.1) are continuable to all of $\mathbb{R}_+$ so that there are no singular solutions. It also gives some useful estimates for $y(t)$ and $y^{[1]}(t)$.

Lemma 2.3.1. *Let y be a solution of* (2.2.1). *Then:*

(i) *y is proper or trivial on $\mathbb{R}_+$;*
(ii) *the estimates*

$$|y(t)| \leq \delta_1 R^{-\frac{p}{(\lambda+2)p+1}}(t)\, F^{\frac{1}{\lambda+1}}(t), \quad \textit{where } \delta_1 = \gamma^{-\frac{1}{\lambda+1}} \tag{2.3.1}$$

and

$$|y^{[1]}(t)| \leq R^{\frac{p}{(\lambda+2)p+1}}(t)\, F^{\frac{p}{p+1}}(t) \tag{2.3.2}$$

hold for all $t \geq 0$;

(iii) *for* $0 \leq \tau < t$, *we have*

$$F(t) = F(\tau) - \alpha g(\tau)y(\tau)y^{[1]}(\tau) + \alpha g(t)y(t)y^{[1]}(t)$$
$$-\alpha \int_{\tau}^{t} g'(\sigma)y(\sigma)y^{[1]}(\sigma)d\sigma. \tag{2.3.3}$$

Proof. Due to the smoothness of the functions a and R, part (i) follows from [13, Theorem 2]; part (ii) follows immediately from the definition of F. To prove (iii), first observe that

$$\int_{\tau}^{t} [F(\sigma) - \alpha g(\sigma)y(\sigma)y^{[1]}(\sigma)]'\, d\sigma$$
$$= F(t) - F(\tau) + \alpha g(\tau)y(\tau)y^{[1]}(\tau) - \alpha g(t)y(t)y^{[1]}(t).$$

However, differentiating first and then integrating, we have

$$\int_{\tau}^{t} [F(\sigma) - \alpha g(\sigma)y(\sigma)y^{[1]}(\sigma)]'d\sigma$$
$$= \int_{\tau}^{t} [-\alpha R^{-\alpha-1}(\sigma)R'(\sigma)|y^{[1]}(\sigma)|^{\delta} + \gamma\beta R^{\beta-1}(\sigma)R'(\sigma)|y(\sigma)|^{\lambda+1}$$
$$-\delta R^{-\alpha}(\sigma)a^{1/p}(\sigma)|y'(\sigma)|r(\sigma)|y(\sigma)|^{\lambda}\operatorname{sgn} y(\sigma)\operatorname{sgn} y'(\sigma)$$
$$+\gamma(\lambda+1)R^{\beta}(\sigma)|y(\sigma)|^{\lambda}y'(\sigma)\operatorname{sgn} y(\sigma) - \alpha g'(\sigma)y(\sigma)y^{[1]}(\sigma)$$
$$-\alpha g(\sigma)y'(\sigma)y^{[1]}(\sigma) + \alpha g(\sigma)r(\sigma)|y(\sigma)|^{\lambda+1}]d\sigma$$
$$= -\alpha \int_{\tau}^{t} g'(\sigma)y(\sigma)y^{[1]}(\sigma)d\sigma.$$

□

Our next lemma gives some bounds on the function F.

Lemma 2.3.2. *Let condition* (2.2.4) *hold. In addition, if* $\lambda = p$, *assume that* $\lim_{t\to\infty} g(t) = 0$. *Then there exist a solution* y *of* (2.2.1), *a constant*

$C_0 > 0$, and $t_0 \in \mathbb{R}_+$ such that

$$0 < \frac{3}{4}C_0 \le F(t) \le \frac{3}{2}C_0 \quad \textit{for} \quad t \ge t_0.$$

Moreover, C_0 can be chosen so that $C_0^{\beta_3}$ is arbitrarily large in case $\lambda \ne p$.

Proof. Let $K = 3\alpha\delta_1 \left(\frac{3}{2}\right)^{\beta_1}$, where δ_1 is given in (2.3.1). Now condition (2.2.4) implies that g is bounded, so we can choose $M > 0$, $t_0 \in \mathbb{R}_+$, and $C_0 > 0$ such that

$$t_0 = 0, \quad |g(t)| \le M \quad \text{for} \quad t \ge t_0, \quad \int_{t_0}^{\infty} |g'(s)|ds \le M, \quad \text{and} \quad C_0^{\beta_3} \ge 4KM$$

if $\lambda \ne p$, and

$$|g(t)| \le M \text{ for } \quad t \ge t_0, \quad \int_{t_0}^{\infty} |g'(s)|ds \le M, \; 4KM < 1, \quad \text{and} \quad C_0 = 1$$

if $\lambda = p$.

Consider a solution y of (2.2.1) such that $F(t_0) = C_0$. First, we will show that

$$F(t) \le \frac{3}{2}C_0 \quad \text{for all } t \ge t_0. \tag{2.3.4}$$

Suppose (2.3.4) does not hold. Then there exist $t_2 > t_1 > t_0$ such that

$$F(t_2) = \frac{3C_0}{2}, \quad F(t_1) = C_0, \quad \text{and} \quad C_0 < F(t) < \frac{3C_0}{2} \quad \text{for}\, t \in (t_1, t_2).$$

Parts (ii) and (iii) of Lemma 2.3.1 with $\tau = t_1$ and $t = t_2$ yield

$$\frac{C_0}{2} \le 3\alpha M \max_{t_1 \le \sigma \le t_2} |y(\sigma)y^{[1]}(\sigma)| \le KMC_0^{\beta_1},$$

so $C_0^{\beta_3} \le 2KM$. This contradicts the choice of C_0. Hence, (2.3.4) holds. Now from parts (ii) and (iii) of Lemma 2.3.1 with $t = t$ and $\tau = t_0$, we obtain

$$|F(t) - C_0| \le 3\alpha M \max_{0 \le \sigma \le t} |y(\sigma)y^{[1]}(\sigma)| \le KMC_0^{\beta_1} \le \frac{1}{4}C_0^{\beta_3+\beta_1} = \frac{C_0}{4}.$$

Hence, the conclusion of the lemma holds. □

Remark 2.3.1. Notice that in Lemma 2.3.2, if $\lambda < p$, then $\beta_2 < 0$, so C_0 needs to be chosen small to ensure that $C_0^{\beta_3}$ is large.

Our next two lemmas give conditions under which the functions G and F, respectively, are bounded. The first one is for the super-half-linear case ($\lambda > p$) and the second is for the sub-half-linear case ($\lambda < p$).

Lemma 2.3.3. *Suppose that* (2.2.4) *holds,* $\lambda > p$, *and* $g(t)$ *is not identically constant on* $\mathbb{R}_+$. *Then for every solution* y *of* (2.2.1), *the function* G *is bounded on* $[t_G, \infty)$.

Proof. Let y be a solution of (2.2.1) and suppose that G is not bounded. Now $G \geq 0$, so there is an increasing sequence $\{t_k\}_{k=1}^{\infty} \to \infty$ as $k \to \infty$ such that $t_1 \geq t_G$ and

$$\lim_{k\to\infty} G(t_k) = \infty. \tag{2.3.5}$$

This implies the existence of sequences $\{\sigma_k\}_{k=1}^{\infty}$ and $\{\tau_k\}_{k=1}^{\infty}$ with $t_k \leq \sigma_k < \tau_k$ such that

$$\frac{1}{2}G(\tau_k) = G(\sigma_k) = G(t_k) \tag{2.3.6}$$

and

$$G(\sigma_k) \leq G(t) \leq G(\tau_k) \quad \text{for } \sigma_k \leq t \leq \tau_k,\ k = 1, 2, \ldots.$$

It is easy to see that there exist an integer k_0 and a constant $M > 0$ such that

$$2|g(0)| \leq M \int_0^{\tau_k} |g'(\sigma)| d\sigma$$

for $k \geq k_0$. From (2.2.3) and (2.3.6), we see that $\max_{\sigma_k \leq t \leq \tau_k} F(t) = F(\tau_k)$. Setting $\tau = \sigma_k$ and $t = \tau_k$ in (2.3.1)–(2.3.3), we obtain

$$F(\tau_k) - F(\sigma_k) \leq \alpha_1 F(\tau_k)^{\beta_1} \left(|g(\sigma_k)| + |g(\tau_k)| + \int_{\sigma_k}^{\tau_k} |g'(\sigma)| d\sigma \right)$$

where $\alpha_1 = \alpha\gamma^{-\frac{1}{\lambda+1}}$. Condition (2.2.4) implies that g is of bounded variation, so

$$|g(\sigma_k)| - |g(0)| \leq |g(\sigma_k) - g(0)| \leq \int_0^{\sigma_k} |g'(\sigma)| d\sigma,$$

and thus

$$F(\tau_k) - F(\sigma_k) \le \alpha_1 F(\tau_k)^{\beta_1} (2|g(0)| + 3 \int_0^{\tau_k} |g'(\sigma)| d\sigma)$$
$$\le \alpha_1 (M+3) F(\tau_k)^{\beta_1} \int_0^{\tau_k} |g'(\sigma)| d\sigma$$

for $k \ge k_0$. Hence,

$$\begin{aligned}(2^{\beta_3} - 2^{-\beta_1}) G^{\beta_3}(\sigma_k) &= G^{\beta_3}(\tau_k) - \frac{G(\sigma_k)}{G^{\beta_1}(\tau_k)} \\ &\le \left(F^{\beta_3}(\tau_k) - \frac{F(\sigma_k)}{F^{\beta_1}(\tau_k)} \right) \left(\int_0^{\tau_k} |g'(\sigma)| d\sigma \right)^{-1} \\ &\le \alpha_1 (M+3).\end{aligned}$$

This contradicts (2.3.5) and completes the proof of the lemma. □

Lemma 2.3.4. *Suppose that (2.2.4) holds, $\lambda < p$, and $\lim_{t\to\infty} g(t) = 0$. If either*

(i) *$g(t)$ is identically a constant on $\mathbb{R}_+$, or*
(ii) *$g(t)$ is not identically a constant on $\mathbb{R}_+$ and*

$$\limsup_{t\to\infty} R^{-\beta}(t) \left(\int_t^\infty |g'(s)| ds \right)^{\beta_2} \times \exp\left\{ - \int_0^t (R^{-1}(s))'_+ R(s) ds \right\} = \infty, \tag{2.3.7}$$

then for every solution y of (2.2.1), the function F is bounded on $\mathbb{R}_+$.

Proof. If $g(t)$ is identically a constant on $\mathbb{R}_+$, then $g(t)$ is identically zero, and it follows from (2.3.3) that F is bounded on $\mathbb{R}_+$. Now assume that $g(t)$ is not identically a constant on $\mathbb{R}_+$ and (2.3.7) holds. Let y be a solution of (2.2.1) and suppose that F is not bounded. Since $F \ge 0$, for any $t_0 \in \mathbb{R}_+$ there exist σ and τ such that $t_0 \le \sigma < \tau$,

$$\frac{1}{2} F(\tau) = F(\sigma) = F(t_0) \tag{2.3.8}$$

and $F(\sigma) \le F(t) \le F(\tau)$ for $\sigma \le t \le \tau$.

Let α_1 be defined as in the proof of Lemma 2.3.3. Since g is of bounded variation and $\lim_{t\to\infty} g(t) = 0$, we see that

$$|g(\sigma)| = |g(\sigma) - g(\infty)| \leq \int_\sigma^\infty |g'(s)|ds.$$

Setting $\tau = \sigma$ and $t = \tau$ in (2.3.1)–(2.3.3), we have

$$F(t_0) = F(\sigma) = F(\tau) - F(\sigma) \leq \alpha_1 F(\tau)^{\beta_1}\left(|g(\sigma)| + |g(\tau)| + \int_\sigma^\tau |g'(s)|ds\right)$$

$$\leq 3\alpha_1 2^{\beta_1} F^{\beta_1}(\sigma) \int_{t_0}^\infty |g'(s)|ds.$$

Thus, (2.3.8) implies

$$F(t_0) \geq M\left(\int_{t_0}^\infty |g'(s)|ds\right)^{\beta_2}, \quad \text{where } M = (3\alpha_1 2^{\beta_1})^{\beta_2}.$$

Since t_0 was arbitrary, we have

$$F(t) \geq M\left(\int_t^\infty |g'(s)|ds\right)^{\beta_2} \tag{2.3.9}$$

for $t \in \mathbb{R}_+$. On the other hand, if $Z(t) = F(t)R^{-\beta}(t)$, then (2.2.2) yields

$$Z'(t) = \left(\frac{1}{R(t)}\right)' |y^{[1]}(t)|^\delta \leq (R^{-1}(t))'_+ R(t) Z(t)$$

for $t \in \mathbb{R}_+$, so

$$Z(t) \leq Z(0)\exp\left\{\int_0^t (R^{-1}(s))'_+ R(s)ds\right\}$$

for $t \in \mathbb{R}_+$. From this and (2.3.9), we have

$$MR^{-\beta}(t)\left(\int_t^\infty |g'(s)|ds\right)^{\beta_2} \leq Z(t) \leq Z(0)\exp\left\{\int_0^t (R^{-1}(s))'_+ R(s)ds\right\},$$

which contradicts (2.3.7). □

The next two lemmas hold for all values of λ and p.

Lemma 2.3.5. *Let* (2.2.4) *hold,* $\lim_{t\to\infty} g(t) \neq 0$, *and*

$$\int_0^\infty \left(\int_s^\infty a^{-1/p}(\sigma)d\sigma \right)^{\beta/\alpha} ds = \infty. \tag{2.3.10}$$

Then

$$\int_0^\infty R^{-\beta}(t)dt = \infty. \tag{2.3.11}$$

Proof. Since g is of bounded variation on $\mathbb{R}_+$, $\lim_{t\to\infty} g(t) = C$ exists and is finite. If $C > 0$, then from the definition of g, we see that R is decreasing for large t, and so (2.3.11) holds. If $C < 0$, then R is increasing, and so $\lim_{t\to\infty} R(t) = R_0 \in (0, \infty]$. Hence, there exists $t_0 \in \mathbb{R}_+$ such that $g(t) \leq C/2$ for $t \geq t_0$. If $R_0 < \infty$, then (2.3.11) holds. So suppose $R_0 = \infty$. Then,

$$\begin{aligned} R^{-\alpha}(t) &= \alpha \int_t^\infty R'(t) R^{-\alpha-1}(t)dt \\ &\geq -\alpha \int_t^\infty g(s)a^{-1/p}(s)ds \geq \frac{\alpha|C|}{2} \int_t^\infty a^{-1/p}(s)ds \end{aligned}$$

for $t \geq t_0$. From this and (2.3.10), we have

$$\int_{t_0}^\infty R^{-\beta}(s)ds \geq \left(\frac{\alpha|C|}{2}\right)^{\beta/\alpha} \int_{t_0}^\infty \left(\int_s^\infty a^{-1/p}(\sigma)d\sigma \right)^{\beta/\alpha} ds = \infty,$$

so again (2.3.11) holds. □

Remark 2.3.2. It is clear that (2.3.10) holds if $\int_0^\infty a^{-1/p}(t)dt = \infty$.

Lemma 2.3.6. *Suppose that* (2.2.4) *holds,*

$$\int_0^\infty R^{-\beta}(t)dt = \infty, \tag{2.3.12}$$

and either

$$\int_0^\infty \frac{|r'(t)|}{r^2(t)} dt < \infty \tag{2.3.13}$$

or

$$\int_0^\infty \frac{|a'(t)|}{a(t)r(t)} dt < \infty. \tag{2.3.14}$$

In addition, if $\lambda = p$, assume that $\lim_{t\to\infty} g(t) = 0$. Then there is a solution y of (2.2.1) such that

$$\int_0^\infty |y(t)|^{\lambda+1} dt = \infty. \tag{2.3.15}$$

Proof. Lemma 2.3.2 yields the existence of a solution y of (2.2.1), $t_0 \in \mathbb{R}_+$, and a constant $C_0 > 0$ such that

$$0 < \frac{3C_0}{4} \le F(t) \le \frac{3C_0}{2} \tag{2.3.16}$$

for $t \ge t_0$. Moreover, C_0 can be chosen so that

$$|g(t)| \le \left(\frac{2}{3}\right)^{\beta_1} \frac{C_0^{\beta_3}}{4\delta_1} \tag{2.3.17}$$

for $t \ge t_0$, where δ_1 is given in (2.3.1).

We will show that (2.3.15) holds. Suppose, to the contrary, that

$$\int_0^\infty |y(t)|^{\lambda+1} dt < \infty. \tag{2.3.18}$$

If we multiply $F(t)$ by $R^{-\beta}(t)$, then, in view of (2.2.2), (2.3.12), (2.3.16), and (2.3.18), there exists $t_1 \ge t_0$ such that

$$\int_0^t \frac{|y^{[1]}(t)|^\delta}{R(t)} dt \ge \frac{1}{2} C_0 \int_0^t R^{-\beta}(s)\, ds \tag{2.3.19}$$

for $t \ge t_1$. Lemma 2.3.1(ii) and (2.3.16) imply

$$|y(t)y^{[1]}(t)| \le M_1 C_0^{\beta_1} \left(\frac{3}{2}\right)^{\beta_1} \tag{2.3.20}$$

for $t \ge t_0$. Define

$$J(t) = \int_0^t \frac{|y^{[1]}(t)|^\delta}{R(t)} dt + \int_0^t \left(\frac{1}{r(\sigma)}\right)' y(\sigma) y^{[1]}(\sigma) d\sigma.$$

Then (2.3.20) yields

$$J(t) \ge \int_0^t \frac{|y^{[1]}(t)|^\delta}{R(t)} dt - M_1 C_0^{\beta_1} \left(\frac{3}{2}\right)^{\beta_1} \int_0^t \left| \left(\frac{1}{r(\sigma)}\right)' \right| d\sigma. \tag{2.3.21}$$

If (2.3.13) holds, then from (2.3.12) and (2.3.19) we have

$$\lim_{t\to\infty} J(t) = \infty. \tag{2.3.22}$$

Now suppose (2.3.14) holds. Then, from (2.3.17),

$$\begin{aligned}
\int_0^t \left|\left(\frac{1}{r(\sigma)}\right)'\right| d\sigma &= \int_0^t \left|\left(\frac{a^{1/p}(\sigma)}{R(\sigma)}\right)'\right| d\sigma \\
&= \int_0^t \left|\left(\frac{a'(\sigma)}{pa(\sigma)r(\sigma)} - \frac{a^{1/p}(\sigma)R'(\sigma)}{R^2(\sigma)}\right)\right| d\sigma \\
&\le \frac{1}{p}\int_0^t \frac{|a'(\sigma)|}{a(\sigma)r(\sigma)}\,d\sigma + \int_0^t \frac{|g(\sigma)|d\sigma}{R^{\beta}(\sigma)} \\
&\le \frac{1}{p}\int_0^t \frac{|a'(\sigma)|}{a(\sigma)r(\sigma)}\,d\sigma + \frac{C_0^{\beta_3}}{4M_1}\left(\frac{2}{3}\right)^{\beta_1}\int_0^t R^{-\beta}(\sigma)d\sigma
\end{aligned}$$

for $t \ge t_0$. From this, (2.3.19), and (2.3.21), we obtain

$$\begin{aligned}
J(t) &\ge \frac{1}{2}C_0\int_0^t R^{-\beta}(\sigma)d\sigma \\
&\quad - \frac{M_1C_0^{\beta_1}}{p}\left(\frac{3}{2}\right)^{\beta_1}\int_0^t \frac{|a'(\sigma)|}{a(\sigma)r(\sigma)}\,d\sigma - \frac{1}{4}C_0\int_0^t R^{-\beta}(\sigma)\,d\sigma
\end{aligned}$$

for $t \ge t_0$. Thus, (2.3.12) and (2.3.14) yield (2.3.22). Hence, if either (2.3.13) or (2.3.14) holds, then we have (2.3.22) holding.

It follows from (2.2.1) and (2.3.18) that

$$\begin{aligned}
\infty > \int_0^\infty |y(t)|^{\lambda+1}\,dt &\ge -\int_0^t \frac{y(\sigma)(y^{[1]}(\sigma))'}{r(\sigma)}\,d\sigma \\
&= -\frac{y(t)y^{[1]}(t)}{r(t)} + \frac{y(0)y^{[1]}(0)}{r(0)} + \int_0^t \left(\frac{y(\sigma)}{r(\sigma)}\right)' y^{[1]}(\sigma)\,d\sigma \\
&= -\frac{y(t)y^{[1]}(t)}{r(t)} + \frac{y(0)y^{[1]}(0)}{r(0)} + J(t)
\end{aligned} \tag{2.3.23}$$

for $t \ge t_0$. If y is oscillatory, let $\{t_k\}_{k=1}^{\infty} \to \infty$ be a sequences of zeros of y. Then letting $t = t_k$ in (2.3.23) contradicts (2.3.22) for large k. Let y be nonoscillatory. Then, in view of (2.2.1), y is either increasing or decreasing for large t, and in case $|y|$ is increasing, it is clear that (2.3.18) does not hold. If $y(t)y'(t) < 0$ for large t, then $y(t)y^{[1]}(t) < 0$, and (2.3.22) and (2.3.23) yield a contradiction. □

Our final lemma also includes the case where g is identically a constant on $\mathbb{R}_+$.

Lemma 2.3.7. *Assume that either*

(i) $\lambda > p$ *and* $g(t) \equiv c$ *on* $\mathbb{R}_+$, *where* c *is a constant, or*
(ii) $\lambda = p$, (2.2.4) *holds, and* $\lim_{t\to\infty} g(t) = 0$.

Then for every solution y *of* (2.2.1), *the function* F *is bounded on* $\mathbb{R}_+$.

Proof. Let y be a solution of (2.2.1) and let $t_0 \in \mathbb{R}_+$ and $c > 0$ be chosen so that

$$t_0 = 0 \quad \text{and} \quad |g(t)| \equiv c$$

in case (i), and

$$c < \gamma^{\frac{1}{\lambda+1}}/6\alpha, \quad |g(t)| \le c \quad \text{for} \quad t \ge t_0, \quad \text{and} \quad \int_{t_0}^{\infty} |g'(s)|ds \le c$$

in case (ii). Suppose that F is not bounded. Since $F \ge 0$, there is an increasing sequence $\{t_k\}_{k=1}^{\infty} \to \infty$ as $k \to \infty$ such that $t_1 \ge t_0$ and

$$\lim_{k\to\infty} F(t_k) = \infty. \tag{2.3.24}$$

This implies the existence of sequences $\{\sigma_k\}_{k=1}^{\infty}$ and $\{\tau_k\}_{k=1}^{\infty}$ with $t_k \le \sigma_k < \tau_k$ such that

$$\frac{1}{2}F(\tau_k) = F(\sigma_k) = F(t_k) \tag{2.3.25}$$

and

$$F(\sigma_k) \le F(t) \le F(\tau_k) \quad \text{for } \sigma_k \le t \le \tau_k,\ k = 1, 2, \ldots.$$

Setting $\tau = \sigma_k$ and $t = \tau_k$ in (2.3.1)–(2.3.3), we obtain

$$\frac{1}{2}F(\tau_k) = F(\tau_k) - F(\sigma_k) \le 3c\alpha_1 F^{\beta_1}(\tau_k)$$

where $\alpha_1 = \alpha\gamma^{-\frac{1}{\lambda+1}}$. From this and (2.3.25), we obtain

$$2^{\beta_3} F^{\beta_3}(t_k) = F^{\beta_3}(\tau_k) \le 6c\alpha_1, \quad k = 1, 2, \ldots.$$

This contradicts (2.3.24) for large k in case (i) and the choice of c in case (ii). This completes the proof of the lemma. □

2.4 Nonlinear Limit-Point/Limit-Circle Results

We begin by considering the super-half-linear case.

Theorem 2.4.1. *Let* $\lambda > p$.

(i) *If* $g(t)$ *is not identically a constant on* $\mathbb{R}_+$ *and*

$$\int_0^\infty R^{-\beta}(t) \left(\int_0^t \left| \left(\frac{a^{1/p}(s) R'(s)}{R^{\alpha+1}(s)} \right)' \right| ds \right)^{\beta_2} dt < \infty,$$

then equation (2.2.1) *is of the nonlinear limit-circle type.*

(ii) *Let*

$$\int_0^\infty \left| \left(\frac{a^{1/p}(s) R'(s)}{R^{\alpha+1}(s)} \right)' \right| ds < \infty \tag{2.4.1}$$

and either

$$\int_0^\infty \frac{|r'(t)|}{r^2(t)} dt < \infty \quad \text{or} \quad \int_0^\infty \frac{|a'(t)|}{a(t) r(t)} dt < \infty \tag{2.4.2}$$

hold. Then (2.2.1) *is of the nonlinear limit-circle type if and only if*

$$\int_0^\infty R^{-\beta}(t)\, dt < \infty. \tag{2.4.3}$$

Proof. (i) According to (2.2.2) and (2.2.3),

$$\int_{t_G}^\infty G(t) R^{-\beta}(t) \left(\int_0^t |g'(\sigma)| d\sigma \right)^{\beta_2} dt \ge \gamma\alpha \int_{t_G}^\infty |y(t)|^{\lambda+1}\, dt,$$

and so the conclusion follows from Lemma 2.3.3.

(ii) Let (2.4.3) hold and suppose that $g(t)$ is not identically a constant on $\mathbb{R}_+$. Then, (2.4.1) ensures that the hypotheses of case (i) are satisfied, and so equation (2.2.1) is of the nonlinear limit-circle type. If (2.4.3) does not hold and $g(t)$ is not identically a constant on $\mathbb{R}_+$, then Lemma 2.3.6 holds, so (2.2.1) is of the nonlinear limit-point type.

Finally, if $g(t)$ is identically a constant on $\mathbb{R}_+$ and (2.4.3) holds, then Lemma 2.3.7(i) implies that F is bounded, and the conclusion follows immediately. If (2.4.3) does not hold and $g(t)$ is identically a constant on $\mathbb{R}_+$, then the conclusion follows from Lemma 2.3.6. □

Corollary 2.4.1. *Let $\lambda > p$, (2.4.1) hold, and let r be nondecreasing for large t. Then equation (2.2.1) is of the nonlinear limit-circle type if and only if (2.4.3) holds.*

Proof. Since r is nondecreasing, the first inequality in (2.4.2) holds. The conclusion then follows from part (ii) of Theorem 2.4.1. □

Applying the above results to the important case $a \equiv 1$, i.e., the equation

$$\left(|y'|^{p-1}y'\right)' + r(t)|y|^{\lambda}\operatorname{sgn} y = 0, \tag{2.4.4}$$

we obtain the following corollary.

Corollary 2.4.2. *Let $\lambda > p$ and*

$$\int_0^\infty \left| \left(\frac{r'(\sigma)}{r^{\frac{(\lambda+3)p+2}{(\lambda+2)p+1}}(\sigma)} \right)' \right| d\sigma < \infty. \tag{2.4.5}$$

Then equation (2.4.4) is of the nonlinear limit-circle type if and only if

$$\int_0^\infty r^{-\frac{(\lambda+1)p}{(\lambda+2)p+1}}(t)\,dt < \infty. \tag{2.4.6}$$

The following theorem is concerned with the sub-half-linear case.

Theorem 2.4.2. *Let $\lambda < p$ and (2.4.1) and (2.4.2) hold.*

(i) *If*

$$\int_0^\infty R^{-\beta}(t)\,dt = \infty, \tag{2.4.7}$$

then equation (2.2.1) is of the nonlinear limit-point type.

(ii) *Assume that $g(t)$ is not identically a constant on $\mathbb{R}_+$ and*

$$\limsup_{t\to\infty} R^{-\beta}(t)\left(\int_t^\infty |g'(\sigma)|\,d\sigma\right)^{\beta_2} \times \exp\{-\int_0^t (R^{-1}(s))'_+ R(s)\,ds\} = \infty.$$

In addition, if

$$\int_0^\infty a^{-1/p}(t)\,dt < \infty, \tag{2.4.8}$$

then let either

$$\int_0^\infty \left(\int_s^\infty a^{-1/p}(\sigma)d\sigma\right)^{\frac{(\lambda+1)p}{p+1}} ds = \infty \quad \text{or} \quad \lim_{t\to\infty} g(t) = 0 \tag{2.4.9}$$

hold. Then equation (2.2.1) *is of the nonlinear limit-circle type if and only if* (2.4.3) *holds.*

Proof. Part (i) follows immediately from Lemma 2.3.6. To prove (ii), let (2.4.3) hold. If (2.4.8) does not hold, then by Remark 2.3.2, (2.4.9), and Lemma 2.3.5, we have $\lim_{t\to\infty} g(t) = 0$. Since the hypotheses of Lemma 2.3.4 are satisfied, the function F is bounded for any solution y of (2.2.1). From (2.2.2), we have

$$\infty > \int_0^\infty F(t)R^{-\beta}(t)\,dt \ge \gamma \int_0^\infty |y(t)|^{\lambda+1}\,dt,$$

and so equation (2.2.1) is of the nonlinear limit-circle type.

If (2.4.7) holds, then by part (i), (2.2.1) is of the nonlinear limit-point type. □

Corollary 2.4.3. *Assume that $\lambda < p$,* (2.4.1) *holds, the functions r and $a^{1/p}r$ are nondecreasing for large t, $g(t)$ is not identically a constant on $\mathbb{R}_+$, and*

$$\limsup_{t\to\infty} R^{-\beta}(t)\left(\int_t^\infty |g'(\sigma)|\right)^{\beta_2} d\sigma = \infty.$$

Moreover, if (2.4.8) *holds, assume that* (2.4.9) *holds. Then equation* (2.2.1) *is of the nonlinear limit-circle type if and only if* (2.4.3) *holds.*

Proof. Since r is nondecreasing, the first inequality in (2.4.2) holds, and since $a^{1/p}r$ is nondecreasing, $(R^{-1}(t))'_+ \equiv 0$. The conclusions then follow from Theorem 2.4.2(ii). □

The following corollary is an immediate consequence of Theorem 2.4.2 since $a \equiv 1$ implies (2.4.2) holds.

Corollary 2.4.4. *Assume that $\lambda < p$, (2.4.5) holds, $r'(t)/r^{\alpha+1}(t)$ is not identically a constant on $\mathbb{R}_+$, and*

$$\limsup_{t\to\infty} r^{-\beta}(t) \left(\int_t^\infty \left| \left(\frac{r'(\sigma)}{r^{\frac{(\lambda+3)p+2}{(\lambda+2)p+1}}(\sigma)} \right)' \right| d\sigma \right)^{\beta_2}$$

$$\times \exp\left\{ -\int_0^t r(s) \left(\frac{1}{r(s)} \right)'_+ ds \right\} = \infty.$$

Then equation (2.4.4) is of the nonlinear limit-circle type if and only if (2.4.6) holds.

The last case to consider is that where $p = \lambda$, i.e., equation (2.2.1) is a half-linear equation. In this case, $\alpha = 1/(\lambda+1)$ and $\beta = \lambda/(\lambda+1)$.

Theorem 2.4.3. *Let $\lambda = p$, $\lim_{t\to\infty} g(t) = 0$, and (2.4.1) and (2.4.2) hold. Then equation (2.2.1) is of the nonlinear limit-circle type if and only if (2.4.3) holds.*

Proof. Let (2.4.3) hold. Then, by Lemma 2.3.7(ii), F is bounded for every solution of (2.2.1), and (2.4.3) and (2.2.2) imply that equation (2.2.1) is of the nonlinear limit-circle type. If (2.4.3) does not hold, then the conclusion follows from Lemma 2.3.6. □

Corollary 2.4.5. *Let $\lambda = p$,*

$$\lim_{t\to\infty} r^{-\frac{\lambda+2}{\lambda+1}}(t) r'(t) = 0, \quad \text{and} \quad \int_0^\infty \left| (r'(\sigma) r^{-\frac{\lambda+2}{\lambda+1}}(\sigma))' \right| d\sigma < \infty.$$

Then equation (2.4.4) is of the nonlinear limit-circle type if and only if

$$\int_0^\infty r^{-\frac{\lambda}{\lambda+1}}(s) ds < \infty.$$

The last theorem in this section is actually an oscillation result for equation (2.2.1). However, it yields a rather interesting nonlinear limit-point result.

Theorem 2.4.4. *If equation* (2.2.1) *is of the nonlinear limit-circle type and*

$$\int_0^\infty \frac{1}{a^{\frac{1}{p}}(s)} ds = \infty, \tag{2.4.10}$$

then all proper solutions of equation (2.2.1) *are oscillatory.*

Proof. Let y be a nonoscillatory solution of (2.2.1), say, $y(t) > 0$ for $t \geq t_1 \geq 0$. From (2.2.1), we have $(a(t)|y'(t)|^{p-1}y'(t))' \leq 0$, so $a(t)|y'(t)|^{p-1}y'(t)$ is decreasing. If y is a nonlinear limit-circle type solution of (2.2.1), then $y'(t)$ must eventually become negative, say $y'(t) < 0$ for $t \geq T \geq t_1$. Thus,

$$a(t)|y'(t)|^{p-1}y'(t) \leq a(T)|y'(T)|^{p-1}y'(T) < 0$$

for $t \geq T$. We then have

$$-|y'(t)|^p = |y'(t)|^{p-1}y'(t) \leq -\frac{K}{a(t)} < 0$$

for some constant $K > 0$ and all $t \geq T$. Hence,

$$|y'(t)|^p \geq \frac{K}{a(t)} > 0,$$

so

$$-y'(t) = |y'(t)| \geq \frac{K_1}{a^{\frac{1}{p}}(t)} > 0,$$

where $K_1 = (K)^{\frac{1}{p}}$. An integration then contradicts the fact that $y(t)$ is eventually positive. □

Remark 2.4.1. It is interesting to note the contrapositive of Theorem 2.4.4, namely, *if equation* (2.2.1) *has a nonoscillatory solution and* (2.4.10) *holds, then equation* (2.2.1) *is of the nonlinear limit-point type.*

If $p = 1$, then Theorem 2.4.4 agrees with Theorem 5.7 in [25]; also see the discussion in [98].

2.5 The Forced Equation

In this section and the next, we consider the forced version of equation (2.2.1), namely,

$$\left(a(t)|y'|^{p-1}y'\right)' + r(t)|y|^\lambda \operatorname{sgn} y = e(t), \tag{2.5.1}$$

where $e \in C(\mathbb{R}_+)$ and a, r, p, and λ are as before. We want to point out here that due to the conditions imposed on the forcing term e, it may be large is some sense (see condition (2.6.1) below).

Definition 2.5.1. A solution y of (2.2.1) is continuable if it is defined on $\mathbb{R}_+$. It is said to be noncontinuable if it is defined on $[0,\tau)$ with $\tau < \infty$, and it cannot be defined at $t = \tau$. A continuable solution y is called proper if it is nontrivial in any neighborhood of ∞.

The definitions of a nonlinear limit-point and a nonlinear limit-circle solution are exactly the same as they are for the unforced equation (see Definition 2.2.3). Next we define what we mean by an oscillatory solution.

Definition 2.5.2. We say that a proper solution $y(t)$ of (2.5.1) is oscillatory if it has a sequence of zeros tending to ∞, and it is nonoscillatory otherwise. A solution $y(t)$ is said to be weakly oscillatory if $y(t)$ is nonoscillatory but $y'(t)$ is oscillatory.

In addition to the constants defined in Section 2.2 above, we set

$$q = \max\left(\frac{1}{p+1}, \frac{(\lambda+2)p+1}{(\lambda+1)(p+1)}\right) \quad \text{and} \quad q_1 = \frac{1}{1-q} \quad \text{for}\, q < 1.$$

We also define the function $H : \mathbb{R}_+ \to \mathbb{R}$ by

$$H(t) = F(t)\left(\int_0^t |g'(s)|ds + 1\right)^{-q_1}. \tag{2.5.2}$$

Notice that $1/(\lambda+1) \leq q$ and that $q < 1$ if and only if $p < \lambda$.

Except for part (iii) with the additional term involving the function e, the content of the following lemma is identical to that of Lemma 2.3.1

Lemma 2.5.1. *Let y be a solution of* (2.5.1). *Then:*

(i) *y is proper or trivial on $\mathbb{R}_+$;*
(ii) *the estimates*

$$|y(t)| \leq \delta_1 R^{-\frac{p}{(\lambda+2)p+1}}(t)\, F^{\frac{1}{\lambda+1}}(t), \quad \textit{where}\, \delta_1 = \gamma^{-\frac{1}{\lambda+1}}, \tag{2.5.3}$$

and

$$|y^{[1]}(t)| \leq R^{\frac{p}{(\lambda+2)p+1}}(t)\, F^{\frac{p}{p+1}}(t) \tag{2.5.4}$$

hold for all $t \geq 0$;

(iii) *for* $0 \le \tau < t$, *we have*

$$F(t) = F(\tau) - \alpha g(\tau)y(\tau)y^{[1]}(\tau) + \alpha g(t)y(t)y^{[1]}(t)$$
$$-\alpha \int_\tau^t g'(s)y(s)y^{[1]}(s)ds$$
$$+ \int_\tau^t [\delta R^{-\alpha}(s)|y^{[1]}(s)|^{\frac{1}{p}} \operatorname{sgn} y^{[1]}(s) - \alpha g(s)y(s)]e(s)ds.$$

Proof. To prove (i), let y be a solution of (2.5.1) that fails to be defined beyond T, that is, $\limsup_{t\to T}[|y(t)+|y'(t)|] = +\infty$. Then, a straight forward calculation gives

$$V'(t) = \left(\frac{1}{R(t)}\right)' |y^{[1]}(t)|^\delta + \delta \frac{e(t)}{r(t)}|y'(t)|\operatorname{sgn} y'(t).$$

Since $p\delta > 1$, the inequality

$$|u| \le |u|^{p\delta} + 1 \tag{2.5.5}$$

holds for all u, so

$$\begin{aligned}
\left|\frac{e(t)}{r(t)}|y'(t)|\operatorname{sgn} y'(t)\right| &\le \frac{|e(t)|}{r(t)}\left[|y'(t)|^{p\delta} + 1\right] \\
&\le \frac{|e(t)|}{a^\delta(t)r(t)}\left|y^{[1]}(t)\right|^\delta + \frac{|e(t)|}{r(t)} = \frac{|e(t)|}{a(t)}\frac{|y^{[1]}(t)|^\delta}{R(t)} + \frac{|e(t)|}{r(t)} \\
&\le \frac{|e(t)|}{a(t)}V(t) + \frac{|e(t)|}{r(t)}.
\end{aligned} \tag{2.5.6}$$

Thus,

$$V'(t) \le \left[\frac{R'_-(t)}{R(t)} + \delta\frac{|e(t)|}{a(t)}\right] V(t) + \delta\frac{|e(t)|}{r(t)}.$$

Integrating, we obtain

$$\begin{aligned}
V(t) &\le V(0) + \int_0^t \left[\frac{R'_-(s)}{R(s)} + \delta\frac{|e(s)|}{a(s)}\right] V(s)ds + \delta \int_0^t \frac{|e(s)|}{r(s)}ds \\
&\le K_1 + \int_0^t \left[\frac{R'_-(s)}{R(s)} + \delta\frac{|e(s)|}{a(s)}\right] V(s)ds,
\end{aligned}$$

where $K_1 = V(0) + \delta \int_0^T [|e(s)|/r(s)]ds$. By Gronwall's inequality, we have

$$V(t) \le K_1 \exp\left\{\int_0^t \left[\frac{R'_-(s)}{R(s)} + \delta\frac{|e(s)|}{a(s)}\right] ds\right\}$$
$$\le K_1 \exp\left\{\int_0^T \left[\frac{R'_-(s)}{R(s)} + \delta\frac{|e(s)|}{a(s)}\right] ds\right\}.$$

This implies that $a(t)|y'(t)|^{p+1}/r(t)$ is bounded on $[0, T]$, and consequently that $y'(t)$ is bounded on $[0, T]$. An integration shows that $y(t)$ is bounded on $[0, T]$ contradicting our assumption. Hence, y is defined on $\mathbb{R}_+$ and it is either trivial or proper according to Lemma 1 and Theorem 4 in [48].

Part (ii) follows immediately from the definition of F. To prove (iii), first observe that

$$\int_\tau^t [F(s) - \alpha g(s)y(s)y^{[1]}(s)]' ds$$
$$= F(t) - F(\tau) + \alpha g(\tau)y(\tau)y^{[1]}(\tau) - \alpha g(t)y(t)y^{[1]}(t).$$

However, differentiating first and then integrating, we have

$$\int_\tau^t [F(s) - \alpha g(s)y(s)y^{[1]}(s)]' ds$$
$$= \int_\tau^t [-\alpha R^{-\alpha-1}(s)R'(s)|y^{[1]}(s)|^\delta + \gamma\beta R^{\beta-1}(s)R'(s)|y(s)|^{\lambda+1}$$
$$-\delta R^{-\alpha}(s)a^{1/p}(s)|y'(s)|r(s)|y(s)|^\lambda \operatorname{sgn} y(s) \operatorname{sgn} y'(s)$$
$$+\gamma(\lambda+1)R^\beta(s)|y(s)|^\lambda y'(s) \operatorname{sgn} y(s) - \alpha g'(s)y(s)y^{[1]}(s)$$
$$-\alpha g(s)y'(s)y^{[1]}(s) + \alpha g(s)r(s)|y(s)|^{\lambda+1}$$
$$+\delta R^{-\alpha}(s)|y^{[1]}(s)|^{\frac{1}{p}} \operatorname{sgn} y^{[1]}(s)e(s) - \alpha g(s)y(s)e(s)]ds$$
$$= -\alpha \int_\tau^t g'(s)y(s)y^{[1]}(s)ds + \int_\tau^t [\delta R^{-\alpha}(s)|y^{[1]}(s)|^{\frac{1}{p}} \operatorname{sgn} y^{[1]}(s)$$
$$-\alpha g(s)y(s)]e(s)ds.$$

□

The next lemma provides information about the behavior of the function H.

Lemma 2.5.2. *Assume that $\lambda > p$ and*

$$\int_0^\infty R^{-\frac{p}{(\lambda+2)p+1}}(t)|e(t)|dt < \infty. \tag{2.5.7}$$

Then H is bounded.

Proof. Let y be a solution of (2.5.1) and suppose that H is not bounded. Since $H(t) \geq 0$, there is a sequence $\{t_k\} \to \infty$ as $k \to \infty$ such that

$$\lim_{k\to\infty} H(t_k) = \infty. \tag{2.5.8}$$

It follows from (2.5.2) that $\lim_{k\to\infty} F(t_k) = \infty$, so we may assume that $F(t_k) \geq 1$ for $k \geq 1$. Thus, there exist two sequences $\{\sigma_k\}_{k=1}^\infty$ and $\{\tau_k\}_{k=1}^\infty$ with $t_k \leq \sigma_k < \tau_k$ such that

$$\frac{1}{2}H(\tau_k) = H(\sigma_k) = H(t_k) \tag{2.5.9}$$

and

$$H(\sigma_k) \leq H(t) \leq H(\tau_k) \text{ for } \sigma_k \leq t \leq \tau_k, \quad k = 1, 2, \ldots.$$

From (2.5.2) and (2.5.9), we see that $\max_{\sigma_k \leq t \leq \tau_k} F(t) = F(\tau_k)$. Now g is locally of bounded variation, so

$$|g(t)| - |g(0)| \leq |g(t) - g(0)| \leq \int_0^{\tau_k} |g'(s)|ds$$

for $t \in [\sigma_k, \tau_k]$. Lemma 2.5.1(ii) yields

$$\begin{aligned}
&\left|\int_{\sigma_k}^{\tau_k} [\delta R^{-\alpha}(s)|y^{[1]}(s)|^{\frac{1}{p}} \operatorname{sgn} y^{[1]}(s) - \alpha g(s)y(s)]e(s)ds\right| \\
&\leq \int_{\sigma_k}^{\tau_k} [\delta R^{-\alpha}(s) R^{\frac{1}{(\lambda+2)p+1}}(s)\, F^{\frac{1}{p+1}}(s) \\
&\quad + \alpha\delta_1|g(s)|R^{-\frac{p}{(\lambda+2)p+1}}(s)\, F^{\frac{1}{\lambda+1}}(s)]e(s)ds \\
&\leq C\int_{\sigma_k}^{\tau_k} R^{-\frac{p}{(\lambda+2)p+1}}(s) \left(F^{\frac{1}{p+1}}(s) + F^{\frac{1}{\lambda+1}}(s)\right) |e(s)|ds \\
&\leq 2CF^q(\tau_k) \int_{\sigma_k}^{\tau_k} R^{-\frac{p}{(\lambda+2)p+1}}(s)|e(s)|ds \\
&\leq \left(C_2 + C_3 \int_0^{\tau_k} |g'(s)|ds\right) F^q(\tau_k),
\end{aligned} \tag{2.5.10}$$

where

$$C = \delta + \alpha\delta_1 \left(|g(0)| + \int_0^{\tau_k} |g'(s)| ds \right),$$

$$C_1 = \int_0^\infty R^{-\frac{p}{(\lambda+2)p+1}}(s)|e(s)| ds,$$

$$C_2 = 2(\delta + \alpha\delta_1|g(0)|)C_1, \quad \text{and} \quad C_3 = 2\alpha\delta_1 C_1.$$

Letting $\tau = \sigma_k$ and $t = \tau_k$ in part (iii) of Lemma 2.5.1 and using (2.5.3), (2.5.4), and (2.5.10), we obtain

$$F(\tau_k) - F(\sigma_k) \le \alpha_1 F^q(\tau_k) \left(|g(\sigma_k)| + |g(\tau_k)| + \int_{\sigma_k}^{\tau_k} |g'(s)| ds \right.$$
$$\left. + \frac{C_2}{\alpha_1} + \frac{C_3}{\alpha_1} \int_0^{\tau_k} |g'(s)| ds \right), \tag{2.5.11}$$

where $\alpha_1 = \alpha\gamma^{-\frac{1}{\lambda+1}}$.

It is easy to see that there exist an integer k_0 and a constant $M > 0$ such that

$$2\alpha_1|g(0)| + C_2 \le M \left(\int_0^{\tau_k} |g'(s)| ds + 1 \right).$$

Thus,

$$\begin{aligned} F(\tau_k) - F(\sigma_k) &\le F^q(\tau_k) \\ &\quad \times \left[2\alpha_1|g(0)| + C_2 + (3\alpha_1 + C_3) \left(\int_0^{\tau_k} |g'(s)| ds + 1 \right) \right] \\ &\le F^q(\tau_k)\,(M + 3\alpha_1 + C_3) \left(\int_0^{\tau_k} |g'(s)| ds + 1 \right) \end{aligned}$$

for $k \ge k_0$. Now $\lambda > p$ implies $q < 1$, so from (2.5.9) we have

$$\begin{aligned} (2^{1-q} - 2^{-q})G^{1-q}(\sigma_k) &= G^{1-q}(\tau_k) - \frac{G(\sigma_k)}{G^q(\tau_k)} \\ &\le \left(F^{1-q}(\tau_k) - \frac{F(\sigma_k)}{F^q(\tau_k)} \right) \left(\int_0^{\tau_k} |g'(s)| ds + 1 \right)^{-1} \\ &\le M + 3\alpha_1 + C_3. \end{aligned}$$

This contradicts (2.5.8) and completes the proof of the lemma. □

The next lemma gives sufficient conditions for the function F to be bounded from below along solutions of (2.5.1).

Lemma 2.5.3. *Assume that $p > \lambda$ and conditions* (2.2.4) *and* (2.5.7) *hold. Then, for any nontrivial solution of equation* (2.5.1), *the function F is bounded from below by a positive constant.*

Proof. Suppose that there is a nontrivial solution of (2.5.1) such that

$$\liminf_{t\to\infty} F(t) = 0. \tag{2.5.12}$$

In view of Lemma 2.5.1, y is proper. Then, for any $t_0 \in \mathbb{R}_+$ such that $F(t_0) < 1$, there exist τ and σ such that $t_0 \le \sigma < \tau$ and

$$2F(\tau) = F(\sigma) = F(t_0) \quad \text{and} \quad F(\tau) \le F(t) \le F(\sigma) < 1 \text{ for } \sigma \le t \le \tau.$$

Recall from the proof of Lemma 2.5.2 that inequality (2.5.10) holds for $p > \lambda$ since $F(t) < 1$ on $[\sigma, \tau]$. Now $q > 1$ since $p > \lambda$, so from (2.5.10) and parts (ii) and (iii) of Lemma 2.5.1, we have

$$\begin{aligned}
\frac{F(\sigma)}{2} &= F(\sigma) - F(\tau) \\
&\le \alpha_1 F^q(\sigma)\left[|g(\tau)| + |g(\sigma)| + \int_\sigma^\tau |g'(s)|ds + \frac{C_2}{\alpha_1} + \frac{C_3}{\alpha_1}\int_0^\tau |g'(s)|ds\right] \\
&\le \alpha_1 F^q(\sigma)\left[|g(\tau)| + |g(\sigma)| + \left(\frac{C_3}{\alpha_1} + 1\right)\int_0^\tau |g'(s)|ds + \frac{C_2}{\alpha_1}\right] \\
&\le C_4 F^q(\sigma),
\end{aligned}$$

where C_2 and C_3 are given in the proof of Lemma 2.5.2 and

$$C_4 = \frac{C_2}{\alpha_1} + 2\sup_{s\in\mathbb{R}_+} |g(s)| + \left(\frac{C_3}{\alpha_1} + 1\right)\int_0^\infty |g'(s)|ds.$$

Hence, $F(\sigma) \ge [1/(2C_4)]^{\frac{1}{q-1}}$. Since t_0 is arbitrary, $q > 1$, and C_4 is a constant, we obtain a contradiction to (2.5.12). □

2.6 Asymptotic Properties of Solutions of the Forced Equation

Our first theorem gives sufficient conditions for equation (2.5.1) to be of the nonlinear limit-circle type.

Theorem 2.6.1. *Assume that $\lambda > p$, (2.5.7) holds, and*

$$\int_0^\infty R^{-\beta}(t)\left(\int_0^t \left|\left(\frac{a^{1/p}(s)R'(s)}{R^{\alpha+1}(s)}\right)'\right| ds + 1\right)^{q_1} dt < \infty.$$

Then (2.5.1) is of the nonlinear limit-circle type.

Proof. Let y be a solution of equation (2.5.1). Then (2.2.2) yields

$$\begin{aligned}\int_0^\infty |y(t)|^{\lambda+1} dt &\le \frac{1}{\gamma}\int_0^\infty \frac{F(t)}{R^\beta(t)} dt \\ &= \frac{1}{\gamma}\int_0^\infty H(t)R^{-\beta}(t)\left(\int_0^t |g'(s)|ds + 1\right)^{q_1} dt < \infty\end{aligned}$$

by Lemma 2.5.2 and the hypotheses of the theorem. □

Remark 2.6.1. Notice that the proof of Theorem 2.6.1 also shows that

$$\int_0^\infty \frac{a(t)|y'(t)|^{p+1}}{r(t)} dt < \infty.$$

The next two theorems give sufficient conditions for the boundedness of all solutions of (2.5.1).

Theorem 2.6.2. *Suppose*

$$\int_0^\infty \frac{|e(s)|}{a(s)} ds < \infty \quad \text{and} \quad \int_0^\infty \frac{|e(s)|}{r(s)} ds < \infty \tag{2.6.1}$$

hold. If

$$\int_0^\infty \frac{R'_-(t)}{R(t)} dt < \infty, \tag{2.6.2}$$

then every solution $y(t)$ of equation (2.5.1) is bounded and so is the quantity $a(t)|y'(t)|^{p+1}/r(t)$.

Proof. Proceeding as in the proof of Lemma 2.5.1, for any solution $y(t)$ of (2.5.1), we obtain

$$V'(t) \le \left[\frac{R'_-(t)}{R(t)} + \delta\frac{|e(t)|}{a(t)}\right] V(t) + \delta\frac{|e(t)|}{r(t)}.$$

Integrating, using Gronwall's inequality, and applying conditions (2.6.1) and (2.6.2) shows that $V(t)$ is bounded, and so the conclusion of the theorem follows. □

Theorem 2.6.3. *If* (2.6.2) *holds and*

$$\int_0^\infty |e(s)|\,ds < \infty, \tag{2.6.3}$$

then every solution $y(t)$ of (2.5.1) *is bounded and so is $a(t)|y'(t)|^{p+1}/r(t)$.*

Proof. In the proof of Theorem 2.6.2, we use the inequality

$$a^{-1/p}|a^{1/p}u| \le a^{-1/p}[|a^{1/p}u|^{p\delta} + 1] \quad \text{for all}\, u$$

in place of (2.5.5). We then obtain

$$V'(t) \le \left[\frac{R'_-(t)}{R(t)} + \delta|e(t)|\right] V(t) + \delta\frac{|e(t)|}{R(t)}.$$

Observing that condition (2.6.2) bounds $R(t)$ from below away from zero, the conclusion then follows from an integration and an application of Gronwall's inequality. □

The advantage of Theorem 2.6.2 over Theorem 2.6.3 is that the forcing term e in condition (2.6.1) may be considerably larger than is allowed by (2.6.3).

Our next result is concerned with the oscillatory and weakly oscillatory solutions.

Theorem 2.6.4. *In addition to conditions* (2.6.1) *and* (2.6.2), *assume that*

$$\int_0^\infty \frac{|r'(t)|a^{\frac{1}{p+1}}(t)}{r^{\frac{p+2}{p+1}}(t)} < \infty, \tag{2.6.4}$$

and let y be an oscillatory or weakly oscillatory solution of (2.5.1). *Then*

$$\int_0^\infty |y(t)|^{\lambda+1}dt = \infty \quad \textit{if and only if} \quad \int_0^\infty \frac{|y^{[1]}(t)|^\delta}{R(t)}dt = \infty. \tag{2.6.5}$$

Proof. For any solution y of (2.5.1), we have

$$\begin{aligned}\int_0^t |y(s)|^{\lambda+1}ds &= -\int_0^t \frac{y(s)(y^{[1]}(s))'}{r(s)}ds + \int_0^t \frac{e(s)y(s)}{r(s)}ds \\ &= -\frac{y(t)y^{[1]}(t)}{r(t)} + \frac{y(0)y^{[1]}(0)}{r(0)} + \int_0^t \frac{|y^{[1]}(s)|^\delta}{R(s)}ds \\ &\quad + \int_0^t \left(\frac{1}{r(s)}\right)' y(s)y^{[1]}(s)ds + \int_0^t \frac{e(s)y(s)}{r(s)}ds.\end{aligned} \tag{2.6.6}$$

By Theorem 2.6.2, $y(t)$ and $|y^{[1]}(t)|/R^{\frac{p}{p+1}}(t)$ are bounded. Now,

$$\begin{aligned}\left|\frac{r'(t)}{r^2(t)}y(t)y^{[1]}(t)\right| &\leq \frac{|r'(t)|}{r^2(t)}|y(t)|\frac{|y^{[1]}(t)|}{R^{\frac{1}{\delta}}(t)}R^{\frac{p}{p+1}}(t) \\ &= \frac{|r'(t)|a^{\frac{1}{p+1}}(t)}{r^{\frac{p+2}{p+1}}(t)}|y(t)|\frac{|y^{[1]}(t)|}{R^{\frac{p}{p+1}}(t)}\end{aligned}$$

and

$$\left|\left(\frac{1}{r(t)}\right)' y(t)y^{[1]}(t)\right| \leq |y(t)|\frac{y^{[1]}(t)}{R^{\frac{p}{p+1}}(t)}\frac{|r'(t)|a^{\frac{1}{p+1}}(t)}{r^{\frac{p+2}{p+1}}(t)},$$

so the expression

$$\frac{y(0)y^{[1]}(0)}{r(0)} + \int_0^t \left(\frac{1}{r(s)}\right)' y(s)y^{[1]}(s)ds + \int_0^t \frac{e(s)y(s)}{r(s)}ds \tag{2.6.7}$$

is bounded on $\mathbb{R}_+$. The conclusion then follows by evaluating (2.6.6) along a sequence of zeros of $y'(t)$. □

The proof of the following theorem is essentially the same as that of Theorem 2.6.4 and will be left to the reader.

Theorem 2.6.5. *Let (2.6.2)–(2.6.4) hold and let y be an oscillatory or weakly oscillatory solution of (2.5.1). Then (2.6.5) holds.*

Remark 2.6.2. If $a(t) \equiv 1$ and $r(t) = t^\sigma$, then

$$\int_0^\infty \frac{|r'(t)|}{r^{\frac{p+2}{p+1}}(t)}dt < \infty \quad \text{if and only if} \quad \sigma \geq 0.$$

Next, we give sufficient conditions that not only show that equation (2.5.1) is of the nonlinear limit-point type, but also in fact that every solution is of the nonlinear limit-point type.

Theorem 2.6.6. *Assume that* $p > \lambda$ *and conditions* (2.2.4), (2.5.7), (2.6.1), (2.6.2), *and* (2.6.4) *hold. If*

$$\int_0^\infty R^{-\beta}(t)\,dt = \infty,$$

then every nontrivial solution of equation (2.5.1) *is of the nonlinear limit-point type.*

Proof. Lemma 2.5.3 implies that for every nontrivial solution of (2.5.1), the function F is bounded from below, so we have

$$\int_0^\infty \frac{|y^{[1]}(t)|^\delta}{R(t)}dt + \gamma\int_0^\infty |y(t)|^{\lambda+1}dt = \int_0^\infty \frac{F(t)}{R^\beta(t)}dt = \infty. \tag{2.6.8}$$

If y is an oscillatory or weakly oscillatory solution, then both integrals on the left-hand side of (2.6.8) diverge by Theorem 2.6.4, so let y be a monotonic solution of (2.5.1). If $y(t)y'(t) > 0$ eventually, then we are done, so assume that $y(t)y'(t) < 0$ for large t. If

$$\int_0^\infty |y(t)|^{\lambda+1}dt < \infty,$$

then we must have

$$\int_0^\infty \frac{|y^{[1]}(t)|^\delta}{R(t)}dt = \infty.$$

In view of (2.6.6), (2.6.7), and the fact that $y(t)y^{[1]}(t) < 0$, we obtain a contradiction. This completes the proof of the theorem. □

Our next result gives a necessary condition for y to be a nonlinear limit-circle type solution of (2.5.1).

Theorem 2.6.7. *Assume that* (2.6.1), (2.6.2), *and* (2.6.4) *hold. If* y *is a nonlinear limit-circle type solution of* (2.5.1), *then*

$$\int_0^\infty \frac{a(t)|y'(t)|^{p+1}}{r(t)}dt < \infty. \tag{2.6.9}$$

Proof. Let y be a nonlinear limit-circle type solution of equation (2.5.1). If y is oscillatory, the results follows from Theorem 2.6.4, so suppose that y is nonoscillatory. If $y(t)y'(t) > 0$ for large t, then y is not a nonlinear limit-circle type solution. If $y(t)y'(t) < 0$ for large t, then the conclusion follows from (2.6.6) and (2.6.7). □

If $p > 1$, we have the following variation of Theorem 2.6.7.

Theorem 2.6.8. *Assume that* $p > 1$ *and conditions* (2.6.1) *and* (2.6.2) *hold. In addition, assume that*

$$\int_0^\infty \frac{(r'(t))^2 a^{\frac{2}{p+1}}(t)}{r^{\frac{2(p+2)}{p+1}}(t)} dt < \infty. \tag{2.6.10}$$

If y is a nonlinear limit-circle type solution of (2.5.1), *then* (2.6.9) *holds.*

Proof. Let y be a nonlinear limit-circle type solution of equation (2.5.1) and multiply (2.5.1) by $y(t)/r(t)$. Observing that

$$(y^{[1]})'y = (y^{[1]}y)' - a|y'|^{p-1}(y')^2, \tag{2.6.11}$$

and then integrating from $T \geq 0$ to t by parts, we obtain

$$\begin{aligned} &\frac{a(t)|y'(t)|^{p-1}y'(t)y(t)}{r(t)} - \frac{a(T)|y'(T)|^{p-1}y'(T)y(T)}{r(T)} \\ &\quad + \int_T^t \frac{a(s)|y'(s)|^{p-1}y'(s)y(s)r'(s)}{r^2(s)} ds \\ &\quad + \int_T^t |y(s)|^{\lambda+1} ds - \int_T^t \frac{a(s)|y'(s)|^{p+1}}{r(s)} ds = \int_T^t \frac{e(s)y(s)}{r(s)} ds. \end{aligned} \tag{2.6.12}$$

By Schwarz's inequality,

$$\begin{aligned} &\left| \int_T^t \frac{a(s)|y'(s)|^{p-1}y'(s)y(s)r'(s)}{r^2(s)} ds \right| \\ &\quad \leq \left| \int_T^t \frac{a^{1/2}(s)|y'(s)|^{(p+1)/2}}{r^{1/2}(s)} \cdot \frac{a^{1/2}(s)|y'(s)|^{(p-1)/2}y(s)|r'(s)|}{r^{3/2}(s)} ds \right| \\ &\quad \leq \left[\int_T^t \frac{a(s)|y'(s)|^{p+1}}{r(s)} ds \right]^{1/2} \left[\int_T^t \frac{a(s)|y'(s)|^{p-1}y^2(s)(r'(s))^2}{r^3(s)} ds \right]^{1/2}. \end{aligned}$$

Now

$$\frac{a|y'|^{p-1}(r')^2}{r^3} = \frac{(r')^2 a^{\frac{2}{p+1}}}{r^{\frac{2(p+2)}{p+1}}}\left[\frac{a^{\frac{p-1}{p+1}}}{r^{\frac{p-1}{p+1}}}|y'|^{p-1}\right] = \frac{(r')^2 a^{\frac{2}{p+1}}}{r^{\frac{2(p+2)}{p+1}}}\left[\frac{a^{\frac{1}{p+1}}}{r^{\frac{1}{p+1}}}|y'|\right]^{p-1}$$

$$\leq \frac{(r')^2 a^{\frac{2}{p+1}}}{r^{\frac{2(p+2)}{p+1}}}\left\{\left\{\left[\frac{a^{\frac{1}{p+1}}}{r^{\frac{1}{p+1}}}|y'|\right]^{p-1}\right\}^{\frac{p+1}{p-1}} + 1\right\}$$

$$= \frac{(r')^2 a^{\frac{2}{p+1}}}{r^{\frac{2(p+2)}{p+1}}}\left[\frac{a|y'|^{p+1}}{r} + 1\right]$$

since the inequality

$$|w| \leq |w|^{\frac{p+1}{p-1}} + 1 \tag{2.6.13}$$

holds for $p > 1$ and all w. By Theorem 2.6.2, both $y(t)$ and $a(t)|y'(t)|^{p+1}/r(t)$ are bounded, say $|y(t)| \leq C_1$ and $a(t)|y'(t)|^{p+1}/r(t) \leq C_2$ for $t \geq 0$ for some constants C_1, $C_2 > 0$. Thus, we have

$$\left|\int_T^t \frac{a(s)|y'(s)|^{p-1}y'(s)y(s)r'(s)}{r^2(s)}ds\right|$$
$$\leq W^{1/2}(t)C_1^2(C_2+1)\int_0^t \frac{(r'(s))^2 a^{\frac{2}{p+1}}(s)}{r^{\frac{2(p+2)}{p+1}}(s)}ds,$$

where

$$W(t) = \int_T^t \frac{a(s)|y'(s)|^{p+1}}{r(s)}ds.$$

Thus, from (2.6.12), condition (2.6.10), and the fact that $y(t)$ is nonlinear limit-circle type solution, we have

$$a(t)|y'(t)|^{p-1}y'(t)y(t)/r(t) + C_3 + C_4 W^{1/2} \geq W(t) \tag{2.6.14}$$

for some constants C_3, $C_4 > 0$. Suppose $W(t) \to \infty$ as $t \to \infty$. If y is not monotonic, let $\{t_n\} \to \infty$ be an increasing sequence of zeros of y'. Then (2.6.14) immediately yields a contradiction. If y is monotonic, then $y(t)y'(t) < 0$ since y is a nonlinear limit-circle type solution, so we again obtain a contradiction from (2.6.14). □

Remark 2.6.3. Notice that the necessary condition (2.6.9) in Theorems 2.6.7 and 2.6.8 also appears as a consequence of the sufficient conditions for equation (2.5.1) to be of the nonlinear limit-circle type in Theorem 2.6.1 (see Remark 2.6.1).

Our next theorem gives sufficient conditions for equation (2.5.1) to be of the nonlinear limit-point type by making use of the necessary conditions obtained in Theorems 2.6.7 and 2.6.8.

Theorem 2.6.9. *In addition to conditions* (2.6.1) *and* (2.6.2), *assume that either* (2.6.4) *holds, or* $p > 1$ *and* (2.6.10) *holds. If*

$$\int_0^\infty \frac{|e(t)|}{r(t)} R^\beta(t) \exp\left\{\int_0^t K(s)ds\right\} dt < \infty \tag{2.6.15}$$

and

$$\int_0^\infty R^{-\beta}(t) \exp\left\{-\int_0^t K(s)ds\right\} dt = \infty, \tag{2.6.16}$$

where

$$K(t) = (1-\beta)\frac{R'_+(t)}{R(t)},$$

then equation (2.5.1) *is of the nonlinear limit-point type.*

Proof. From (2.5.6), we obtain

$$\begin{aligned}
F'(t) &= R^\beta(t)\left[V'(t) + \beta\frac{R'(t)}{R(t)}V(t)\right] \\
&= R^\beta(t)\left[\left(\frac{1}{R(t)}\right)'|y^{[1]}(t)|^\delta + \delta\frac{e(t)}{r(t)}|y'(t)|\operatorname{sgn} y'(t) + \beta\frac{R'(t)}{R(t)}V(t)\right] \\
&\geq R^\beta(t)\left[-\frac{R'_+(t)}{R(t)}V(t) - \delta\frac{|e(t)|}{a(t)}V(t) - \delta\frac{|e(t)|}{r(t)} + \beta\frac{R'(t)}{R(t)}V(t)\right] \\
&= R^\beta(t)\left[-\frac{R'_+(t)}{R(t)}V(t) - \delta\frac{|e(t)|}{a(t)}V(t) - \delta\frac{|e(t)|}{r(t)}\right. \\
&\quad \left. + \beta\frac{R'_+(t) - R'_-(t)}{R(t)}V(t)\right] \\
&= \left\{-(1-\beta)\frac{R'_+(t)}{R(t)} - \delta\frac{|e(t)|}{a(t)} - \beta\frac{R'_-(t)}{R(t)}\right\}F(t) - \delta\frac{|e(t)|}{r(t)}R^\beta(t).
\end{aligned}$$

Thus,

$$F'(t) + K_1(t)F(t) \geq -\delta \frac{|e(t)|}{r(t)} R^{\beta}(t),$$

where

$$K_1(t) = (1-\beta)\frac{R'_+(t)}{R(t)} + \beta \frac{R'_-(t)}{R(t)} + \delta \frac{|e(t)|}{a(t)}.$$

Observe that condition (2.6.1) and (2.6.2) imply that

$$e^{\int_0^t K(s)ds} \leq e^{\int_0^t K_1(s)ds} \leq k_1 e^{\int_0^t K(s)ds}$$

for some constant $k_1 > 0$.

Now, let $y(t)$ be a solution of (2.5.1) such that

$$F(0) \geq \delta k_1 \int_0^{\infty} \frac{|e(t)|}{r(t)} R^{\beta}(t) e^{\int_0^t K(s)ds} + 1.$$

Then,

$$(F(t)e^{\int_0^t K_1(s)ds})' \geq -\delta k_1 \frac{|e(t)|}{r(t)} R^{\beta}(t) e^{\int_0^t K(s)ds},$$

so integrating, we obtain

$$V(t) \geq R^{-\beta}(t) e^{-\int_0^t K_1(s)ds} \geq \frac{1}{k_1} R^{-\beta}(t) e^{-\int_0^t K(s)ds}.$$

Integrating and applying condition (2.6.16), we obtain that

$$\int_0^t V(s)ds = \int_0^t \left[\frac{a(s)}{r(s)} |y'(s)|^{p+1} + \gamma |y(s)|^{\lambda+1}\right] ds \to \infty \qquad (2.6.17)$$

as $t \to \infty$. In view of either Theorem 2.6.7 or 2.6.8, the first integral on the right-hand side of (2.6.17) converges, and the conclusion follows. □

2.7 Examples and Discussion

If we apply our results in Section 2.4 to the Emden–Fowler equation

$$(a(t)y')' + r(t)|y|^{\lambda}\operatorname{sgn} y = 0, \tag{E–F}$$

we obtain the following corollary.

Corollary 2.7.1. *Let*

$$\int^{\infty} \left| \left(\frac{(a(t)r(t))'}{a^{\frac{2}{\lambda+3}}(t) r^{\frac{\lambda+5}{\lambda+3}}(t)} \right)' \right| dt < \infty, \tag{2.7.1}$$

(2.4.2) *hold, and let* $\lim_{t\to\infty} \frac{(a(t)r(t))'}{a^{1/2}(t)r^{3/2}(t)} = 0$ *if* $\lambda = 1$. *In addition, if* $\lambda < 1$, *let* $g(t)$ *with* $p = 1$ *not be identically a constant on* $\mathbb{R}_+$ *and*

$$\limsup_{t\to\infty} (a(t)r(t))^{-\frac{\lambda+1}{\lambda+3}} \left(\int_t^{\infty} \left| \left(\frac{(a(t)r(t))'}{a^{\frac{2}{\lambda+3}}(t) r^{\frac{\lambda+5}{\lambda+3}}(t)} \right)' \right| dt \right)^{\frac{2(\lambda+1)}{\lambda-1}}$$
$$\times \exp\left\{ -\int_0^t a(\sigma)r(\sigma) \left(\frac{1}{a(\sigma)r(\sigma)} \right)'_+ d\sigma \right\} = \infty.$$

Moreover, if $\lambda < 1$ *and* $\int_0^{\infty} \frac{dt}{a(t)} < \infty$, *let either*

$$\int_0^{\infty} \left(\int_s^{\infty} a^{-1}(\sigma) d\sigma \right)^{\frac{\lambda+1}{2}} dt = \infty$$

or

$$\lim_{t\to\infty} \frac{(a(t)r(t))'}{a^{\frac{2}{\lambda+3}}(t) r^{\frac{\lambda+5}{\lambda+3}}(t)} = 0$$

hold. Then equation (E–F) *is of the nonlinear limit-circle type if and only if*

$$\int_0^{\infty} \frac{dt}{(a(t)r(t))^{\frac{\lambda+1}{\lambda+3}}} < \infty. \tag{2.7.2}$$

The best known results on the linear limit-point/limit-circle problem are the following ones due to Dunford and Schwartz.

Theorem 2.7.1. **([72, p. 1410]).** *Assume that*

$$\int_0^{\infty} \left| \left[\frac{(a(u)r(u))'}{a^{1/2}(u)r^{3/2}(u)} \right]' + \frac{\{[a(u)r(u)]'\}^2}{4a^{3/2}(u)r^{5/2}(u)} \right| du < \infty. \tag{2.7.3}$$

If

$$\int_0^\infty [1/(a(u)r(u))^{1/2}]du < \infty, \tag{2.7.4}$$

then the equation

$$(a(t)y')' + r(t)y = 0 \tag{L}$$

is of the (linear) limit-circle type, i.e., every solution $y(t)$ satisfies

$$\int_0^\infty y^2(u)du < \infty.$$

If

$$\int_0^\infty [1/(a(u)r(u))^{1/2}]du = \infty, \tag{2.7.5}$$

then equation (L) *is of the (linear) limit-point type, i.e., there is a solution $y(t)$ of* (L) *such that*

$$\int_0^\infty y^2(u)du = \infty.$$

It is interesting to observe how some of the hypotheses used here for the nonlinear equation (2.2.1) compare to those above of Dunford and Schwartz. Notice, for example, that the integrand in (2.7.1) with $\lambda = 1$ is exactly what appears in the first term of the integrand in (2.7.3). Also, if $p = 1 = \lambda$, then conditions (2.4.3) and (2.4.7) are exactly conditions (2.7.4) and (2.7.5), respectively, of Dunford and Schwartz above. The same is true for (2.7.2) if $\lambda = 1$.

Example 2.7.1. Consider the equation

$$\left(e^{at}|y'|^{\lambda-1}y'\right)' + e^{bt}|y|^\lambda \operatorname{sgn} y = 0, \quad t \in [0,\infty), \tag{2.7.6}$$

where a and b are constants. If $b > 0$ and $a < b$, then by Theorem 2.4.3, equation (2.7.6) is of the nonlinear limit-circle type if and only if $a > -b\lambda$.

Example 2.7.2. Consider the equation

$$\left(t^{a}|y'|^{\lambda-1}y'\right)' + t^{b}|y|^\lambda \operatorname{sgn} y = 0, \quad t \in [1,\infty), \tag{2.7.7}$$

where a and b are constants. If $b > 0$ and $a - b < \lambda + 1$, then Theorem 2.4.3 implies that equation (2.7.7) is of the nonlinear limit-circle type if and only if $a + b\lambda > \lambda + 1$. Now consider the special case of (2.7.7) with $a = 0$, i.e., the equation

$$\left(|y'|^{\lambda-1}y'\right)' + t^{\sigma}|y|^{\lambda}\operatorname{sgn} y = 0, \quad t \in [1, \infty). \tag{2.7.8}$$

By Corollary 2.4.5, if $\sigma > -1 - \lambda$, then equation (2.7.8) is of the nonlinear limit-circle type if and only if

$$\sigma > 1 + \frac{1}{\lambda}. \tag{2.7.9}$$

Now suppose that λ is an odd integer, say, $\lambda = 2k - 1$ for some integer $k > 1$. Then (2.7.9) becomes

$$\sigma > 1 + \frac{1}{2k-1}.$$

It is known (see [25, 98, 104]) that the equation

$$y'' + t^{\sigma_1}y^{2k-1} = 0$$

is of the nonlinear limit-circle type if and only if

$$\sigma_1 > 1 + \frac{1}{k}.$$

Notice that

$$1 + \frac{1}{k} - \left(1 + \frac{1}{2k-1}\right) = \frac{k-1}{k(2k-1)},$$

and in a sense, this is a measure of the difference between the limit-circle behavior of Emden–Fowler equations and half-linear equations.

Example 2.7.3. Consider the equation

$$\left(|y'|^{p-1}y'\right)' + t^{\sigma}|y|^{\lambda}\operatorname{sgn} y = 0, \tag{2.7.10}$$

where $p > 0$, $\lambda > 0$, and $\sigma \in [0, 1]$. The hypotheses of Theorem 2.6.9 are satisfied (in fact, if $p > 1$, then both possibilities hold), so equation (2.7.10) is of the nonlinear limit-point type. If $p = \lambda$, so that equation (2.7.10) is half-linear, then we see that Theorem 2.6.6 does not apply. On the other hand, if $p = 2$, $\lambda = 1$, and $\sigma = 3/2$, then Theorem 2.6.6 holds but Theorem 2.6.9 does not since (2.6.16) fails. Hence, Theorems 2.6.6 and 2.6.9 are independent of each other.

Chapter 3

Strong Limit-Point/Limit-Circle Properties

3.1 Introduction

In this chapter, we introduce the reader to what we have called the strong limit-point and strong limit-circle properties of solutions. Once these are presented here, they will be used in subsequent chapters in this book. For clarity and simplicity, in Section 3.2, we begin by describing the strong limit-point property for the simple case of the Emden–Fowler equation. After giving some lemmas in Section 3.3, strong nonlinear limit-point results are presented in Section 3.4. In Section 3.5, we initiate a discussion of the strong limit-point and strong limit-circle properties for sub-half-linear equations. The needed lemmas are given in Section 3.6 and the actual results given in Section 3.7. Section 3.8 contains strong limit-point and limit-circle results for super-half-linear equations. In Section 3.9, we consider the forced equation while Section 3.10 contains additional results on the unforced equation. Section 3.11 discusses the case where a certain integral is required to be convergent. Sections 3.12–3.14 are devoted to Thomas–Fermi equations. The final section in the chapter contains illustrative examples.

3.2 The Strong Limit-Point Property

Consider the Emden–Fowler equation

$$y'' + r(t)|y|^{\lambda} \operatorname{sgn} y = 0, \tag{3.2.1}$$

where $\lambda > 0$, $r \in AC^1_{\text{loc}}(\mathbb{R}_+)$, and $r > 0$. Under these conditions it is known that all solutions of (3.2.1) are either proper or trivial on $\mathbb{R}_+$, i.e., there are no singular solutions (see, for example, [121, Theorem 9.4] or the results in [100–102]).

We begin by defining what it will mean to say that a solution of (3.2.1) is of the strong nonlinear limit-point type.

Definition 3.2.1. A solution y of equation (3.2.1) will be said to be of the strong nonlinear limit-point type if

$$\int_0^\infty |y(t)|^{\lambda+1} dt = \infty$$

and

$$\int_0^\infty \frac{(y'(t))^2}{r(t)} dt = \infty.$$

An equation will be said to be of the strong nonlinear limit-point type if every nontrivial solution is of the strong nonlinear limit-point type.

For convenience, we also define what it will mean to say that an equation is of the *weak nonlinear limit-point type.*

Definition 3.2.2. An equation will be said to be of the weak nonlinear limit-point type if every nontrivial solution is of the nonlinear limit-point type but at least one nontrivial solution fails to be of the strong nonlinear limit-point type.

To see that this classification of limit-point solutions and equations is not empty, consider the equation

$$y'' + \frac{1}{4} t^{-\frac{\lambda+3}{2}} |y|^\lambda \operatorname{sgn} y = 0, \quad t \geq 1.$$

Here, $y(t) = t^{\frac{1}{2}}$ is a limit point solution and

$$\int_1^\infty \frac{(y'(t))^2}{r(t)} dt = \frac{1}{4} \int_1^\infty t^{\frac{\lambda+1}{2}} dt = \infty,$$

so y is a strong nonlinear limit-point type solution. On the other hand, $y(t) = (t^2-1)/t^2 \to 1$ is a limit-point solution of the equation

$$y'' + 6 \frac{1}{t^2(t^2-1)} |y| \operatorname{sgn} y = 0, \quad t \geq 2,$$

but

$$\int_2^\infty \frac{(y'(t))^2}{r(t)} dt = \frac{2}{3} \int_2^\infty \frac{t^2-1}{t^4} dt < \infty.$$

Remark 3.2.1. The terminology *strong* and *weak limit-point* was introduced by Everitt, Giertz, and Weidmann [86] (also see [80, 83]) in the study of linear differential operators. The way they are defined in this paper does not correspond to how they are used for such operators in the linear case.

We define the function $g : \mathbb{R}_+ \to \mathbb{R}$ by

$$g(t) = -\frac{r'(t)}{r^{\frac{\lambda+5}{\lambda+3}}(t)}.$$

In the remainder of this section, as well as in Sections 3.3 and 3.4, the condition

$$\int_0^\infty |g'(t)|dt < \infty \tag{3.2.2}$$

holds without further mention.

For any solution y of (3.2.1), we set

$$F(t) = r^{\frac{\lambda+1}{\lambda+3}}(t)\left[\frac{(y'(t))^2}{r(t)} + \frac{2}{\lambda+1}|y(t)|^{\lambda+1}\right]. \tag{3.2.3}$$

We will continue to use the notation that for any continuous function $h : \mathbb{R}_+ \to \mathbb{R}$, $h(t)_+ = \max\{h(t), 0\}$ and $h(t)_- = \max\{-h(t), 0\}$ so that $h(t) = h(t)_+ - h(t)_-$.

We state the following theorem for reference purposes.

Theorem 3.2.1. *Assume that either*

(i) $\lambda = 1$ *and* $\lim_{t\to\infty} g(t) = 0$,
(ii) $\lambda > 1$, *or*
(iii) $\lambda < 1$ *and*

$$\liminf_{t\to\infty} r^{\frac{\lambda+1}{\lambda+3}}(t)\left(\int_t^\infty |g'(s)|ds\right)^{\frac{2(\lambda+1)}{1-\lambda}} \exp\left\{\int_0^t (r^{-1}(s))'_+ r(s)ds\right\} = 0. \tag{3.2.4}$$

Then (3.2.1) *is of the nonlinear limit-circle type if and only if*

$$\int_0^\infty r^{-\frac{\lambda+1}{\lambda+3}}(t)dt < \infty. \tag{3.2.5}$$

Proof. The conclusions follow from Theorem 2.4.3, Corollary 2.4.2, and Theorem 2.4.2(ii), respectively. □

In the next section, we presents some lemmas that are needed in the sequel.

3.3 Some Lemmas

Recall that condition (3.2.2) holds throughout this section. The first lemma describes the behavior of the function g.

Lemma 3.3.1. *The function g satisfies $\lim_{t\to\infty} g(t) \in [0, \infty)$.*

Proof. Since g is of bounded variation on $[s_0, \infty)$, $\lim_{s\to\infty} g(s) = C$ exists and is finite. Suppose $C < 0$. Then, there exists $s_1 \geq s_0$ such that $g(s) < C/2$ for $s \geq s_1$. Hence,

$$-\frac{\lambda+3}{2r^{2/(\lambda+3)}(s_1)} \leq \frac{\lambda+3}{2}\left(\frac{1}{r^{2/(\lambda+3)}(s)} - \frac{1}{r^{2/(\lambda+3)}(s_1)}\right)$$
$$= \int_{s_1}^{s} g(s)ds \leq \frac{C}{2}(s - s_1) \to -\infty$$

as $s \to \infty$, which is a contradiction. □

The following two lemmas tell us about the behavior of the function F.

Lemma 3.3.2. (i) *Any nontrivial solution y of* (3.2.1) *satisfies*

$$|y(t)y'(t)| \leq \left(\frac{\lambda+1}{2}\right)^{\frac{1}{\lambda+1}} F^{\frac{\lambda+3}{2(\lambda+1)}}(t). \tag{3.3.1}$$

(ii) *If $\lambda < 1$, then for any nontrivial solution y of* (3.2.1), *F is bounded from below by a positive constant.*

Proof. Part (i) follows immediately from the definition of F in (3.2.3), and part (ii) is a special case of Lemma 2.5.3. □

Lemma 3.3.3. *Assume that $\lim_{t\to\infty} g(t) = 0$.*

(i) *If $\lambda = 1$, then for every nontrivial solution y of* (3.2.1), *we have*

$$\lim_{t\to\infty} F(t) = C \in (0, \infty). \tag{3.3.2}$$

(ii) *If $\lambda > 1$, then for every nontrivial solution y of* (3.2.1), *either* (3.3.2) *holds, or*

$$0 < F(t) \leq M\left(\int_t^{\infty} |g'(s)|ds\right)^{\frac{2(\lambda+1)}{\lambda-1}}, \tag{3.3.3}$$

where $M = (\frac{8}{\lambda+3})^{\frac{2(\lambda+1)}{\lambda-1}}(\lambda+1)^{\frac{2}{\lambda-1}}$.

Proof. First note that in part (ii), apart from the values of λ, the behaviors of F in (3.3.2) and (3.3.3) are mutually exclusive. Part (i) follows from [138, Theorem 17.7] and part (ii) from [122, Theorem 20.1]. $\square$

Our next lemma includes a sufficient condition for (3.3.3) not to hold.

Lemma 3.3.4. *Suppose* $\lim_{t\to\infty} g(t) = 0$. *If either* $\lambda < 1$ *and* (3.2.4) *hold, or* $\lambda > 1$ *and*

$$\liminf_{t\to\infty} r^{-\frac{\lambda+1}{\lambda+3}}(t) \left(\int_t^\infty |g'(s)|ds\right)^{\frac{2(\lambda+1)}{\lambda-1}} \exp\left(\int_0^t (r^{-1}(s))'_- r(s)\, ds\right) = 0, \tag{3.3.4}$$

then (3.3.2) *holds for every nontrivial solution of* (3.2.1).

Proof. If $\lambda < 1$ and (3.2.4) hold, then the conclusion follows from Lemma 2.3.4. Let $\lambda > 1$, (3.3.4) hold, and let y be a nontrivial solution of (3.2.1). Suppose, to the contrary, that (3.3.3) holds, namely,

$$F(t) \le M\left(\int_t^\infty |g'(s)|ds\right)^{\frac{2(\lambda+1)}{\lambda-1}} \tag{3.3.5}$$

where $M > 0$ is a constant. Furthermore, if $Z(t) = F(t) r^{-\frac{\lambda+1}{\lambda+3}}(t)$ for $t \in R_+$, then (3.2.3) yields

$$Z'(t) = \left(\frac{1}{r(t)}\right)'(y'(t))^2 \ge -(r^{-1}(t))'_- r(t) Z(t) \tag{3.3.6}$$

for $t \in R_+$, so

$$Z(t) \ge Z(0)\exp\left\{-\int_0^t (r^{-1}(s))'_-\, r(s)ds\right\}$$

for $t \in R_+$. From this and (3.3.5), we have

$$Z(0)\exp\left\{-\int_0^t (r^{-1}(s))'_-\, r(s)ds\right\}$$

$$\le Z(t) \le M r^{-\frac{\lambda+1}{\lambda+3}}(t)\left(\int_t^\infty |g'(s)|ds\right)^{\frac{2(\lambda+1)}{\lambda-1}}$$

for $t \in R_+$, which contradicts (3.3.4). $\square$

The next two lemmas are also important in what follows.

Lemma 3.3.5.

(i) *If* $\lambda \ne 1$, *then equation* (3.2.1) *has an oscillatory solution.*

(ii) *For* $\lambda > 1$*, equation* (3.2.1) *has a nonoscillatory solution if and only if*

$$\int_0^\infty \int_t^\infty r(s)dsdt < \infty. \tag{3.3.7}$$

(iii) *If* $\lambda \neq 1$*, equation* (3.2.1) *has a nonoscillatory solution* y *with* $\lim_{t\to\infty} |y(t)| = C \in (0, \infty)$ *if and only if*

$$\int_0^\infty \operatorname{tr}(t)dt < \infty.$$

Proof. Part (i) follows from [15, Theorem 2], part (ii) from [138, Corollary 11.3], and part (iii) from [122, Theorem 16.9] with $k = 1$. □

Lemma 3.3.6. *If either*

$$\lambda < 1 \quad \textit{and} \quad \lim_{t\to\infty} r(t) = 0, \tag{3.3.8}$$

$$\lambda = 1, \quad \lim_{t\to\infty} r(t) = 0, \quad \textit{and} \quad \lim_{t\to\infty} g(t) = 0, \tag{3.3.9}$$

or

$$\lambda > 1, \quad r(t) \le M < \infty, \quad \textit{and} \quad \int_0^\infty \int_t^\infty r(s)ds\,dt < \infty, \tag{3.3.10}$$

then for every nontrivial solution y *of* (3.2.1)*, we have*

$$\int_0^\infty r^{-\frac{\lambda+1}{\lambda+3}}(s)ds = \infty \quad \textit{and} \quad \int_0^\infty |y(s)|^{\lambda+1}ds = \infty. \tag{3.3.11}$$

Moreover, for any nonoscillatory solution y *of* (3.2.1) *satisfying* $\lim_{t\to\infty} y(t) = C \neq 0$*, we have*

$$\int_0^\infty \frac{(y'(t))^2}{r(t)}dt = \infty \quad \textit{if and only if} \quad \int_0^\infty \frac{1}{r(s)}\left(\int_s^\infty r(\sigma)d\sigma\right)^2 ds = \infty. \tag{3.3.12}$$

For all other nontrivial solutions, we have

$$\int_0^\infty \frac{(y'(t))^2}{r(t)}dt = \infty. \tag{3.3.13}$$

Proof. It is clear that in all cases there exists M_0 such that

$$|r(t)| \le M_0 \tag{3.3.14}$$

for all $t \in \mathbb{R}_+$. Let y be an oscillatory solution of (3.2.1). We will show that

$$y \text{ is unbounded on } \mathbb{R}_+. \tag{3.3.15}$$

First, suppose that (3.3.8) or (3.3.9) holds. Then by [15, Lemma 2],

$$y(t) = r^{-\frac{1}{\lambda+3}}(t)F^{\frac{1}{\lambda+1}}(t)w(\varphi(t)),$$

where w is the periodic solution of the problem

$$w'' + |w|^{\lambda} \operatorname{sgn} w = 0, \quad w(0) = \left(\frac{2}{\lambda+1}\right)^{-\frac{1}{\lambda+1}}, \quad w'(0) = 0,$$

having $\varphi \in C^1(\mathbb{R}_+)$ and $\lim_{t\to\infty} \varphi(t) = \infty$. In view of part (ii) of Lemma 3.3.2 and part (i) of Lemma 3.3.3 above, F is bounded from below by a positive constant, so (3.3.15) follows from the fact that $\lim_{t\to\infty} r(t) = 0$.

Next, suppose (3.3.10) holds, and assume to the contrary that $|y(t)| \le M_1$ for some $M_1 > 0$ and all $t \in \mathbb{R}_+$. Then, y is a solution of the equation

$$v'' + r(t)|y(t)|^{\lambda-1}v = 0, \tag{3.3.16}$$

and so (3.3.16) is oscillatory. Furthermore, the last condition in (3.3.10) and Lemma 3.3.5(ii) yield the existence of a nonoscillatory solution of (3.2.1). Note that the absolute value of any nonoscillatory solution of (3.2.1) is increasing for large t. Thus, if all nonoscillatory solutions have finite limits as $t \to \infty$, then according to Theorem 16.9 in [122] with $k = 1$, there exists a nonoscillatory solution $\bar{y}$ of equation (3.2.1) such that

$$|\bar{y}(t)| \ge M_1^{\lambda-1} \quad \text{for large } t. \tag{3.3.17}$$

Hence, in any case, there is a nonoscillatory solution $\bar{y}$ of (3.2.1) for which (3.3.17) holds. Moreover, $\bar{y}$ is a solution of the nonoscillatory equation

$$v'' + r(t)|\bar{y}(t)|^{\lambda-1}v = 0. \tag{3.3.18}$$

But since $|y(t)|^{\lambda-1} \le M_1^{\lambda-1} \le |\bar{y}(t)|^{\lambda-1}$ for large t, the Sturm Comparison Theorem applied to equations (3.3.16) and (3.3.18) gives us a contradiction, and so (3.3.15) holds.

The first relation in (3.3.11) follows from (3.3.14). To prove that the second expression in (3.3.11) holds, let y be a solution of (3.2.1) such that

$$\int_0^{\infty} |y(t)|^{\lambda+1} dt < \infty. \tag{3.3.19}$$

Clearly, y must be oscillatory. From (3.3.14), we have

$$\int_0^{\infty} |y''(t)|^{\frac{\lambda+1}{\lambda}} dt = \int_0^{\infty} r^{\frac{\lambda+1}{\lambda}}(t)|y(t)|^{\lambda+1} dt \le M_0^{\frac{\lambda+1}{\lambda}} \int_0^{\infty} |y(t)|^{\lambda+1} dt < \infty,$$

so $y \in L^{\lambda+1}(\mathbb{R}_+)$ and $y'' \in L^{\frac{\lambda+1}{\lambda}}(\mathbb{R}_+)$. Then [49, Chapter V, Theorem 2] implies

$$y' \in L^s(\mathbb{R}_+), \quad s = \max\{\lambda+1, (\lambda+1)/\lambda\},$$

and it follows from [49, Chapter V, Theorem 1] that y is bounded on $\mathbb{R}_+$. This contradicts (3.3.15) and so (3.3.11) holds.

To prove (3.3.12), let y be a nonoscillatory solution of (3.2.1) for which $\lim_{t\to\infty} y(t) = C > 0$. Then, (3.2.1) implies that $\lim_{t\to\infty} y'(t) = 0$, and we see that there exists $t_0 \in \mathbb{R}_+$ such that

$$C/2 \leq y(t) \leq C \quad \text{on } [t_0, \infty).$$

Integrating equation (3.2.1) yields

$$\left(\frac{C}{2}\right)^{\lambda} \int_t^\infty r(s)ds \leq y'(t) = \int_t^\infty r(s)y^\lambda(s)ds \leq C^\lambda \int_t^\infty r(s)ds, \quad t \geq t_0,$$

so squaring, dividing by $r(t)$, and integrating again, we obtain

$$\left(\frac{C}{2}\right)^{2\lambda} \int_{t_0}^\infty r^{-1}(s)\left(\int_s^\infty r(\sigma)d\sigma\right)^2 ds$$

$$\leq \int_{t_0}^\infty \frac{(y'(s))^2}{r(s)} ds \leq C^{2\lambda} \int_{t_0}^\infty r^{-1}(s)\left(\int_s^\infty r(\sigma)d\sigma\right)^2 ds.$$

Hence, (3.3.12) holds. If $C < 0$, the proof is similar.

Finally, let y be either an oscillatory solution or a nonoscillatory solution satisfying $\lim_{t\to\infty} |y(t)| = \infty$. Then (3.3.15) implies

$$y \text{ is unbounded on } \mathbb{R}_+. \tag{3.3.20}$$

By Lemma 3.3.1, there exists a constant $C_1 \in (0, \infty)$ such that $|g(t)| \leq C_1$ for $t \in \mathbb{R}_+$. Suppose (3.3.13) does not hold, say

$$\int_0^\infty \frac{(y'(t))^2}{r(t)} dt = C_2 < \infty \tag{3.3.21}$$

for some $C_2 > 0$. If we define $Z(t) = F(t)r^{-\frac{\lambda+1}{\lambda+3}}(t)$ as above, then (3.3.20) implies

$$Z \text{ is unbounded on } \mathbb{R}_+. \tag{3.3.22}$$

On the other hand, in view of (3.3.6), (3.3.14), and (3.3.21), we have

$$|Z(t)| = \left|Z(0) + \int_0^t Z'(t)dt\right| \leq |Z(0)| + \int_0^t |(r^{-1}(s))|' \, (y'(s))^2 \, ds$$

$$= |Z(0)| + \int_0^t |g(s)| r^{\frac{2}{\lambda+3}}(s) \frac{(y'(s))^2}{r(s)} \, ds \leq |Z(0)| + C_1 M_0^{\frac{2}{\lambda+3}} C_2$$

for $t \in \mathbb{R}_+$. This contradicts (3.3.22) and so (3.3.13) holds. This completes the proof of the lemma. □

Lemma 3.3.7. *Let* $\lambda > 0$, $\lim_{t\to\infty} g(t) = 0$,

$$\int_0^\infty r^{-\frac{\lambda+1}{\lambda+3}} \, dt = \infty, \tag{3.3.23}$$

and let y *be a solution of* (3.2.1) *such that* $\lim_{t\to\infty} F(t) = C \in (0, \infty)$. *Then*

$$\int_0^\infty |y(t)|^{\lambda+1} \, dt = \int_0^\infty \frac{(y'(t))^2}{r(t)} \, dt = \infty. \tag{3.3.24}$$

Proof. There exists $t_0 \in \mathbb{R}_+$ such that

$$\frac{C}{2} \leq F(t) \leq 2C \quad \text{for } t \geq t_0. \tag{3.3.25}$$

From this and (3.2.3), we have

$$\int_{t_1}^t \frac{(y'(s))^2}{r(s)} \, ds + \frac{2}{\lambda+1} \int_{t_1}^t |y(s)|^{\lambda+1} \, ds$$

$$= \int_{t_1}^t F(s) r^{-\frac{\lambda+1}{\lambda+3}}(s) \, ds \geq \frac{C}{2} \int_{t_1}^t r^{-\frac{\lambda+1}{\lambda+3}}(s) \, ds \tag{3.3.26}$$

for $t > t_1$ for any $t_1 \geq t_0$. Now (3.3.26) and (3.3.23) imply that either

$$\int_0^\infty \frac{(y'(s))^2}{r(s)} \, ds = \infty \quad \text{or} \quad \int_0^\infty |y(s)|^{\lambda+1} \, ds = \infty. \tag{3.3.27}$$

Furthermore,

$$\int_{t_1}^{t} |y(s)|^{\lambda+1}\,ds = -\int_{t_1}^{t} \frac{y(s)y''(s)}{r(s)}\,ds$$
$$= -\frac{y(t)y'(t)}{r(t)} + \frac{y(t_1)y'(t_1)}{r(t_1)} + \int_{t_1}^{t} \frac{(y'(s))^2}{r(s)}\,ds$$
$$+ \int_{t_1}^{t} g(s)r^{-\frac{\lambda+1}{\lambda+3}}(s)y(s)y'(s)\,ds \tag{3.3.28}$$

for $t \geq t_1$. Lemma 3.3.2(i), (3.3.1), and (3.3.25) imply

$$|y(t)y'(t)| \leq \left(\frac{\lambda+1}{2}\right)^{\frac{1}{\lambda+1}} (2C)^{\frac{\lambda+3}{2(\lambda+1)}} := C_1 \tag{3.3.29}$$

for $t \geq t_0$. Suppose that

$$\int_0^{\infty} \frac{(y'(s))^2}{r(s)}\,ds < \infty.$$

Let t_2 and t_3 be such that $t_3 \geq t_2 \geq t_0$,

$$|g(t)| \leq \frac{\lambda+1}{16}\frac{C}{C_1} \quad \text{for } t \geq t_2,$$

and

$$\frac{2}{\lambda+1}\int_{t_0}^{t} |y(s)|^{\lambda+1}\,ds \geq \int_{t_0}^{t} \frac{(y'(s))^2}{r(s)}\,ds \quad \text{for } t \geq t_3.$$

These choices of t_2 and t_3 are possible because (3.3.27) holds and the fact that $\lim_{t\to\infty} g(t) = 0$. Then, in view of (3.3.26), (3.3.28), and (3.3.29), we obtain

$$\frac{C}{2}\int_{t_2}^{t} r^{-\frac{\lambda+1}{\lambda+3}}(s)\,ds \leq \frac{4}{\lambda+1}\int_{t_2}^{t} |y(s)|^{\lambda+1}\,ds$$
$$\leq C_2 - \frac{4}{\lambda+1}\frac{y(t)y'(t)}{r(t)} + \frac{C}{4}\int_{t_2}^{t} r^{-\frac{\lambda+1}{\lambda+3}}(s)\,ds$$

for $t \geq t_3$, where

$$C_2 = \frac{4}{\lambda+1}\left[\frac{y(t_2)y'(t_2)}{r(t_2)} + \int_{t_2}^{\infty} \frac{(y'(s))^2}{r(s)}\,ds\right].$$

Thus,

$$C_2 - \frac{4}{\lambda+1}\frac{y(t)y'(t)}{r(t)} \geq \frac{C}{4}\int_{t_2}^{t} r^{-\frac{\lambda+1}{\lambda+3}}\,ds \tag{3.3.30}$$

for $t \geq t_3$. Now (3.3.23) implies that the right-hand side of (3.3.30) tends to ∞ as $t \to \infty$, and we immediately have a contradiction if y is oscillatory and we consider a sequence of zeros of y. If y is nonoscillatory, then $y(t)y'(t) > 0$ for large t, and this contradicts (3.3.30), so the second relation in (3.3.24) holds.

Now suppose

$$\int_0^\infty |y(s)|^{\lambda+1}\,ds < \infty. \tag{3.3.31}$$

Let t_4 and t_5 be such that $t_5 \geq t_4 \geq t_0$,

$$|g(t)| \leq \frac{C}{4C_1} \quad \text{for } t \geq t_4,$$

and

$$\int_{t_4}^{t} \frac{(y'(s))^2}{r(s)}\,ds \geq \frac{C}{3}\int_{t_4}^{t} r^{-\frac{\lambda+1}{\lambda+3}}(s)\,ds \quad \text{for } t \geq t_5.$$

These choices of t_4 and t_5 exist due to (3.3.26) (with $t_4 = t_1$), (3.3.23), (3.3.31), and the fact that $\lim_{t\to\infty} g(t) = 0$. Hence, for $t \geq t_5$, (3.3.28), (3.3.29), and (3.3.31) imply

$$\begin{aligned}\infty &> \int_{t_4}^{\infty} |y(s)|^{\lambda+1}\,ds \\ &\geq -\frac{y(t)y'(t)}{r(t)} + \frac{y(t_4)y'(t_4)}{r(t_4)} \\ &\quad + \frac{C}{3}\int_{t_4}^{t} r^{-\frac{\lambda+1}{\lambda+3}}(s)ds - \frac{C}{4}\int_{t_4}^{t} r^{-\frac{\lambda+1}{\lambda+3}}(s)\,ds.\end{aligned} \tag{3.3.32}$$

If y is nonoscillatory, then $|y|$ is increasing for large t, so (3.3.31) does not hold. If y is oscillatory, then (3.3.23) contradicts (3.3.31) at large zeros of y. This completes the proof of the lemma. □

Lemma 3.3.8. *Assume that $\lambda \neq 1$ and $\lim_{t\to\infty} g(t) = K \neq 0$. Then there are positive constants K_1 and K_2 such that*

$$K_1 t^{-\frac{\lambda+3}{2}} \leq r(t) \leq K_2 t^{-\frac{\lambda+3}{2}}$$

for $t \in \mathbb{R}_+$, and the conclusions of Lemma 3.3.6 hold.

Proof. By Lemma 3.3.1, the function g has a finite limit $K \in [0, \infty)$ and so $K > 0$. We will estimate the function r. Let $t_0 \in \mathbb{R}_+$ be such that

$$2K \geq g(t) = -r^{-\frac{\lambda+5}{\lambda+3}}(t)r'(t) \geq \frac{K}{2}$$

for $t \geq t_0$. Integrating, we obtain

$$2K(t - t_0) \geq \frac{\lambda+3}{2}[r^{-\frac{2}{\lambda+3}}(t) - r^{-\frac{2}{\lambda+3}}(t_0)] \geq \frac{K}{2}(t - t_0)$$

for $t \geq t_0$. From this we see that there exists $t_1 \geq t_0$ such that

$$\frac{K}{2(\lambda+3)}t \leq r^{-\frac{2}{\lambda+3}}(t) \leq \frac{6K}{\lambda+3}t$$

for $t \geq t_1$, or

$$\left(\frac{6K}{\lambda+3}\right)^{-\frac{\lambda+3}{2}} t^{-\frac{\lambda+3}{2}} \leq r(t) \leq \left(\frac{K}{2(\lambda+3)}\right)^{-\frac{\lambda+3}{2}} t^{-\frac{\lambda+3}{2}}$$

for $t \geq t_1$. Setting

$$K_1 = \left(\frac{6K}{\lambda+3}\right)^{-\frac{\lambda+3}{2}} \quad \text{and} \quad K_2 = \left(\frac{K}{2(\lambda+3)}\right)^{-\frac{\lambda+3}{2}},$$

we easily see that (3.3.8) and (3.3.10) hold. □

3.4 Strong Nonlinear Limit-Point Results for Emden–Fowler Equations

We are now ready to prove some strong nonlinear limit-point results for equation (3.2.1). We begin with a theorem that gives some conditions equivalent to equation (3.2.1) being of the nonlinear limit-circle type. Although not explicitly stated, condition (3.2.2) is holding throughout this section.

Theorem 3.4.1. *Assume that* $\lim_{t\to\infty} g(t) = 0$. *Let either* $\lambda = 1$, *or* $\lambda > 1$ *and* (3.3.4) *hold, or* $\lambda < 1$ *and* (3.2.4) *hold. Then the following conclusions are equivalent:*

(i) *Equation* (3.2.1) *is of the nonlinear limit-circle type.*
(ii)

$$\int_0^\infty r^{-\frac{\lambda+1}{\lambda+3}}(t)dt < \infty. \tag{3.4.1}$$

(iii)

$$\int_0^\infty \frac{(y'(t))^2}{r(t)}\,dt < \infty \quad \textit{for all solutions of } (3.2.1). \tag{3.4.2}$$

Proof. Conditions (i) and (ii) are equivalent by Theorem 3.2.1. By Lemma 3.3.4, if $\lambda \neq 1$ and Lemma 3.3.3(i), if $\lambda = 1$, we see that $\lim_{t\to\infty} F(t) = C \in (0,\infty)$, and so F is bounded from below and from above by positive constants for every nontrivial solution y of (3.2.1).

Let (ii) hold and let y be a solution of (3.2.1). Then, from (3.2.3) and (3.4.1), we have

$$\infty > \int_0^\infty F(t) r^{-\frac{\lambda+1}{\lambda+3}}(t)\,dt = \int_0^\infty \frac{(y'(t))^2}{r(t)}\,dt + \frac{2}{\lambda+1}\int_0^\infty |y(t)|^{\lambda+1}\,dt. \tag{3.4.3}$$

Hence, (3.4.2) holds.

Finally, to show that (iii) implies (ii), suppose, to the contrary, that

$$\int_0^\infty r^{-\frac{\lambda+1}{\lambda+3}}(t)dt = \infty.$$

Then all the hypotheses of Lemma 3.3.7 are satisfied for any nontrivial solution y of (3.2.1), and we see that (3.3.24) contradicts (3.4.2). This completes the proof of the theorem. □

Remark 3.4.1. If $\lambda \neq 1$ and $\lim_{t\to\infty} g(t) = K \in R\setminus\{0\}$, then, according to Lemma 3.3.8, (3.3.11) holds for every nontrivial solution y of (3.2.1), and so (3.2.1) is not of the nonlinear limit-circle type. Thus, the assumption $\lim_{t\to\infty} g(t) = 0$ is really not restrictive when studying the limit-circle case.

The next theorem gives a necessary and sufficient condition for equation (3.2.1) to be of the strong nonlinear limit-point type.

Theorem 3.4.2. *Let $\lim_{t\to\infty} g(t) = 0$ and assume that either $\lambda = 1$, or $\lambda > 1$ and (3.3.4) hold, or $\lambda < 1$ and (3.2.4) hold. Then the following conclusions are equivalent:*

(i) *Every nontrivial solution of (3.2.1) is of the nonlinear limit-point type.*

(ii)

$$\int_0^\infty r^{-\frac{\lambda+1}{\lambda+3}}(t)dt = \infty. \tag{3.4.4}$$

(iii)

$$\int_0^\infty \frac{(y'(t))^2}{r(t)}\,dt = \infty \quad \textit{for all nontrivial solutions } y \textit{ of } (3.2.1). \tag{3.4.5}$$

That is, equation (3.2.1) *is of the strong nonlinear limit-point type if and only if* (3.4.4) *holds.*

Proof. If $\lambda \neq 1$, then Lemma 3.3.4 implies $\lim_{t\to\infty} F(t) = C \in (0,\infty)$ for every nontrivial solution of (3.2.1), and the same result follows from Lemma 3.3.3(i) if $\lambda = 1$. Let (3.4.4) hold; then the hypotheses of Lemma 3.3.7 are satisfied, so (3.3.24) implies (i) and (iii) hold.

Let either (i) or (iii) hold; we will prove indirectly that (3.4.4) holds. Suppose that

$$\int_0^\infty r^{-\frac{\lambda+1}{\lambda+3}}(t)dt < \infty.$$

Since F is bounded from above, (3.4.3) holds, and this contradicts either (i) or (iii). □

Next, we prove a result giving sufficient conditions for equation (3.2.1) to be of the strong nonlinear limit-point type. Here, we do not require $\lim_{t\to\infty} g(t) = 0$.

Theorem 3.4.3. *Assume that either* (3.3.8), (3.3.9), *or* (3.3.10) *holds. Furthermore, for* $\lambda > 1$, *let*

$$\int_0^\infty \frac{1}{r(s)}\left(\int_s^\infty r(\sigma)d\sigma\right)^2 ds = \infty \textit{ if } \int_0^\infty tr(t)dt < \infty. \tag{3.4.6}$$

Then (3.4.4) *holds, and for every nontrivial solution* y *of* (3.2.1), *we have*

$$\int_0^\infty |y(t)|^{\lambda+1}\,dt = \int_0^\infty \frac{(y'(t))^2}{r(t)}\,dt = \infty. \tag{3.4.7}$$

That is, (3.2.1) *is of the strong nonlinear limit-point type.*

Proof. For the case $\lambda \le 1$, if we can prove that equation (3.2.1) has no solution y with a finite nonzero limit as $t \to \infty$ that also satisfies

$$\int_0^\infty \frac{(y'(t))^2}{r(t)}\, dt = \infty,$$

then the result would follow from Lemma 3.3.6. To the contrary, suppose y is a solution of (3.2.1) with $\lim_{t\to\infty} |y(t)| = C \in (0,\infty)$ and

$$\int_0^\infty \frac{(y'(t))^2}{r(t)}\, dt = M_1 < \infty. \tag{3.4.8}$$

Then from (3.3.8), (3.3.9), and Lemma 3.3.1, there exist constants M_2 and M_3 such that

$$r(t) \le M_2 \quad \text{and} \quad |g(t)| \le M_3 \quad \text{on } \mathbb{R}_+. \tag{3.4.9}$$

Letting $Z(t) = F(t) r^{-\frac{\lambda+1}{\lambda+3}}$, we have $Z'(t) = g(t) r^{\frac{2}{\lambda+3}}(t)(y'(t))^2/r(t)$. From this, (3.4.8), and (3.4.9), we obtain

$$\int_0^t Z(s)\, ds = \int_0^t \left(\int_0^s Z'(\sigma)\, d\sigma + Z(0) \right) ds \le [M_3 M_2^{\frac{2}{\lambda+3}} M_1 + Z(0)]t.$$

Since $\lim_{t\to\infty} r(t) = 0$, an application of l'Hôpital's rule shows that the left-hand side of the above inequality tends to ∞ as $t \to \infty$, which is a contradiction.

Now suppose $\lambda > 1$. Then, by Lemma 3.3.6, (3.4.4) holds and the first integral in (3.4.7) diverges. If y is a nonoscillatory solution of (3.2.1) such that $\lim_{t\to\infty} y(t) = C \ne 0$, then we consider two cases. First, if

$$\int_0^\infty \mathrm{tr}(t) dt < \infty, \tag{3.4.10}$$

then the second integral in (3.4.7) diverges due to (3.4.6) and (3.3.12). Second, if (3.4.10) fails to hold, then by Lemma 3.3.5, there are no solutions tending to a nonzero constant, and so the conclusion again follows from Lemma 3.3.6. □

Thus far, our results have ensured that equation (3.2.1) is of the strong nonlinear limit-point type. Our next result is concerned with the weak nonlinear limit-point property.

Theorem 3.4.4. *Let $\lambda > 1$, (3.3.10) hold,*

$$\int_0^\infty tr(t)dt < \infty, \quad \text{and} \quad \int_0^\infty \frac{1}{r(s)} \left(\int_s^\infty r(\sigma)d\sigma \right)^2 ds < \infty. \tag{3.4.11}$$

Then, (3.4.4) holds and every nontrivial solution of (3.2.1) is of the nonlinear limit-point type. Furthermore,

$$\int_0^\infty \frac{(y'(t))^2}{r(t)} dt < \infty \quad \text{if and only if } y \text{ satisfies } \lim_{t\to\infty} y(t) = K \in R \setminus \{0\}, \tag{3.4.12}$$

and

$$\int_0^\infty \frac{(y'(t))^2}{r(t)} dt = \infty \quad \text{for all other nontrivial solutions,} \tag{3.4.13}$$

i.e., equation (3.2.1) is of the weak nonlinear limit-point type. In addition, both sets of solutions satisfying (3.4.12) and (3.4.13) are not empty.

Proof. By parts (i) and (iii) of Lemma 3.3.5, oscillatory solutions as well as a solution y tending to a nonzero finite limit exist. The conclusion then follows from Lemma 3.3.6. □

Remark 3.4.2. If $\lim_{t\to\infty} g(t) = K \neq 0$, then in view of Lemma 3.3.8, we see that (3.3.13) and (3.4.12) hold if $\lambda > 3$. Hence, Theorem 3.4.4 is not empty.

In our final theorem in this section, we consider the case where the function r is monotonic.

Theorem 3.4.5. *Assume that $\lim_{t\to\infty} g(t) = 0$.*

(i) *If $\lambda \leq 1$ and $r' < 0$ on $\mathbb{R}_+$, then equation (3.2.1) is of the strong nonlinear limit-point type.*
(ii) *Let $r' > 0$ on $\mathbb{R}_+$ and let*

$$g' \text{ be bounded on } \mathbb{R}_+. \tag{3.4.14}$$

If

$$\int_0^\infty r^{-\frac{\lambda+1}{\lambda+3}}(t)\, dt < \infty,$$

then

$$\int_0^\infty |y(t)|^{\lambda+1} dt < \infty \quad \text{and} \quad \int_0^\infty \frac{(y'(t))^2}{r(t)} dt < \infty$$

for every solution y of (3.2.1), i.e., equation (3.2.1) is of the nonlinear limit-circle type. If (3.4.4) holds, then equation (3.2.1) is of the strong nonlinear limit-point type.

Proof. Note first that $g \neq 0$ on $\mathbb{R}_+$.

(i) We have $\lim_{t\to\infty} r(t) = r_0 \in [0,\infty)$ and r is bounded on $\mathbb{R}_+$, so (3.4.4) holds. If $r_0 > 0$, then (3.2.4) holds for $\lambda < 1$, and the conclusion follows from Theorem 3.4.2. If $r_0 = 0$, then (3.3.8) and (3.3.9) hold, and the conclusion follows from Theorem 3.4.3.

(ii) Since r is increasing, $\lim_{t\to\infty} r(t) = r_0 \in (0,\infty]$. If $r_0 \in (0,\infty)$, then (3.4.4) holds, as do (3.2.4) and (3.3.4) for $\lambda < 1$ and $\lambda > 1$, respectively. The result then follows from Theorem 3.4.2.

Let $r_0 = \infty$. We first consider the case $\lambda < 1$. Then (3.2.4) takes the form

$$\liminf_{t\to\infty} r^{\frac{\lambda+1}{\lambda+3}}(t)\left(\int_t^\infty |g'(s)|\,ds\right)^{\frac{2(\lambda+1)}{1-\lambda}} = 0.$$

Furthermore, since $\lim_{t\to\infty} g(t) = 0$, by l'Hôpital's rule,

$$\lim_{t\to\infty}\left(\int_t^\infty |g'(s)|\,ds\right)^{\frac{3\lambda+1}{1-\lambda}} g^{-1}(t)$$
$$= -\frac{3\lambda+1}{1-\lambda}\lim_{t\to\infty}\left(\int_t^\infty |g'(s)|\,ds\right)^{\frac{4\lambda}{1-\lambda}}\frac{|g'(t)|}{g'(t)} = 0,$$

and hence, in view of (3.4.14),

$$\lim_{t\to\infty} r^{\frac{\lambda+1}{\lambda+3}}(t)\left(\int_t^\infty |g'(s)|\,ds\right)^{\frac{2(\lambda+1)}{1-\lambda}}$$
$$= \frac{2(\lambda+3)}{1-\lambda}\lim_{t\to\infty}\frac{\left(\int_t^\infty |g'(s)|\,ds\right)^{\frac{3\lambda+1}{1-\lambda}}|g'(t)|}{r^{\frac{1-\lambda}{\lambda+3}}(t)g(t)} = 0.$$

Thus, (3.2.4) holds, and the results follow from Theorems 3.4.1 and 3.4.2 if $\lambda \leq 1$.

Finally, in case $\lambda > 1$, we see that (3.3.4) has the form

$$\liminf_{t\to\infty} r^{\frac{2}{\lambda+3}}(t)\left(\int_t^\infty |g'(s)|\,ds\right)^{\frac{2(\lambda+1)}{\lambda-1}} = 0.$$

Since $\lim_{t\to\infty} g(t) = 0$, we can again use l'Hôpital's rule to obtain

$$\lim_{t\to\infty}\left(\int_t^\infty |g'(s)|\,ds\right)^{\frac{\lambda+3}{\lambda-1}} g^{-1}(t)$$

$$= \frac{\lambda+3}{\lambda-1}\lim_{t\to\infty}\left(\int_t^\infty |g'(s)|\,ds\right)^{\frac{4}{\lambda-1}} \frac{|g'(t)|}{g'(t)} = 0.$$

Hence, from l'Hôpital's rule and (3.4.14), we have

$$\lim_{t\to\infty} r^{\frac{2}{\lambda+3}}(t)\left(\int_t^\infty |g'(s)|\,ds\right)^{\frac{2(\lambda+1)}{\lambda-1}}$$

$$= -\frac{(\lambda+1)(\lambda+3)}{\lambda-1}\lim_{t\to\infty}\frac{\left(\int_t^\infty |g'(s)\,ds\right)^{\frac{\lambda+3}{\lambda-1}}|g'(t)|}{g(t)} = 0,$$

so (3.3.4) holds. The conclusions then follow from Theorems 3.4.1 and 3.4.2. □

The following corollary is an immediate consequence of Theorem 3.4.5.

Corollary 3.4.1. *Let $r' > 0$ and (3.4.14) hold. Then equation (3.2.1) is of the strong nonlinear limit-point type if and only if (3.4.4) holds.*

3.5 Sub-Half-Linear and Super-Half-Linear Equations

Beginning with this section we will be discussing equations with the p-Laplacian. Now that we have introduced the strong nonlinear limit-point property of solutions, we can now introduce the strong nonlinear limit-circle property. So consider the equation

$$\left(a(t)|y'|^{p-1}y'\right)' + r(t)|y|^\lambda \operatorname{sgn} y = 0, \tag{3.5.1}$$

where $p > 0$, $\lambda > 0$, $a \in C^1(\mathbb{R}_+)$, $a^{1/p}r \in AC^1_{\text{loc}}(\mathbb{R}_+)$, $a(t) > 0$, and $r(t) > 0$. As was the case with equation (2.2.1), all nontrivial solutions of (3.5.1) are defined on $\mathbb{R}_+$ and are nontrivial in any neighborhood of ∞ (see, for example, [13, Theorem 1], or Lemma 2.3.1 above).

Because of the form of equation (3.5.1), we let

$$y^{[1]}(t) = a(t)|y'(t)|^{p-1}y'(t),$$

define the function $R : \mathbb{R}_+ \to \mathbb{R}_+$ by

$$R(t) = a^{1/p}(t)r(t),$$

and let δ denote the constant

$$\delta = \frac{p+1}{p}.$$

The definition of a strong nonlinear limit-point solution takes the following form for equation (3.5.1).

Definition 3.5.1. A solution y of (3.5.1) is said to be of the strong nonlinear limit-point type if

$$\int_0^\infty |y(\sigma)|^{\lambda+1}\, d\sigma = \infty$$

and

$$\int_0^\infty \frac{|y^{[1]}(\sigma)|^\delta}{R(\sigma)}\, d\sigma = \infty.$$

Equation (3.5.1) is said to be of the strong nonlinear limit-point type if every nontrivial solution is of the strong nonlinear limit-point type.

As is noted above, every nontrivial solution of (3.5.1) is proper and so such solutions do exist. We now introduce the following definition for a strong nonlinear limit-circle solution of equation (3.5.1).

Definition 3.5.2. A solution y of (3.5.1) is said to be of the strong nonlinear limit-circle type if

$$\int_0^\infty |y(\sigma)|^{\lambda+1}\, d\sigma < \infty$$

and

$$\int_0^\infty \frac{|y^{[1]}(\sigma)|^\delta}{R(\sigma)}\, d\sigma < \infty.$$

Equation (3.5.1) is said to be of the strong nonlinear limit-circle type if every solution is of the strong nonlinear limit-circle type.

It will be convenient to define the following constants:

$$\alpha = \frac{p+1}{(\lambda+2)p+1}, \quad \beta = \frac{(\lambda+1)p}{(\lambda+2)p+1}, \quad \gamma = \frac{p+1}{p(\lambda+1)},$$

$$\beta_1 = \frac{(\lambda+2)p+1}{(\lambda+1)(p+1)}, \quad \omega = \frac{(\lambda+1)(p+1)}{\lambda-p} \quad \text{for } \lambda \neq p, \quad \text{and}$$

$$\omega_1 = \frac{1}{\omega} = \frac{\lambda-p}{(\lambda+1)(p+1)}.$$

Note that $\alpha = 1-\beta$. We define the function $g : \mathbb{R}_+ \to \mathbb{R}_+$ by

$$g(t) = -\frac{a^{1/p}(t)R'(t)}{R^{\alpha+1}(t)},$$

and for any solution $y : \mathbb{R}_+ \to \mathbb{R}_+$ of (3.5.1), we let

$$\begin{aligned} F(t) &= R^{\beta}(t)\left[\frac{a(t)}{r(t)}|y'(t)|^{p+1} + \gamma|y(t)|^{\lambda+1}\right] \\ &= R^{\beta}(t)\left(\frac{|y^{[1]}(t)|^{\delta}}{R(t)} + \gamma|y(t)|^{\lambda+1}\right). \end{aligned} \tag{3.5.2}$$

Note that $F > 0$ on $\mathbb{R}_+$ for every nontrivial solution of (3.5.1) (see [13, Theorem 1]). In the remainder of this section, as well as in Sections 3.6–3.8, and without additional mention, we assume that

$$\lim_{t\to\infty} g(t) = 0, \quad \int_0^{\infty} |g'(\sigma)|\, d\sigma < \infty. \tag{3.5.3}$$

Also in the remainder of this section, and in Sections 3.6–3.8, we assume that

$$\int_0^{\infty} \left(a^{-\frac{1}{p}}(\sigma) + r(\sigma)\right) d\sigma = \infty. \tag{3.5.4}$$

It should not really come as a surprise that an integral condition like (3.5.3) must hold when discussing the limit-point/limit-circle behavior of solutions. In fact, if we let $p = 1$, $\lambda = 1$, and $a(t) \equiv 1$, then

$$\int_0^{\infty} |g'(\sigma)|\, d\sigma = \int_0^{\infty} \left|\frac{r''(\sigma)}{r^{3/2}(\sigma)} - \frac{3}{2}\frac{(r'(\sigma))^2}{r^{5/2}(\sigma)}\right| d\sigma < \infty,$$

which is essentially the well-known condition of Dunford and Schwartz [72, p. 1414]

$$\int_0^{\infty} \left|\frac{r''(\sigma)}{r^{3/2}(\sigma)} - \frac{5}{4}\frac{(r'(\sigma))^2}{r^{5/2}(\sigma)}\right| d\sigma < \infty$$

for second-order linear equations. For a discussion of the relationship between the linear and nonlinear limit-point/limit-circle properties, we refer the reader to the book of Bartušek, Došlá, and Graef [25].

3.6 Preliminary Lemmas

In this section, we gather together some lemmas that are needed in the proofs of our main results in the remainder of this chapter. Some of them also provide insight into the behavior of solutions of equation (3.5.1). Recall that conditions (3.5.3) and (3.5.4) are assumed to hold here.

Note: It is important to note here that unless specifically indicated in their hypotheses, none of the lemmas in this section depend on a relationship between λ and p, that is, they holds in both the sub-half-linear ($\lambda \leq p$) and super-half-linear ($\lambda \geq p$) cases.

Lemma 3.6.1.

(i) *If* $\lambda = p$, *then for every nontrivial solution* y *of* (3.5.1), *we have*

$$\lim_{t\to\infty} F(t) = C \in (0,\infty). \tag{3.6.1}$$

(ii) *If* $\lambda < p$, *then for every nontrivial solution* y *of* (3.5.1), *either* (3.6.1) *holds, or* $g(t) \not\equiv 0$ *on* $\mathbb{R}_+$ *and*

$$F(t) \geq M \left(\int_t^\infty |g'(\sigma)|\, d\sigma \right)^{\omega}, \tag{3.6.2}$$

where M *is a positive constant depending on* λ *and* p.

Proof. Parts (i) and (ii) follow from Theorems 17.6 and 17.5 in [138], respectively. □

Apart from the values of λ, the behaviors of F in (3.6.1) and (3.6.2) are mutually exclusive.

Lemma 3.6.2. *Let* $\lambda > p$ *and* y *be a nontrivial solution of equation* (3.5.1).

(i) *Either*

$$\lim_{t\to\infty} F(t) = C \in (0,\infty), \tag{3.6.3}$$

or

$$0 < F(t) \leq M \left(\int_t^\infty |g'(\sigma)|\, d\sigma \right)^{\omega}, \tag{3.6.4}$$

where M *is a positive constant depending on* λ *and* p.

(ii) *Moreover, if*

$$\liminf_{t\to\infty} R^{-\beta}(t)\left(\int_t^\infty |g'(\sigma)|\,d\sigma\right)^{\omega}\exp\left\{\int_0^t \left(R^{-1}(\sigma)\right)'_{-} R(\sigma)\,d\sigma\right\}=0, \tag{3.6.5}$$

then in fact (3.6.3) *holds.*

Proof. Part (i) is a consequence of [138, Theorem 17.4]. To prove (ii), let y be a nontrivial solution of (3.5.1) and suppose, to the contrary, that (3.6.4) holds. Let $Z(t) = F(t)R^{-\beta}(t)$ for $t \in R_+$; then (3.5.2) implies

$$Z'(t) = \left(\frac{1}{R(t)}\right)'\left(y^{[1]}(t)\right)^{\delta} \geq -\left(R^{-1}(t)\right)'_{-} R(t)\,Z(t)$$

for $t \in \mathbb{R}_+$. Hence,

$$Z(t) \geq Z(0)\exp\left\{-\int_0^t \left(R^{-1}(\sigma)\right)'_{-} R(\sigma)\,d\sigma\right\}$$

for $t \in \mathbb{R}_+$. From this and (3.6.4), we have

$$Z(0)\exp\left\{-\int_0^t \left(R^{-1}(\sigma)\right)'_{-} R(\sigma)\,d\sigma\right\}$$
$$\leq Z(t) \leq MR^{-\beta}(t)\left(\int_t^\infty |g'(\sigma)|\,d\sigma\right)^{\omega}$$

for $t \in \mathbb{R}_+$, which contradicts (3.6.5). □

Remark 3.6.1. Similar to what we had with the previous lemma, the behaviors of F in (3.6.3) and (3.6.4) are mutually exclusive.

Our next lemma actually follows from Lemma 2.3.4.

Lemma 3.6.3. *If* $\lambda < p$ *and*

$$\liminf_{t\to\infty} R^{\beta}(t)\left(\int_t^\infty |g'(\sigma)|\,ds\right)^{-\omega}\exp\left\{\int_0^t \left(R^{-1}(\sigma)\right)'_{+} R(\sigma)\,d\sigma\right\}=0, \tag{3.6.6}$$

then (3.6.1) *holds for every nontrivial solution* y *of* (3.5.1).

In addition to yielding useful expressions for y and $y^{[1]}$, the following lemma gives a characterization of the oscillatory solutions of (3.5.1).

Lemma 3.6.4. *For every nontrivial solution y of equation* (3.5.1), *there exists a positive function $\varphi \in C^1(\mathbb{R}_+)$ such that*

$$\begin{aligned} y(t) &= R^{-\frac{\beta}{\lambda+1}}(t)F^{\frac{1}{\lambda+1}}(t)w\left(\varphi(t)\right), \\ y^{[1]}(t) &= R^{\frac{\beta}{\lambda+1}}(t)F^{\frac{p}{p+1}}(t)w^{[1]}\left(\varphi(t)\right), \end{aligned} \tag{3.6.7}$$

where $w^{[1]}(s) = |w'(s)|^{p-1}w'(s)$ and w is a periodic solution of the problem

$$\left(|w'|^{p-1}w'\right)' + |w|^{\lambda+1}w = 0, \quad w(0) = \gamma^{\frac{-1}{\lambda+1}}, \quad w'(0) = 0.$$

Moreover,

$$\varphi'(t) = a^{-\frac{1}{p}}(t)R^{\alpha}(t)\left[F^{\omega_1}(t) - \frac{1}{\lambda+1}g(t)w\left(\varphi(t)\right)w^{[1]}\left(\varphi(t)\right)\right], \tag{3.6.8}$$

and y is oscillatory if and only if $\lim_{t\to\infty}\varphi(t) = \infty$.

This result follows from [15, Lemma 2 and Theorem 1]. Moreover, it follows from (14) in [15] that

$$\max_{t\in\mathbb{R}_+}|w(t)| = \gamma^{-\frac{1}{\lambda+1}} \tag{3.6.9}$$

and

$$|w^{[1]}(t)| = 1 \text{ at all relative extrema of } w^{[1]}. \tag{3.6.10}$$

The next lemma shows that equation (3.5.1) always has an oscillatory solution and provides additional information about the behavior of the function F for nonoscillatory solutions.

Lemma 3.6.5. (i) *Equation* (3.5.1) *has an oscillatory solution.*
(ii) *If $\lambda \le p$, then every nonoscillatory solution y of* (3.5.1) *satisfies*

$$\lim_{t\to\infty} F(t) = \infty.$$

Proof. Part (i) follows from [15, Theorem 2]. To prove (ii), notice that Lemma 3.6.1(ii) implies $\lim_{t\to\infty} F(t) \in (0,\infty]$. Let y be a nonoscillatory solution and suppose, to the contrary, that

$$\lim_{t\to\infty} F(t) = C \in (0,\infty).$$

In view of the first part of (3.5.3), we then see that $\varphi'(t)$ is eventually positive, and since y is nonoscillatory, by Lemma 3.6.4 we have that $\varphi(t)$

is bounded. Since (3.5.4) holds, Lemma 6 in [15] implies

$$\int_0^\infty a^{-\frac{1}{p}}(\sigma)RR^\alpha(\sigma)\,d\sigma = \infty.$$

Thus, by l'Hôpital's rule, we have

$$\begin{aligned}
0 &= \lim_{t\to\infty}\frac{\varphi(t)}{\int_0^t a^{-1/p}(\sigma)R^\alpha(\sigma)\,d\sigma} = \lim_{t\to\infty}\frac{\varphi'(t)}{a^{-1/p}(t)R^\alpha(t)}\\
&= \lim_{t\to\infty}\left[F^{\omega_1}(t) - \frac{1}{\lambda+1}g(t)w\left(\varphi(t)\right)w^{[1]}\left(\varphi(t)\right)\right]\\
&= C^{\omega_1} \neq 0,
\end{aligned}$$

and this contradiction completes the proof of the lemma. □

Lemma 3.6.6. *Assume that*

$$\lim_{t\to\infty}\frac{a'(t)}{a^{1-\frac{\beta}{p}}(t)r^\alpha(t)} = 0, \tag{3.6.11}$$

$$\int_0^\infty R^{-\beta}(\sigma)\,d\sigma = \infty, \tag{3.6.12}$$

and let y be a nontrivial solution of (3.5.1) such that $\lim_{t\to\infty} F(t) = C \in (0,\infty)$. Then

$$\int_0^\infty |y(\sigma)|^{\lambda+1}\,d\sigma = \int_0^\infty \frac{|y^{[1]}(\sigma)|^\delta}{R(\sigma)}\,d\sigma = \infty. \tag{3.6.13}$$

Proof. Let y be a nontrivial solution of (3.5.1) such that $\lim_{t\to\infty} F(t) = C \in (0,\infty)$. By Lemma 3.6.5(ii), y is oscillatory. Moreover, there exists $t_0 \in \mathbb{R}_+$ such that

$$\frac{C}{2} \leq F(t) \leq 2C \quad \text{for } t \geq t_0. \tag{3.6.14}$$

From this and (3.5.2), we have

$$\begin{aligned}
\int_{t_0}^t \frac{|y^{[1]}(\sigma)|^\delta}{R(\sigma)}\,d\sigma + \gamma\int_{t_0}^t |y(\sigma)|^{\lambda+1}\,d\sigma &= \int_{t_0}^t F(\sigma)R^{-\beta}(\sigma)\,d\sigma\\
&\geq \frac{C}{2}\int_{t_0}^t R^{-\beta}(\sigma)\,d\sigma
\end{aligned} \tag{3.6.15}$$

for $t \geq t_0$. Now, (3.6.12) and (3.6.15) imply that either

$$\int_0^\infty |y(\sigma)|^{\lambda+1}\,d\sigma = \infty \quad \text{or} \quad \int_0^\infty \frac{|y^{[1]}(\sigma)|^\delta}{R(\sigma)}\,d\sigma = \infty. \tag{3.6.16}$$

Furthermore,

$$\int_{t_0}^t |y(\sigma)|^{\lambda+1}\,d\sigma = -\int_{t_0}^t \frac{y(\sigma)(y^{[1]}(\sigma))'}{r(\sigma)}\,d\sigma = -\frac{y(t)y^{[1]}(t)}{r(t)} + \frac{y(t_0)y^{[1]}(t_0)}{r(t_0)}$$
$$+\int_{t_0}^t \frac{|y^{[1]}(\sigma)|^\delta}{R(\sigma)}\,d\sigma + \int_{t_0}^t \left(\frac{1}{r(\sigma)}\right)' y(\sigma)y^{[1]}(\sigma)\,d\sigma. \tag{3.6.17}$$

In view of (3.6.7), (3.6.9), and (3.6.10), we have

$$\left|y(t)y^{[1]}(t)\right| \leq \gamma^{-\frac{1}{\lambda+1}} F^{\beta_1}(t) \leq \gamma^{-\frac{1}{\lambda+1}}(2C)^{\beta_1} := C_1, \quad t \geq t_0. \tag{3.6.18}$$

Moreover,

$$\int_{t_0}^t \left|\left(\frac{1}{r(\sigma)}\right)'\right| d\sigma = \int_{t_0}^t \left|\left(\frac{a^{\frac{1}{p}}(\sigma)}{R(\sigma)}\right)'\right| d\sigma$$
$$\leq \int_{t_0}^t \frac{|a'(\sigma)|}{p\,a(\sigma)r(\sigma)}\,d\sigma + \int_{t_0}^t |g(\sigma)|R^{-\beta}(\sigma)\,d\sigma, \tag{3.6.19}$$

and using l'Hôpital rule, (3.6.11), and (3.6.12), we obtain

$$\lim_{t\to\infty} \int_{t_0}^t \frac{|a'(\sigma)|}{a(\sigma)r(\sigma)}\,d\sigma \left(\int_{t_0}^t R^{-\beta}(\sigma)\,d\sigma\right)^{-1} = \lim_{t\to\infty} \frac{|a'(t)|}{a(t)r(t)} R^\beta(t)$$
$$= \lim_{t\to\infty} \frac{|a'(t)|}{a^{1-\frac{\beta}{p}}(t)r^\alpha(t)} = 0. \tag{3.6.20}$$

We will now prove that the first integral in (3.6.13) is divergent, so suppose

$$\int_0^\infty |y(\sigma)|^{\lambda+1}\,d\sigma < \infty. \tag{3.6.21}$$

Then (3.6.16) implies

$$\int_0^\infty \frac{|y^{[1]}(\sigma)|^\delta}{R(\sigma)}\,d\sigma = \infty. \tag{3.6.22}$$

From the fact that $\lim_{t\to\infty} g(t) = 0$ and that (3.6.15), (3.6.20), (3.6.21), and (3.6.22) hold, we can choose $T \geq t_0$ such that

$$|g(t)| \leq \frac{C}{4C_1}, \quad \int_{t_0}^{t} \frac{|y^{[1]}(\sigma)|^{\delta}}{R(\sigma)}\, d\sigma \geq \frac{C}{3}\int_{t_0}^{t} R^{-\beta}(\sigma)\, d\sigma$$

$$\text{and} \quad \frac{C_1}{p}\int_{t_0}^{t} \frac{|a'(\sigma)|}{a(\sigma)r(\sigma)}\, d\sigma \leq \frac{C}{24}\int_{t_0}^{t} R^{-\beta}(\sigma)\, d\sigma$$

for $t \geq T$. Hence, (3.6.17), (3.6.18) and (3.6.19) yield

$$\begin{aligned}\int_{t_0}^{t} |y(\sigma)|^{\lambda+1}\, d\sigma &\geq C_2 - \frac{y(t)y^{[1]}(t)}{r(t)} + \int_{t_0}^{t} \frac{|y^{[1]}(\sigma)|^{\delta}}{R(\sigma)}\, d\sigma \\ &\quad - \frac{C_1}{p}\int_{t_0}^{t} \frac{|a'(\sigma)|}{a(\sigma)r(\sigma)}\, d\sigma - \frac{C}{4}\int_{t_0}^{t} R^{-\beta}(\sigma)\, d\sigma \\ &\geq C_2 - \frac{y(t)y^{[1]}(t)}{r(t)} + \frac{C}{24}\int_{t_0}^{t} R^{-\beta}(\sigma)\, d\sigma \qquad (3.6.23)\end{aligned}$$

for $t \geq T$ and some constant C_2. Since y is oscillatory, if we take an increasing sequence $\{t_n\}_1^{\infty}$ of zeros of y with $t_1 \geq T$, then (3.6.23) contradicts (3.6.12). Thus,

$$\int_0^{\infty} |y(\sigma)|^{\lambda+1}\, dt = \infty.$$

We next prove the second integral in (3.6.13) is divergent. To the contrary, suppose that

$$\int_0^{\infty} \frac{|y^{[1]}(\sigma)|^{\delta}}{R(\sigma)}\, d\sigma < \infty; \qquad (3.6.24)$$

then, the first integral in (3.6.13) diverges. Similar to what we did above, in view of (3.6.13), (3.6.14), (3.6.20), (3.6.24), and the equality in (3.6.15), we can choose $T_1 \geq t_0$ so that

$$\begin{aligned}&|g(t)| \leq \frac{C}{12C_1}, \quad \int_{t_0}^{t} |y(\sigma)|^{\lambda+1}\, d\sigma \geq \frac{C}{3}\int_{t_0}^{t} R^{-\beta}(\sigma)\, d\sigma, \quad \text{and} \\ &\frac{C_1}{p}\int_{t_0}^{t} \frac{|a'(\sigma)|}{a(\sigma)r(\sigma)}\, d\sigma \leq \frac{C}{12}\int_{t_0}^{t} R^{-\beta}(\sigma)\, d\sigma\end{aligned} \qquad (3.6.25)$$

for $t \geq T_1$. Then (3.6.17), (3.6.18), (3.6.19), and (3.6.25) yield

$$\begin{aligned}\frac{C}{3}\int_{t_0}^{t} R^{-\beta}(\sigma)\,d\sigma &\leq \int_{t_0}^{t} |y(\sigma)|^{\lambda+1}\,d\sigma \\ &\leq C_3 - \frac{y(t)y^{[1]}(t)}{r(t)} \\ &\quad + \frac{C_1}{p}\int_{t_0}^{t}\frac{|a'(\sigma)|}{a(\sigma)r(\sigma)}\,d\sigma + \frac{C}{12}\int_{t_0}^{t} R^{-\beta}(\sigma)\,d\sigma \\ &\leq C_3 - \frac{y(t)y^{[1]}(t)}{r(t)} + \frac{C}{6}\int_{t_0}^{t} R^{-\beta}(\sigma)\,d\sigma \qquad (3.6.26)\end{aligned}$$

for $t \geq T_1$ and some constant C_3. Again taking an increasing sequence of zeros of y tending to ∞, (3.6.26) contradicts (3.6.12). □

Remark 3.6.2. A version of part (ii) of Lemma 3.6.5 can be proved for $\lambda > p$, but in that case, $\lim_{t\to\infty} F(t) = 0$.

Remark 3.6.3. From the proof of Lemma 3.6.6, we see that we could replace condition (3.6.11) by asking instead that

$$\lim_{t\to\infty}\int_{t_0}^{t}\frac{|a'(\sigma)|}{a(\sigma)r(\sigma)}\,d\sigma\left(\int_{t_0}^{t} R^{-\beta}(\sigma)\,d\sigma\right)^{-1} = 0. \qquad (3.6.27)$$

An easy modification of the proof of Lemma 3.6.6 also shows that (3.6.11) can be replaced by the condition

$$\int_0^{\infty}\left|\left(\frac{1}{r(\sigma)}\right)'\right|\,d\sigma < \infty. \qquad (3.6.28)$$

The following lemma gives some properties of the nonoscillatory solutions of equation (3.5.1).

Lemma 3.6.7. (i) *If*

$$\int_0^{\infty} a^{-\frac{1}{p}}(\sigma)\,d\sigma = \infty, \qquad (3.6.29)$$

then any nonoscillatory solution y of (3.5.1) satisfies $y(t)y'(t) > 0$ for all large t.

(ii) *If*

$$\int_0^{\infty} a^{-\frac{1}{p}}(\sigma)\,d\sigma < \infty, \qquad (3.6.30)$$

then there exists $R_0 > 0$ such that $R_0 \leq R(t)$ for $t \in \mathbb{R}_+$.

(iii) *If* $\lambda < p$, (3.6.2) *and* (3.6.30) *hold, and*

$$\lim_{t\to\infty} R^{\beta}(t)\left(\int_t^{\infty} |g'(\sigma)|\, d\sigma\right)^{-\omega} = 0, \tag{3.6.31}$$

then no nonoscillatory solution y *of* (3.5.1) *satisfies* $\lim_{t\to\infty} y(t) = C \in (-\infty, \infty)$.

Proof. Part (i) is straightforward. To prove (ii), first note that $\lim_{t\to\infty} g(t) = 0$ implies that there exists $t_0 \in \mathbb{R}_+$ such that

$$-1 \le -g(t) = \frac{a^{1/p}(t)R'(t)}{R^{1+\alpha}(t)},$$

for $t \ge t_0$. Thus,

$$-a^{-1/p}(t) \le \frac{R'(t)}{R^{1+\alpha}(t)}$$

for $t \ge t_0$, and so

$$-\infty < -\int_0^{\infty} a^{-1/p}(\sigma)\, d\sigma \le \frac{1}{\alpha}\left(R^{-\alpha}(t_0) - R^{-\alpha}(t)\right) \tag{3.6.32}$$

for $t \ge t_0$. If $\liminf_{t\to\infty} R(t) = 0$, then (3.6.32) gives us a contradiction. Hence, there exists $R_0 > 0$ such that $R_0 \le R(t)$ for $t \in \mathbb{R}_+$.

Finally, to prove (iii), suppose that y satisfies

$$\lim_{t\to\infty} |y(t)| = C \in [0, \infty). \tag{3.6.33}$$

By Lemma 3.6.4 and inequality (3.6.2),

$$\begin{aligned}|y(t)|^{\lambda+1} &= R^{-\beta}(t)F(t)\left|w^{\lambda+1}(\varphi(t))\right| \\ &\ge MR^{-\beta}(t)\left(\int_t^{\infty} |g'(\sigma)|\, d\sigma\right)^{\omega}\left|w^{\lambda+1}(\varphi(t))\right|.\end{aligned}$$

Hence, (3.6.33) and the assumptions in this part of the lemma imply $\lim_{t\to\infty} w^{\lambda+1}(\varphi(t)) = 0$.

From Lemma 3.6.5, we have that

$$\lim_{t\to\infty} F(t) = \infty.$$

In view of (3.5.3), (3.6.9), and (3.6.10), we see from (3.6.8) that $\varphi(t)$ is increasing. Since y is nonoscillatory, $\varphi(t)$ is bounded. It then follows from the differential equation for w that $w(\varphi(t)) \to 0$ implies $w^{[1]}(\varphi(t))$ approaches a relative extrema, so (3.6.10) implies

$$\lim_{t\to\infty} |w'(\varphi(t))| = 1. \tag{3.6.34}$$

By part (ii) of this lemma, $R(t) \geq R_0 > 0$ for $t \in \mathbb{R}_+$, so (3.6.34), the second equality in (3.6.7), and (3.6.2) imply

$$\begin{aligned} |y'(t)|^{p+1} &= a^{-\frac{p+1}{p}}(t)R^{\alpha}(t)F(t)\,|w'(\varphi(t))|^{p+1} \\ &\geq a^{-\frac{p+1}{p}}(t)R^{-\beta}(t)\left(\int_t^{\infty} |g'(\sigma)|\,d\sigma\right)^{\omega} |w'(\varphi(t))|^{p+1} R_0 M. \end{aligned}$$

Let $t_0 \in \mathbb{R}_+$ be such that $y \neq 0$ and $|w'(\varphi(t))| \geq \frac{1}{2}$ on $[t_0, \infty)$. Then,

$$\begin{aligned} |y(t) - y(t_0)| \geq \left(\frac{R_0 M}{2}\right)^{\frac{1}{p+1}} & \min_{t_0 \leq t \leq \infty} \left[\left(R^{-\beta}(t)\left(\int_t^{\infty} |g'(\sigma)|\,d\sigma\right)^{\omega}\right)^{\frac{1}{p+1}}\right] \\ & \times \int_{t_0}^{t} a^{-1/p}(\sigma)\,d\sigma \to \infty \end{aligned}$$

as $t \to \infty$. This contradiction to (3.6.33) completes the proof of the lemma. □

The next two lemmas require that R be small in some sense.

Lemma 3.6.8. *Assume that $\lambda < p$ and that (3.6.2) and (3.6.31) hold. Then any oscillatory solution y of equation (3.5.1) is unbounded. Moreover, if*

$$a^{-1/p}(t)R^{\frac{1}{p+1}}(t) \leq B_1 < \infty \tag{3.6.35}$$

for $t \in \mathbb{R}_+$, then

$$\int_0^{\infty} |y(\sigma)|^{\lambda+1}\,d\sigma = \infty.$$

Proof. Let y be an oscillatory solution of (3.5.1). Then from (3.6.2) and Lemma 3.6.4, a function φ exists such that, if $\{t_k\}_1^{\infty}$ is an increasing sequence of relative extrema of y, then

$$\begin{aligned} \left|y^{\lambda+1}(t_k)\right| &= R^{-\beta}(t_k)F(t_k)\left|w^{\lambda+1}(\varphi(t_k))\right| \\ &\geq MR^{-\beta}(t_k)\left(\int_{t_k}^{\infty} |g'(\sigma)|\,d\sigma\right)^{\omega}\left|w^{\lambda+1}(\varphi(t_k))\right|. \end{aligned} \tag{3.6.36}$$

At the same time, (3.6.7) implies that y has a local extrema if and only if w has a local extrema, and by (3.6.9), we have $w^{\lambda+1}(\varphi(t_k)) = \gamma^{-1}$. From this and (3.6.31), we obtain that $\lim_{k\to\infty}|y(t_k)| = \infty$, and so y is unbounded.

Now let $\{\tau_k\}_1^\infty$ be a sequence such that $\tau_k < t_k < \tau_{k+1}$, $y(\tau_k) = 0$, and $y(t) > 0$ on (τ_k, t_k); note that $y'(t_k) = 0$, $k = 1, 2, \ldots$. From (3.6.2), we have

$$F^{\omega_1}(t) \le M^{\omega_1} \int_t^\infty |g'(\sigma)|\, d\sigma,$$

so from (3.6.8), we obtain

$$\varphi'(t) \le a^{-\frac{1}{p}}(t) R^\alpha(t) \left[C_1 \int_t^\infty |g'(\sigma)|\, d\sigma - \frac{1}{\lambda+1} g(t) w(\varphi(t))\, w^{[1]}(\varphi(t)) \right],$$

where $C_1 = M^{\omega_1}$. Now (3.5.3) implies that g is of bounded variation on R_+, and since $\lim_{t\to\infty} g(t) = 0$, we see that $|g(t)| \le \int_t^\infty |g'(\sigma)|\, d\sigma$ for $t \in \mathbb{R}_+$. From this, (3.6.9), and (3.6.10), we have

$$\varphi'(t) \le C_2 a^{-1/p}(t) R^\alpha(t) \int_t^\infty |g'(\sigma)|\, d\sigma, \quad C_2 = C_1 + \gamma^{-\frac{1}{\lambda+1}}/(\lambda+1). \tag{3.6.37}$$

Then, from (3.6.31), (3.6.35), (3.6.36) with t_k replaced by t, and (3.6.37), we have

$$\begin{aligned}
&\int_{\tau_k}^{t_k} |y(\sigma)|^{\lambda+1}\, d\sigma \\
&\ge M \max_{\tau_k \le t \le t_k} \left[R^{-\beta}(t) \left(\int_t^\infty |g'(\sigma)|\, d\sigma \right)^\omega a^{\frac{1}{p}}(t) R^{-\alpha}(t) \right. \\
&\quad \left. \times \left(\int_t^\infty |\, g'(\sigma)\, d\sigma \right)^{-1} \right] \int_{\tau_k}^{t_k} a^{-\frac{1}{p}}(s) R^\alpha(s) \int_s^\infty |g'(\sigma)|\, d\sigma w^{\lambda+1}(\varphi(s))\, ds \\
&\ge \frac{M}{C_2} \max_{\tau_k \le t \le t_k} \left\{ \left[R^{-\beta}(t) \left(\int_t^\infty |g'(\sigma)|\, d\sigma \right)^\omega \right]^{1-\frac{1}{\omega}} a^{\frac{1}{p}}(t) R^{-\frac{1}{p+1}} \right\} \\
&\quad \times \int_{\tau_k}^{t_k} w^{\lambda+1}(\varphi(\sigma))\, \varphi'(\sigma)\, d\sigma \to \infty
\end{aligned}$$

as $k \to \infty$ since $\int_{\varphi(\tau_k)}^{\varphi(t_k)} w^{\lambda+1}(z)\,dz = \text{const.} > 0$ for any $k \in \{1, 2, \ldots\}$ due to the periodicity of w. This completes the proof of the lemma. □

Next is a lemma that is concerned with unbounded solutions of equation (3.5.1).

Lemma 3.6.9. *Let y be an unbounded solution of* (3.5.1) *and assume that there is a positive constant B_2 such that*

$$R'_{-}(t)/R(t) < B_2 \tag{3.6.38}$$

for $t \in \mathbb{R}_+$. Then

$$\int_0^\infty \frac{|y^{[1]}(\sigma)|^\delta}{R(\sigma)}\,d\sigma = \infty.$$

Proof. Let $Z(t) = F(t)R^{-\beta}(t)$. Then (3.5.2) and the fact that y is unbounded imply Z is unbounded as well. On the other hand, suppose that

$$\int_0^\infty \frac{|y^{[1]}(\sigma)|^\delta}{R(\sigma)}\,d\sigma = C < \infty.$$

Then, it is easy to see that $Z'(t) = -\frac{R'(t)}{R^2(t)}\left|y^{[1]}(t)\right|^\delta$ and

$$Z(t) = Z(0) + \int_0^\infty Z'(s)\,ds = Z(0) - \int_0^t \frac{R'(\sigma)}{R^2(\sigma)}\left|y^{[1]}(\sigma)\right|^\delta d\sigma$$

$$\leq Z(0) + \int_0^t \frac{R'_{-}(\sigma)}{R^2(\sigma)}\left|y^{[1]}(\sigma)\right|^\delta d\sigma \leq Z(0) + B_2C < \infty.$$

This contradiction proves the lemma. □

The next lemma gives a useful representation for the function F.

Lemma 3.6.10. *Let y be a nontrivial solution of* (3.5.1). *Then*

$$F(t) = R^{-\alpha}(t)\left[K_1 + K_2\int_0^t R'(\sigma)\,|y(\sigma)|^{\lambda+1}\,d\sigma\right]$$

with $K_1 = R^\alpha(0)F(0) > 0$ and $K_2 = \alpha(1+\gamma) > 0$.

Proof. Let $C_1 = 1 + \gamma$; then (3.5.2) yields

$$F'(t) = \beta R^{-\alpha}(t)R'(t)\left(\frac{|y^{[1]}(t)|^{\delta}}{R(t)} + \gamma|y(t)|^{\lambda+1}\right) + R^{\beta}(t)\left(R^{-1}(t)\right)'\left|y^{[1]}(t)\right|^{\delta}$$

$$= -\alpha\frac{R'(t)}{R^{\alpha}(t)}\left(\frac{|y^{[1]}(t)|^{\delta}}{R(t)} - |y(t)|^{\lambda+1}\right)$$

$$= -\alpha\frac{R'(t)}{R^{\alpha}(t)}\left(\frac{F(t)}{R^{\beta}(t)} - C_1|y(t)|^{\lambda+1}\right)$$

$$= -\alpha\frac{R'(t)}{R(t)}F(t) + \alpha C_1\frac{R'(t)}{R^{\alpha}(t)}\,|y(t)|^{\lambda+1}.$$

Integrating, we obtain

$$F(t) = \exp\left\{-\alpha\int_0^t \frac{R'(\sigma)}{R(\sigma)}\,d\sigma\right\}\left[\int_0^t \alpha C_1\frac{R'(\sigma)}{R^{\alpha}(\sigma)}\,|y(\sigma)|^{\lambda+1}\right.$$
$$\left.\times\exp\left\{\alpha\int_0^{\sigma}\frac{R'(s)}{R(s)}\,ds\right\}\,d\sigma + F(0)\right],$$

or

$$F(t) = R^{-\alpha}(t)\left[K_1 + K_2\int_0^t R'(\sigma)|y(t)|^{\lambda+1}\,d\sigma\right]$$

for $t \in \mathbb{R}_+$. □

Our final lemma is taken from [89].

Lemma 3.6.11 ([89, Theorem 7]). *If* (3.6.29) *holds, then there exists a nontrivial solution* y *of* (3.5.1) *tending to a finite nonzero limit as* $t \to \infty$ *if and only if*

$$\int_0^{\infty}\left(\frac{1}{a(\sigma)}\int_{\sigma}^{\infty} r(s)ds\right)^{\frac{1}{p}}\,d\sigma < \infty. \tag{3.6.39}$$

3.7 Strong Nonlinear Limit-Point/Limit-Circle Results for Sub-Half-Linear Equations

In this section we prove strong nonlinear limit-point and strong nonlinear limit-circle results for sub-half-linear equations. Again, throughout this section, conditions (3.5.3) and (3.5.4) are assumed to hold.

Theorem 3.7.1. *Suppose that (3.6.11) holds and that either $\lambda = p$, or $\lambda < p$ and (3.6.6) holds.*

(a) *The following statements are equivalent:*

(i) *Equation (3.5.1) is of the nonlinear limit-circle type.*

(ii)

$$\int_0^\infty R^{-\beta}(\sigma)\,d\sigma < \infty. \tag{3.7.1}$$

(iii) $\int_0^\infty \frac{|y^{[1]}(\sigma)|^\delta}{R(\sigma)}\,d\sigma < \infty$ *for all solutions of (3.5.1).*

That is, equation (3.5.1) is of the strong nonlinear limit-circle type if and only if (3.7.1) holds.

(b) *The following statements are equivalent:*

(iv) *Every nontrivial solution of (3.5.1) is of the nonlinear limit-point type.*

(v)

$$\int_0^\infty R^{-\beta}(\sigma)\,d\sigma = \infty. \tag{3.7.2}$$

(vi) $\int_0^\infty \frac{|y^{[1]}(\sigma)|^\delta}{R(\sigma)}\,d\sigma = \infty$ *for every nontrivial solutions of (3.5.1).*

That is, equation (3.5.1) is of the strong nonlinear limit-point type if and only if (3.7.2) holds.

Proof. (a) Lemmas 3.6.1 and 3.6.3 imply that (3.6.1) holds for every nontrivial solution of (3.5.1). Therefore, F is bounded, say

$$C_1 < F(t) < C_2 \tag{3.7.3}$$

for some constants $C_1, C_2 > 0$ and all $t \in R_+$. From (3.5.2) and (3.7.3), we have

$$\infty > C_2 \int_0^\infty R^{-\beta}(\sigma)\,d\sigma \ge \int_0^\infty F(\sigma)R^{-\beta}(\sigma)\,d\sigma$$
$$= \int_0^\infty \frac{|y^{[1]}(\sigma)|^\delta}{R(\sigma)}\,d\sigma + \gamma\int_0^\infty |y(\sigma)|^{\lambda+1}\,d\sigma \ge C_1\int_0^\infty R^{-\beta}(\sigma)\,d\sigma.$$

From this and Lemma 3.6.6, we see that the three conclusions in part (a) are equivalent.

(b) If (3.7.2) holds, then Lemma 3.6.6 implies that every nontrivial solution y of (3.5.1) satisfies

$$\int_0^\infty |y(\sigma)|^{\lambda+1}\, d\sigma = \infty \quad \text{and} \quad \int_0^\infty \frac{|y^{[1]}(\sigma)|^\delta}{R(\sigma)}\, d\sigma = \infty.$$

Hence, (v) implies both (iv) and (vi).

Suppose that either (iv) or (vi) holds. Then there exists a nontrivial solution y of (3.5.1) such that either

$$\int_0^\infty |y(\sigma)|^{\lambda+1}\, d\sigma = \infty \quad \text{or} \quad \int_0^\infty \frac{|y^{[1]}(\sigma)|^\delta}{R(\sigma)}\, d\sigma = \infty.$$

But from part (a), this implies (3.7.1) fails. This completes the proof of the theorem. □

Theorem 3.7.2. *Assume that* (3.6.11), (3.6.29), (3.6.35), *and* (3.6.38) *hold and there is a positive constant* R_1 *such that*

$$R(t) \le R_1 \tag{3.7.4}$$

for $t \in \mathbb{R}_+$. *Then every nontrivial solution of* (3.5.1) *satisfies*

$$\int_0^\infty |y(\sigma)|^{\lambda+1}\, d\sigma = \infty \quad \text{and} \quad \int_0^\infty \frac{|y^{[1]}(\sigma)|^\delta}{R(\sigma)}\, d\sigma = \infty, \tag{3.7.5}$$

that is, equation (3.5.1) *is of the strong nonlinear limit-point type.*

Proof. First note that the hypotheses of the theorem imply that conditions (3.6.12) and (3.6.31) hold. Let y be a nontrivial solution of (3.5.1). Then, by Lemma 3.6.1, either (3.6.1) or (3.6.2) holds. If (3.6.1) holds, the conclusion follows from Lemma 3.6.6.

Suppose (3.6.2) holds. In view of Lemma 3.6.1, we have $\lambda < p$. If y is an oscillatory solution, then the conclusion follows from Lemmas 3.6.8 and 3.6.9. If y is nonoscillatory, then Lemma 3.6.7 (i) implies $|y|$ is increasing for large t. If $\lim_{t\to\infty} |y(t)| = \infty$, then clearly the first integral in (3.7.5) diverges, and the rest of the statement follows from Lemma 3.6.9. If $\lim_{t\to\infty} |y(t)| = C \in (0,\infty)$, then again the first integral in (3.7.5) diverges, and (3.5.2) and (3.6.2) imply

$$\gamma C^{\lambda+1} + \frac{|y^{[1]}(t)|^\delta}{R(t)} \geqq F(t)R^{-\beta}(t) \ge R_1^{-\beta} M \left(\int_t^\infty |g'(\sigma)|\, d\sigma\right)^\omega \to \infty$$

as $t \to \infty$. Hence, $\lim_{t\to\infty} |y^{[1]}(t)|^\delta / R(t) = \infty$ and so the second integral in (3.7.5) diverges. This completes the proof of the theorem. □

Remark 3.7.1. It should be clear from the proof of Theorem 3.7.2 that condition (3.7.4) can be removed if we instead require that conditions (3.6.12) and (3.6.31) hold.

Our final theorem on sub-half-linear equations is also a strong nonlinear limit-point result.

Theorem 3.7.3. *Assume that* $\lambda < p$, (3.6.11), (3.6.30), *and* (3.6.38) *hold, and there is a positive constant* B_3 *such that*

$$R'(t) \le B_3 \tag{3.7.6}$$

for $t \in \mathbb{R}_+$. *If* (3.6.31) *holds, then any nontrivial solution* y *of* (3.5.1) *satisfies*

$$\int_0^\infty |y(\sigma)|^{\lambda+1}\, d\sigma = \infty \quad \text{and} \quad \int_0^\infty \frac{|y^{[1]}(\sigma)|^\delta}{R(\sigma)}\, d\sigma = \infty, \tag{3.7.7}$$

that is, equation (3.5.1) *is of the strong nonlinear limit-point type.*

Proof. First observe that (3.7.6) implies that (3.6.12) holds. Let y be a nontrivial solution of (3.5.1); then, by Lemma 3.6.1, either (3.6.1) or (3.6.2) holds.

If (3.6.1) holds, then the statement follows from Lemma 3.6.6, so assume (3.6.2) holds. If y is oscillatory, then Lemmas 3.6.8 and 3.6.9 imply the second integral in (3.7.7) diverges. To prove that the first integral in (3.7.7) diverges, first observe that Lemma 3.6.7(ii) implies $R(t) \ge R_0$ for $t \in \mathbb{R}_+$ and some $R_0 > 0$. Then, from Lemma 3.6.10 and inequality (3.6.2), we have

$$\begin{aligned} R^{-\beta}(t)\left(\int_t^\infty |g'(\sigma)|\, d\sigma\right)^\omega &\le R^{-\beta}(t)F(t)/M \\ &= R^{-1}(t)\left[K_1 + K_2\int_0^t R'(\sigma)\,|y(\sigma)|^{\lambda+1}\, d\sigma\right]/M \\ &\le R_0^{-1}\left[K_1 + K_2 B_3\int_0^t |y(\sigma)|^{\lambda+1}\, d\sigma\right]/M, \end{aligned}$$

and the desired conclusion follows from (3.6.31). Finally, if y is nonoscillatory, Lemma 3.6.7 (iii) implies $\lim_{t\to\infty}|y(t)| = \infty$, and the divergence of the first integral in (3.7.7) is immediate. The divergence of the second integral follows from Lemma 3.6.9. This completes the proof of the theorem. □

Remark 3.7.2. Alternate versions of Theorems 3.7.1, 3.7.2, and 3.7.3 hold if we replace condition (3.6.11) with either (3.6.27) or (3.6.28).

3.8 Strong Nonlinear Limit-Point/Limit-Circle Results for Super-Half-Linear Equations

In this section we are again considering equation (3.5.1), i.e.,

$$\left(a(t)|y'|^{p-1}y'\right)' + r(t)|y|^{\lambda} \operatorname{sgn} y = 0, \tag{3.8.1}$$

but this time with $\lambda \geq p$. We are also assuming that conditions (3.5.3) and (3.5.4) hold here.

Our first theorem concerns both the strong nonlinear limit-point and strong nonlinear limit-circle properties of solutions of equation (3.8.1)

Theorem 3.8.1. *Assume that* (3.6.5) *and* (3.6.11) *hold.*

(a) *The following statements are equivalent:*

(i) *Equation* (3.8.1) *is of the nonlinear limit-circle type.*

(ii)

$$\int_0^\infty R^{-\beta}(\sigma)\, d\sigma < \infty. \tag{3.8.2}$$

(iii) $\int_0^\infty \frac{|y^{[1]}(\sigma)|^\delta}{R(\sigma)}\, d\sigma < \infty$ *for all solutions of* (3.8.1).

That is, equation (3.8.1) *is of the strong nonlinear limit-circle type if and only if* (3.8.2) *holds.*

(b) *The following statements are equivalent:*

(iv) *Every nontrivial solution of* (3.8.1) *is of the nonlinear limit-point type.*

(v)

$$\int_0^\infty R^{-\beta}(\sigma)\, d\sigma = \infty. \tag{3.8.3}$$

(vi) $\int_0^\infty \frac{|y^{[1]}(\sigma)|^\delta}{R(\sigma)}\, d\sigma = \infty$ *for every nontrivial solutions of* (3.8.1).

That is, equation (3.8.1) *is of the strong nonlinear limit-point type if and only if* (3.8.3) *holds.*

Proof. (a) By Lemma 3.6.2, (3.6.3) holds for every nontrivial solution of (3.8.1). Therefore, F is bounded, say

$$C_1 < F(t) < C_2 \tag{3.8.4}$$

for some constants C_1, $C_2 > 0$ and all $t \in R_+$. Now (3.5.2) and (3.8.4) imply

$$\begin{aligned}\infty > C_2 \int_0^\infty R^{-\beta}(\sigma)\,d\sigma &\geq \int_0^\infty F(\sigma) R^{-\beta}(\sigma)\,d\sigma \\ &= \int_0^\infty \frac{|y^{[1]}(\sigma)|^\delta}{R(\sigma)}\,d\sigma + \gamma \int_0^\infty |y(\sigma)|^{\lambda+1}\,d\sigma \geq C_1 \int_0^\infty R^{-\beta}(\sigma)\,d\sigma.\end{aligned} \tag{3.8.5}$$

Lemma 3.6.6 then shows that the three conclusions in part (a) are equivalent.

(b) If (3.8.3) holds, then, by Lemma 3.6.6, every nontrivial solution y of (3.8.1) satisfies

$$\int_0^\infty |y(\sigma)|^{\lambda+1}\,d\sigma = \infty \quad \text{and} \quad \int_0^\infty \frac{|y^{[1]}(\sigma)|^\delta}{R(\sigma)}\,d\sigma = \infty.$$

Hence, (v) implies both (iv) and (vi).

Finally, suppose that either (iv) or (vi) holds. Then there exists a nontrivial solution y of (3.8.1) such that either

$$\int_0^\infty |y(\sigma)|^{\lambda+1}\,d\sigma = \infty \quad \text{or} \quad \int_0^\infty \frac{|y^{[1]}(\sigma)|^\delta}{R(\sigma)}\,d\sigma = \infty.$$

But from part (a), this implies that (3.8.2) fails. This completes the proof of the theorem. □

From the proof of Theorem 3.8.1, it is clear that we have the following sufficient condition for equation (3.8.1) to be of the strong nonlinear limit-circle type.

Theorem 3.8.2. *If* (3.8.2) *holds, then* (3.8.1) *is of the strong nonlinear limit-circle type.*

Proof. Part (i) of Lemma 3.6.2 shows that F is bounded from above, so the conclusion follows from (3.8.5). □

In the following corollary, we give criteria for the conclusions of Theorem 3.8.1 to hold that are somewhat easier to verify in that we replace condition (3.6.5) by conditions (3.8.6) and (3.8.7) below.

Corollary 3.8.1. *In addition to condition* (3.6.11), *assume that*

$$a^{\frac{1}{p}}(t)g'(t) \text{ is bounded}, \quad R'(t) \geq 0 \text{ on } \mathbb{R}_+, \tag{3.8.6}$$

and

$$p > \frac{\lambda - 1}{\lambda + 3}. \tag{3.8.7}$$

Then the conclusions of Theorem 3.8.1 *hold.*

Proof. It suffices to show that conditions (3.8.6) and (3.8.7) imply that condition (3.6.5) holds. Since R is nondecreasing, $\lim_{t\to\infty} R(t) = R_0 \in (0, \infty]$. Condition (3.6.5) then takes the form

$$\liminf_{t\to\infty} R^{-\beta}(t) \left(\int_t^\infty |g'(\sigma)|\, d\sigma \right)^\omega R(t) = 0. \tag{3.8.8}$$

If $R_0 < \infty$, then (3.6.5) holds, so suppose that $R_0 = \infty$. Note that $g(t)$ is nontrivial in any neighborhood of ∞. Since $\lim_{t\to\infty} g(t) = 0$ and condition (3.8.7) guarantees that $\omega - 2 > 0$, we can use l'Hôpital's rule to obtain

$$\lim_{t\to\infty} \frac{\left(\int_t^\infty |g'(\sigma)|\, d\sigma\right)^{\omega-1}}{g(t)} = -\lim_{t\to\infty} \frac{(\omega - 1)\left(\int_t^\infty |g'(\sigma)|\, d\sigma\right)^{\omega-2} |g'(t)|}{g'(t)} = 0. \tag{3.8.9}$$

l'Hôpital's rule, (3.8.6), and (3.8.9) then yield

$$\begin{aligned} &\lim_{t\to\infty} R^\alpha(t) \left(\int_t^\infty |g'(\sigma)|\, d\sigma \right)^\omega \\ &\quad = -\frac{\omega}{\alpha} \lim_{t\to\infty} \frac{|g'(t)| a^{\frac{1}{p}}(t)}{g(t)} \left(\int_t^\infty |g'(\sigma)|\, d\sigma \right)^{\omega-1} = 0, \end{aligned} \tag{3.8.10}$$

so (3.6.5) holds. □

Remark 3.8.1. For the important case where the functions a and r are powers of t, condition (3.6.11) is not restrictive in that it becomes exactly the same condition needed for the covering condition (3.5.3) to hold.

Remark 3.8.2. If we replace condition (3.6.11) with either (3.6.27) or (3.6.28), then alternate versions of Theorem 3.8.1 and Corollary 3.8.1 are obtained. In fact, if we use condition (3.6.28) in place of (3.6.11), then Theorem 3.8.1(a) includes a part of Theorem 2.4.1(ii) as a special case.

The following strong nonlinear limit point result is relatively easy to apply.

Corollary 3.8.2. *If $r'(t) \geq 0$ and $r''(t) \leq 0$, then the equation*

$$\left(|y'|^{p-1}y'\right)' + r(t)|y|^{\lambda} \operatorname{sgn} y = 0 \tag{3.8.11}$$

is of the strong nonlinear limit-point type.

Proof. It is easy to see that $g'(t) \geq 0$. Moreover, condition (3.6.5) becomes

$$\liminf_{t\to\infty} r^{-\beta}(t)|g(t)|^{\omega} r(t) = r^{\alpha}(t)\left(\frac{r'(t)}{r^{1+\alpha}(t)}\right)^{\omega}$$

$$= \left(\frac{r'(t)}{r^{1+\alpha}(t)}\right)^{\omega-1}\frac{r'(t)}{r(t)} = (-g(t))^{\omega-1}\frac{r'(t)}{r(t)} = 0$$

since $r'(t)/r(t)$ is bounded from above, $\omega - 1 > 0$, and (3.5.3) holds. The conclusion then follows from Theorem 3.8.1(b). □

Theorem 3.8.3. *Assume that* (3.6.29), (3.6.38), *and* (3.6.39) *hold,*

$$\lim_{t\to\infty} R(t) = 0, \tag{3.8.12}$$

and there is a constant $A_2 > 0$ such that

$$a(t) \geq A_2 \tag{3.8.13}$$

for $t \in \mathbb{R}_+$.

(a) *Every nontrivial solution y of* (3.8.1) *satisfies*

$$\int_0^{\infty} \left|y(\sigma)\right|^{\lambda+1} d\sigma = \infty. \tag{3.8.14}$$

(b) *In addition:*

(i) *if y is a bounded nonoscillatory solution of* (3.8.1), *then*

$$\int_0^{\infty} \frac{|y^{[1]}(\sigma)|^{\delta}}{R(\sigma)} d\sigma < \infty \tag{3.8.15}$$

if and only if

$$\int_0^{\infty} R^{-1}(\sigma)\left(\int_{\sigma}^{\infty} r(s)\, ds\right)^{\delta} d\sigma < \infty; \tag{3.8.16}$$

(ii) *for all other nontrivial solutions,*

$$\int_0^\infty \frac{|y^{[1]}(\sigma)|^\delta}{R(\sigma)} d\sigma = \infty.$$

Proof. Let y be an oscillatory solution of (3.8.1). We will first show that

$$y \text{ is unbounded on } \mathbb{R}_+. \tag{3.8.17}$$

Suppose, to the contrary, that $|y(t)| \le B_1$ for some $B_1 > 0$ and all $t \in \mathbb{R}_+$. Then y is a solution of the equation

$$\left(a(t)|v'|^{p-1}v\right)' + r(t)|y(t)|^{\lambda-p}|v|^p \operatorname{sgn} v = 0. \tag{3.8.18}$$

This is a half-linear equation and all of its nontrivial solutions are oscillatory by [71, Theorem 1.2.3].

Now Lemma 3.6.11 implies that equation (3.8.1) has a nonoscillatory solution $\bar{y}$ and, by Lemma 3.6.7 (i), $|\bar{y}|$ is eventually increasing; hence

$$|\bar{y}(t)| \ge B_1^{\lambda-p} \quad \text{for large } t. \tag{3.8.19}$$

Moreover, $\bar{y}$ is a solution of the nonlinear equation

$$\left(a(t)|v'|^{p-1}v\right)' + r(t)\,|\bar{y}(t)|^{\lambda-p}\,|v|^p \operatorname{sgn} v = 0. \tag{3.8.20}$$

But since

$$|y(t)|^{\lambda-p} \le B_1^{\lambda-p} \le |\bar{y}(t)|^{\lambda-p} \quad \text{for large } t,$$

a comparison result of Došlý and Řehák [71, Theorem 2.3.1] applied to equations (3.8.18) and (3.8.20) shows that equation (3.8.18) is nonoscillatory. This contradicts the fact that y is oscillatory and proves that (3.8.17) holds.

In view of (3.8.17) and Lemma 3.6.9, we have that $\int_0^\infty \frac{|y^{[1]}(\sigma)|^\delta}{R(\sigma)} d\sigma = \infty$ if y is either an oscillatory solution or an unbounded nonoscillatory one. So let us suppose that y is a bounded nonoscillatory solution on $\mathbb{R}_+$. Since $|y|$ is increasing for large t, we have

$$\lim_{t\to\infty} y(t) = B_2 \in (0, \infty)$$

for some constant B_2. (The case where $B_2 < 0$ can be handled in a similar fashion.) Then, there exists $t_0 \in \mathbb{R}_+$ such that

$$\frac{B_2}{2} \le y(t) \le B_2$$

for $t \ge t_0$ and $\lim_{t\to\infty} y'(t) = 0$. Since (3.8.12) holds, the second expression in (3.6.7) and part (i) of Lemma 3.6.2 imply $\lim_{t\to\infty} y^{[1]}(t) = 0$.

Integrating equation (3.8.1), we have

$$\left(\frac{B_2}{2}\right)^{\lambda} \int_t^{\infty} r(\sigma)\, d\sigma \le y^{[1]}(t) = \int_t^{\infty} r(\sigma) y^{\lambda}(\sigma)\, d\sigma \le B_2^{\lambda} \int_t^{\infty} r(\sigma)\, d\sigma$$

for $t \ge t_0$. Raising this inequality to the power δ, dividing by $R(t)$, and integrating again, we obtain

$$\left(\frac{B_2}{2}\right)^{\lambda\delta} \int_{t_0}^{\infty} R^{-1}(\sigma) \left(\int_{\sigma}^{\infty} r(s)\, ds\right)^{\delta} d\sigma \le \int_{t_0}^{\infty} \frac{|y^{[1]}(\sigma)|^{\delta}}{R(\sigma)}\, d\sigma$$

$$\le B_2^{\lambda\delta} \int_{t_0}^{\infty} R^{-1}(\sigma) \left(\int_{\sigma}^{\infty} r(s)\, ds\right)^{\delta} d\sigma.$$

Hence, (3.8.15) holds, and so part (b) is proved.

To prove part (a), first let y be a nonoscillatory solution. Then $|y|$ is eventually increasing by Lemma 3.6.7 (i) and so (3.8.14) holds in this case. Finally, let y be an oscillatory solution. Then, Lemmas 3.6.1 and 3.6.4, the boundedness of $w^{[1]}(\varphi(t))$, and (3.8.12) imply $\lim_{t\to\infty} y^{[1]}(t) = 0$. In view of (3.8.13), we have $\lim_{t\to\infty} y'(t) = 0$. Now let $\{\tau_k\}_{k=1}^{\infty}$ and $\{t_k\}_{k=1}^{\infty}$ be increasing sequences such that $\tau_k < t_k < \tau_{k+1}$, $y(\tau_k) = 0$, $y'(t_k) = 0$, and $y(t) > 0$ and $y'(t) > 0$ for $t \in (\tau_k, t_k)$, $k = 1, 2, \ldots$. Since $y'(t) \to 0$ as $t \to \infty$, y' is bounded on $\mathbb{R}_+$, and so there is a constant $B_3 > 0$ such that $|y'(t)| \le B_3$ on $\mathbb{R}_+$. Hence,

$$\int_{\tau_k}^{t_k} |y(\sigma)|^{\lambda+1}\, d\sigma \ge \int_{\tau_k}^{t_k} \frac{y^{\lambda+1}(\sigma) y'(\sigma)}{B_3}\, d\sigma = \frac{y^{\lambda+2}(t_k)}{B_3} \to \infty$$

as $t \to \infty$ since oscillatory solutions are unbounded. Thus, (3.8.14) holds, and this completes the proof of the theorem. □

Remark 3.8.3. Notice that Theorem 3.8.3(b) provides us with conditions under which equation (3.8.1) is of the nonlinear limit-point type but not of the strong nonlinear limit-point type.

It seems reasonable to ask about the relationship between conditions (3.6.39) and (3.8.16). If $a(t) = t^a$ and $b(t) = t^b$, then (3.6.39) holds if and only if $b < a - (p+1)$ while (3.8.16) holds if and only if $b < a - (2p+1)$. Thus, if $a - (2p+1) < b < a - (p+1)$, then (3.6.39) holds but (3.8.16) does not. On the other hand, if $a(t) = t^p e^{-t^2}$ and $r(t) = te^{-t^2}$, then (3.8.16) holds and (3.6.39) fails. That is, (3.6.39) and (3.8.16) are independent conditions.

Remark 3.8.4. If (3.6.30) holds, then Lemma 3.6.7(ii) implies $R(t) \geq R_0 > 0$ on $\mathbb{R}_+$. Thus, in view of condition (3.8.12), (3.6.30) cannot replace (3.6.29) in Theorem 3.8.3.

3.9 The Forced Equation

In this section, we consider the forced version of equation (3.5.1) above (also see (3.8.1)), namely,

$$\left(a(t)|y'|^{p-1}y'\right)' + r(t)|y|^\lambda \operatorname{sgn} y = e(t), \tag{3.9.1}$$

where $e \in C(\mathbb{R}_+)$, and $p > 0$, $\lambda > 0$, $a \in C^1(\mathbb{R}_+)$, $a^{\frac{1}{p}} r \in AC_{\text{loc}}(\mathbb{R}_+)$, $a(t) > 0$, and $r(t) > 0$ as before. As was the case for equation (3.5.1), the functions a, r, and e are smooth enough so that all nontrivial solutions are defined on $\mathbb{R}_+$ and are proper (for example, see Lemma 2.5.1). Moreover, if $e(t) \equiv 0$, all nontrivial solutions of (3.9.1) are nontrivial in any neighborhood of ∞ (for example, see [138, Theorem 9.4]).

The definitions of strong nonlinear limit-point and strong nonlinear limit-circle solutions of equation (3.9.1) are the same as they are for the unforced equation (3.5.1). In addition to the strong nonlinear limit-point and strong nonlinear limit-circle properties, we will also discuss other asymptotic properties of solutions of equation (3.9.1) such as boundedness, oscillation, and convergence to a limit. The strong nonlinear limit-point/limit-circle results in this section do not require that condition (3.5.4) holds, i.e., we do not ask that

$$\int_0^\infty (a^{-\frac{1}{p}}(\sigma) + r(\sigma))d\sigma = \infty \tag{3.9.2}$$

as we did in our discussion of the sub-half-linear and super-half linear unforced equations above.

A solution y of equation (3.9.1) is *proper* if it is nontrivial in any neighborhood of ∞. And a proper solution y of equation (3.9.1) is *oscillatory* if it has a sequence of zeros tending to ∞, and it is called *nonoscillatory* otherwise. As indicated at the beginning of this section, every nontrivial solution of (3.9.1) is proper.

In addition to the constants α, β, γ, β_1, ω, and ω_1 and the functions R and g defined in Section 3.5, we will need the constants

$$\beta_2 = \frac{p}{(\lambda+2)p+1}, \quad \gamma_1 = \alpha\gamma^{-\frac{1}{\lambda+1}}, \quad \delta_1 = \gamma^{-\frac{1}{\lambda+1}},$$
$$q = \max\left\{\frac{1}{p+1}, \beta_1\right\} < 1, \quad q_1 = \frac{1}{1-q}.$$

Notice that since $\lambda > p > 0$, we have $\beta_1 < 1$ here.

To simplify some of the notation in what follows, for any solution $y : \mathbb{R}_+ \to \mathbb{R}$ of (3.9.1), we let

$$V(t) = \frac{|y^{[1]}(t)|^\delta}{R(t)} + \gamma|y(t)|^{\lambda+1} = \frac{a(t)}{r(t)}|y'(t)|^{p+1} + \gamma|y(t)|^{\lambda+1}$$

so that

$$F(t) = R^\beta(t)\, V(t). \tag{3.9.3}$$

We also define

$$A_1(t) = \left(\max_{0\le s\le t}|g(s)| + 1\right)\int_0^t R^{-\beta_2}(s)|e(s)|\, ds, \tag{3.9.4}$$

$$A(t) = \int_0^t |g'(s)|\, ds + 1 + \left(\max_{0\le s\le t}|g(s)| + 1\right)\int_0^t R^{-\beta_2}(s)|e(s)|\, ds, \tag{3.9.5}$$

and

$$G(t) = F(t)\, A^{-q_1}(t). \tag{3.9.6}$$

We will be making use of the conditions

$$\int_0^\infty R^{-\beta_2}(\sigma)|e(\sigma)|\, d\sigma < \infty, \tag{3.9.7}$$

and

$$\int_t^\infty R^{-\beta_2}(\sigma)|e(\sigma)|\, d\sigma \le K\int_t^\infty |g'(\sigma)|\, d\sigma \quad \text{for large } t, \tag{3.9.8}$$

where $K > 0$ is a constant. Observe that condition (3.9.8) implies (3.9.7).

In order to deal with the forced equation, we will need to begin with a couple of lemmas.

Lemma 3.9.1. *Let y be a solution of equation* (3.9.1). *Then*

$$|y(t)| \le \gamma^{-\frac{1}{\lambda+1}} R^{-\beta_2}(t) F^{\frac{1}{\lambda+1}}(t) \quad \text{and} \quad |y^{[1]}(t)| \le R^{\beta_2}(t) F^{\frac{p}{p+1}}(t) \tag{3.9.9}$$

for $t \in \mathbb{R}_+$, and for $0 \le \tau < t$, we have

$$\begin{aligned} F(t) &= F(\tau) - \alpha g(\tau) y(\tau) y^{[1]}(\tau) + \alpha g(t) y(t) y^{[1]}(t) \\ &\quad - \alpha \int_\tau^t g'(s) y(s) y^{[1]}(s)\, ds \\ &\quad + \int_\tau^t \left[\delta R^{-\alpha}(s) |y^{[1]}(s)|^{\frac{1}{p}} \operatorname{sgn} y^{[1]}(s) - \alpha g(s) y(s)\right] e(s)\, ds \end{aligned} \tag{3.9.10}$$

and

$$\begin{aligned} &\left| \int_\tau^t \left[\delta R^{-\alpha}(s) |y^{[1]}(s)|^{\frac{1}{p}} \operatorname{sgn} y^{[1]}(s) - \alpha g(s) y(s)\right] e(s)\, ds \right| \\ &\qquad \le K_1 \int_\tau^t R^{-\beta_2}(s) \left(F^{\frac{1}{p+1}}(s) + F^{\frac{1}{\lambda+1}}(s)\right) |e(s)|\, ds, \end{aligned} \tag{3.9.11}$$

with $K_1 = \delta + \gamma_1 \sup_{t \in \mathbb{R}_+} |g(t)|$.

Proof. Inequalities (3.9.9) are immediate, (3.9.10) was proved as Lemma 2.5.1(iii) and (3.9.11) follows from inequality (2.5.10). □

Although this is for the forced equation, notice the similarities between Lemmas 3.6.1 and 3.6.2 and the following lemma.

Lemma 3.9.2. *Let* (3.5.3) *and* (3.9.8) *hold and let y be a solution of equation* (3.9.1).

(i) *If $\lambda > p$, then either there is a constant $C \in (0, \infty)$ such that*

$$\lim_{t \to \infty} F(t) = C, \tag{3.9.12}$$

or

$$F(t) \le M \left(\int_t^\infty |g'(\sigma)|\, d\sigma\right)^{\frac{\lambda+1}{\lambda}} \quad \text{for large } t, \tag{3.9.13}$$

where M is a positive constant depending on λ and p.

(ii) *Let $\lambda < p$. Then either* (3.9.12) *holds, or*

$$F(t) \le M_1 \left(\int_t^\infty |g'(\sigma)|\, d\sigma\right)^{\delta} \quad \text{for large } t, \tag{3.9.14}$$

or

$$F(t) \ge M_2 \left(\int_t^\infty |g'(\sigma)|\, d\sigma \right)^\omega \quad \textit{for large } t, \tag{3.9.15}$$

where the positive constants M_1 and M_2 depend on λ and p.

(iii) *Let $\lambda = p$. Then either* (3.9.12) *or* (3.9.13) *holds.*

Proof. First, we prove that

$$\lim_{t\to\infty} F(t) = C \in [0, \infty]. \tag{3.9.16}$$

Suppose that $\limsup_{t\to\infty} F(t) = C_2$, $\liminf_{t\to\infty} F(t) = C_1$, and $0 \le C_1 < C_2 \le \infty$. Then there exist sequences $\{t_k\}_1^\infty$ and $\{s_k\}_1^\infty$ such that $0 \le t_k < s_k < t_{k+1}$, $k = 1, 2, \ldots$, $\lim_{k\to\infty} t_k = \infty$, $F(t_k) = C_3$, $F(s_k) = C_4$, and $C_3 \le F(t) \le C_4$ for $t \in [t_k, s_k]$, where $C_3 = C_1 + \frac{1}{3}(C_2 - C_1)$ and $C_4 = C_1 + \frac{2}{3}(C_2 - C_1)$ if $C_2 < \infty$, and $C_3 = C_1 + 1$ and $C_4 = C_1 + 2$ if $C_2 = \infty$. It then follows from Lemma 3.9.1 and (3.9.8) that

$$\begin{aligned} 0 < C_4 - C_3 &= F(s_k) - F(t_k) \\ &\le \gamma_1 \left[|g(s_k)| + |g(t_k)| + \int_{t_k}^\infty |g'(s)|\, ds \right] C_4^{\beta_1} \\ &\quad + KK_1 \left(C_4^{\frac{1}{p+1}} + C_4^{\frac{1}{\lambda+1}} \right) \int_{t_k}^\infty |g'(\sigma)|\, d\sigma. \end{aligned} \tag{3.9.17}$$

In view of (3.5.3), the right-hand side of (3.9.17) tends to zero as $k \to \infty$. This contradiction proves (3.9.16).

If $C \in (0, \infty)$, then (3.9.12) holds, so next let $C = 0$. Then for every large t_0, there exist τ and σ such that $t_0 \le \sigma < \tau$, $2F(\tau) = F(\sigma) = F(t_0) \le 1$, and $F(\tau) \le F(t) \le F(\sigma)$ for $\sigma \le t \le \tau$.

Let $\lambda \ge p$. From Lemma 3.9.1, we obtain

$$\begin{aligned} \frac{F(\sigma)}{2} = F(\sigma) - F(\tau) &\le \gamma_1 F^{\beta_1}(\sigma) \left[|g(\tau)| + |g(\sigma)| + \int_\tau^\infty |g'(s)|\, ds \right] \\ &\quad + 2K_1 F^{\frac{1}{\lambda+1}}(\sigma) \int_\tau^\infty R^{-\beta_2}(s)\, |e(s)|\, ds. \end{aligned} \tag{3.9.18}$$

Since $1/(\lambda + 1) < \beta_1$, from (3.5.3), (3.9.8), and (3.9.18), we have

$$|g(\tau)| \le \int_\sigma^\infty |g'(s)|\, ds, \quad |g(\sigma)| \le \int_\sigma^\infty |g'(s)|\, ds,$$

and

$$F(\sigma) \le [6\gamma_1 + 4K_1K]F^{\frac{1}{\lambda+1}}(\sigma)\int_\sigma^\infty |g'(s)|\,ds.$$

Thus,

$$F(t_0) = F(\sigma) \le [6\gamma_1 + 4K_1K]^{\frac{\lambda+1}{\lambda}}\left(\int_\sigma^\infty |g'(s)|\,ds\right)^{\frac{\lambda+1}{\lambda}}.$$

Since t_0 is arbitrary, (3.9.13) holds.

If $\lambda < p$, the proof of (3.9.14) is similar; we only need to replace $\frac{1}{\lambda+1}$ by $\frac{1}{p+1}$ in (3.9.18) since, in this case, $\frac{1}{p+1} < \frac{1}{\lambda+1}$.

Let $C = \infty$. In view of Lemma 2.5.2, this case is impossible if $\lambda > p$. Hence, let $\lambda \le p$. Then, for large fixed $t_0 \in \mathbb{R}_+$, there exist τ and σ such that $t_0 \le \sigma < \tau$, $1 \le \frac{1}{2}F(\tau) = F(\sigma) = F(t_0)$, and $F(\sigma) \le F(t) \le F(\tau)$ for $t \in [\sigma, \tau]$.

If $\lambda < p$, then Lemma 3.9.1 and the fact that $\beta_1 \ge 1$ imply

$$\begin{aligned}
F(\sigma) &= 2F(\sigma) - F(\sigma) = F(\tau) - F(\sigma) \\
&\le \gamma_1\,|g(\sigma)|\,F^{\beta_1}(\sigma) + \gamma_1\,|g(\tau)|\,F^{\beta_1}(\tau) + \gamma_1\int_\sigma^\tau |g'(s)|\,F^{\beta_1}(s)\,ds \\
&\quad + K_1\int_\sigma^\tau R^{-\beta_2}(s)[F^{\frac{1}{p+1}}(s) + F^{\frac{1}{\lambda+1}}(s)]\,|e(s)|\,ds \\
&\le \gamma_1 F^{\beta_1}(\tau)\left[|g(\sigma)| + |g(\tau)| + \int_\sigma^\tau |g'(s)|\,ds\right] \\
&\quad + K_1\int_\sigma^\tau R^{-\beta_2}(s)[2F^{\frac{1}{\lambda+1}}(s)]\,|e(s)|\,ds \\
&\le 2\gamma_1 F^{\beta_1}(\sigma)\left[|g(\sigma)| + |g(\tau)| + \int_\sigma^\tau |g'(s)|\,ds\right. \\
&\quad \left. + \frac{2K_1K}{\gamma_1}\int_\sigma^\infty |g'(s)|\,ds\right] \\
&\le F^{\beta_1}(\sigma)\,[6\gamma_1 + 4K_1K]\int_\sigma^\infty |g'(s)|\,ds.
\end{aligned}$$

We then have

$$F^{\omega_1}(\sigma) \le [6\gamma_1 + 4K_1K]\left(\int_\sigma^\infty |g'(s)|\,ds\right), \tag{3.9.19}$$

and so

$$F(t_0) = F(\sigma) \geq [6\alpha_1 + 4K_1K]^{\omega} \left(\int_{\sigma}^{\infty} |g'(s)| \, ds \right)^{\omega}.$$

Since t_0 is arbitrary, (3.9.15) holds.

Finally, let $\lambda = p$; then $\omega_1 = 0$ and (3.9.19) yields a contradiction if σ is large enough. Hence, this case is impossible. □

Lemma 3.9.3. *Let* (3.5.3) *and* (3.9.8) *hold. Then there is a solution* y *of* (3.9.1) *such that* $\lim_{t\to\infty} F(t) \in (0, \infty)$.

Proof. Let $N_1 = \left(\frac{3}{2}\right)^{\frac{1}{p+1}} + \left(\frac{3}{2}\right)^{\frac{1}{\lambda+1}}$. In view of conditions (3.5.3), (3.9.7), and (3.9.8) (recall that (3.9.8) implies (3.9.7)), there exist $0 < N \leq \frac{1}{4}[3\gamma_1(\frac{3}{2})^{\beta_1} + K_1N_1]^{-1}$ and $t_0 \in \mathbb{R}_+$ such that

$$|g(t)| \leq N \quad \text{for } t \geq t_0,$$

$$\int_{t_0}^{\infty} |g'(\sigma)| \, d\sigma \leq N, \quad \text{and} \quad \int_{t_0}^{\infty} R^{-\beta_2}(\sigma)|e(\sigma)| \, d\sigma \leq N.$$

Consider a solution y of (1) with $F(t_0) = 1$. First, we will show that

$$F(t) \leq \frac{3}{2} \quad \text{for all } t \geq t_0. \tag{3.9.20}$$

Suppose that (3.9.20) does not hold. Then there exist $t_2 > t_1 \geq t_0$ such that

$$F(t_2) = \frac{3}{2}, \quad F(t_1) = 1, \quad \text{and} \quad 1 < F(t) < \frac{3}{2} \quad \text{for } t \in (t_1, t_2).$$

Then, (3.9.9)–(3.9.11) with $\tau = t_1$ and $t = t_2$ yield

$$\frac{1}{2} < 3\alpha N \max_{t_1 \leq \sigma \leq t_2} \left| y(\sigma)y^{[1]}(\sigma) \right| + K_1N_1 \int_{t_1}^{t_2} R^{-\beta_2}(\sigma)|e(\sigma)| \, d\sigma$$

$$\leq 3N\gamma_1 \left(\frac{3}{2}\right)^{\beta_1} + K_1NN_1 \leq \frac{1}{4}.$$

This contradiction proves that (3.9.20) holds.

Now from (3.9.9)–(3.9.11) with $t = t$ and $\tau = t_0$, we obtain

$$|F(t) - 1| \leq 3\alpha N \sup_{t \in [t_0, \infty)} \left| y(\sigma)y^{[1]}(\sigma) \right| + K_1N_1 \int_{t_0}^{t} R^{-\beta_2}(\sigma)|e(\sigma)| \, d\sigma$$

$$\leq 3\gamma_1 N \left(\frac{3}{2}\right)^{\beta_1} + K_1NN_1 \leq \frac{1}{4}.$$

Hence, $\frac{3}{4} \leq F(t) \leq \frac{5}{4} < \frac{3}{2}$ for $t \in [t_0, \infty)$. By Theorem 3.9.2, the function F has a limit as $t \to \infty$, so the conclusion of the lemma follows. □

It is interesting to compare the conclusions in Lemmas 3.6.1 and 3.6.2 for unforced equations to those in Lemma 3.9.2 for forced equations. If $\lambda > p$, i.e., the super-half-linear case, notice that the exponent $(\lambda+1)/\lambda$ in (3.9.13) is replaced by ω in (3.6.4). In the sub-half-linear case, $\lambda < p$, we see that (3.9.15) and (3.6.2) are the same, but the alternate behavior in (3.9.14) does not occur. Finally, if $\lambda = p$, i.e., the half-linear case, then alternate (3.9.13) in Theorem 3.9.2 does not occur for the unforced equation.

In our next lemma we show that the function G is bounded.

Lemma 3.9.4. *For any solution y of* (3.9.1) *with $\lambda > p$, the function G is bounded.*

Proof. Let y be a solution of (3.9.1). Since $G(t) \geq 0$ on $\mathbb{R}_+$, we only need to show that G is bounded from above. So suppose that this is not the case; then there is a sequence $\{t_k\}_{k=1}^{\infty}$ such that $\lim_{k\to\infty} t_k = \infty$ and

$$\lim_{k\to\infty} G(t_k) = \infty. \tag{3.9.21}$$

It follows from (3.9.6) that $\lim_{k\to\infty} F(t_k) = \infty$, so we may assume that $F(t_k) \geq 1$ for $k \geq 1$. Hence, there exist two sequences $\{\sigma_k\}_{k=1}^{\infty}$ and $\{\tau_k\}_{k=1}^{\infty}$ with $t_k \leq \sigma_k \leq \tau_k$ such that

$$G(t_k) = G(\sigma_k) = \frac{1}{2} G(\tau_k) \tag{3.9.22}$$

and

$$G(\sigma_k) \leq G(t) \leq G(\tau_k) \quad \text{for } \sigma_k \leq t \leq \tau_k, \quad k = 1, 2, \ldots. \tag{3.9.23}$$

From (3.9.6) and (3.9.23), we see that

$$\max_{\sigma_k \leq t \leq \tau_k} F(t) = F(\tau_k) \quad \text{and} \quad F(t) \geq 1 \quad \text{on } [\sigma_k, \tau_k]. \tag{3.9.24}$$

Now, g is locally of bounded variation, so

$$|g(t)| - |g(0)| \leq |g(t) - g(0)| \leq \int_0^{\tau_k} |g'(s)|\, ds \tag{3.9.25}$$

for $t \in [\sigma_k, \tau_k]$. Furthermore, Lemma 3.9.1 and (3.9.24) imply the existence of $k_0 \in \{1, 2, \dots\}$ and $C > 0$ (not depending on k) such that

$$\begin{aligned}
|I(\tau_k, \sigma_k)| &\le \int_{\sigma_k}^{\tau_k} R^{-\beta_2}(s)\left(\delta F^{1/(p+1)}(s) + \alpha\delta_1 |g(s)| F^{1/(\lambda+1)}(s)\right)|e(s)|\,ds \\
&\le \left[\delta + \alpha\delta_1 \max_{0\le s\le \tau_k} |g(s)|\right] \int_{\sigma_k}^{\tau_k} R^{-\beta_2}(s) F^{1/(p+1)}(s)|e(s)|\,ds \\
&\le C\left(\max_{0\le s\le\tau_k} |g(s)| + 1\right) F^{1/(p+1)}(\tau_k) \int_{\sigma_k}^{\tau_k} R^{-\beta_2}(s)|e(s)|\,ds \\
&\le CA_1(\tau_k)F^{1/(p+1)}(\tau_k) \qquad (3.9.26)
\end{aligned}$$

for $k \ge k_0$.

Letting $\tau = \sigma_k$ and $t = \tau_k$ in (3.9.10) and using (3.9.9), (3.9.24), (3.9.25), and (3.9.26), we obtain

$$|y(t)y^{[1]}(t)| \le \delta_1 F^{\beta_1}(\tau_k) \quad \text{for } t \in [\sigma_k, \tau_k]$$

and

$$\begin{aligned}
F(\tau_k) - F(\sigma_k) &\le \gamma_1 F^{\beta_1}(\tau_k)\left(|g(\sigma_k)| + |g(\tau_k)| + \int_0^{\tau_k} |g'(s)|\,ds\right) \\
&\quad + CA_1(\tau_k)F^{1/(p+1)}(\tau_k) \\
&\le F^q(\tau_k)\left\{2\gamma_1|g(0)| + 3\gamma_1 \int_0^{\tau_k} |g'(s)|\,ds + CA_1(\tau_k)\right\} \\
&\qquad (3.9.27)
\end{aligned}$$

for $k \ge k_0$. It is easy to see that there exist an integer $k_1 \ge k_0$ and a constant $M > 0$ such that

$$2\alpha_1|g(0)| \le M\left(\int_0^{\tau_k} |g'(s)|\,ds + 1\right) \quad \text{for } k \ge k_1.$$

From (3.9.4), (3.9.5), and (3.9.27), we obtain

$$F(\tau_k) - F(\sigma_k) \le M_1 A(\tau_k) F^q(\tau_k)$$

for $k \ge k_1$, where $M_1 = \max\{C, M + 3\gamma_1\}$. From this, (3.9.6), and (3.9.22), we have

$$\begin{aligned}
\frac{1}{2}A^{q_1}(\tau_k)G(\tau_k) &= A^{q_1}(\tau_k)[G(\tau_k) - G(\sigma_k)] \le A^{q_1}(\tau_k)G(\tau_k) - A^{q_1}(\sigma_k)G(\sigma_k) \\
&= F(\tau_k) - F(\sigma_k) \le M_1 A^{1+qq_1}(\tau_k) G^q(\tau_k),
\end{aligned}$$

or

$$G^{1-q}(\tau_k) \le 2M_1$$

for $k \ge k_1$. This contradicts (3.9.21) and completes the proof of the lemma. □

By combining Lemmas 3.9.4 and 3.9.1 we can obtain some global estimates on solutions of equation (3.9.1) and their derivatives.

Theorem 3.9.1. *Let $\lambda > p$ and y be a solution of* (3.9.1). *Then there are positive constants C and C_1 such that*

$$|y(t)| \le CR^{-\omega}(t)A^{q_1/(\lambda+1)}(t)$$

and

$$|y'(t)| \le C_1 a^{-\frac{1}{p}}(t)R^{\alpha/(p+1)}(t)A^{q_1/(p+1)}(t)$$

on $\mathbb{R}_+$.

The following result is well known in the case $m = 2$, however we will need it for other not necessarily integer values.

Lemma 3.9.5. *Let $m > 0$, $f \in C^1(\mathbb{R}_+)$, f' be bounded on $\mathbb{R}_+$, and*

$$\int_0^\infty |f(t)|^m \, dt < \infty. \tag{3.9.28}$$

Then

$$\lim_{t\to\infty} f(t) = 0. \tag{3.9.29}$$

Proof. Let $M > 0$ be such that

$$|f'(t)| \le M < \infty \quad \text{on } \mathbb{R}_+, \tag{3.9.30}$$

$\limsup_{t\to\infty} f(t) = N$, and $\liminf_{t\to\infty} f(t) = N_1$, for some N and N_1 with $-\infty \le N_1 \le N \le \infty$. Suppose that (3.9.29) does not hold. Then, in view of (3.9.28), we see that $-\infty \le N_1 \le 0$, $0 \le N \le \infty$, and we do not have $N = N_1 = 0$.

If $N > 0$, then there are sequences $\{t_n\}_{n=1}^\infty$ and $\{\bar{t}_n\}_{n=1}^\infty$ such that

$$t_n < \bar{t}_n < t_{n+1}, \quad f(t_n) = \frac{N}{2}, \quad \text{and} \quad \bar{t}_n = t_n + \frac{N}{2M} \quad \text{for } n = 1, 2, \ldots.$$

Now (3.9.30) implies $-M \le f'(t)$, so the function f lies above the line passing through the points $[t_n, N/2]$ and $[\bar{t}_n, 0]$, i.e.,

$$f(t) \ge \frac{N}{2} - M(t - t_n) \ge 0 \quad \text{for } t \in [t_n, \bar{t}_n].$$

Hence,

$$\int_0^\infty |f(t)|^m \, dt \ge \sum_{n=1}^\infty \int_{t_n}^{\bar{t}_n} [f(t)]^m \, dt \ge \sum_{n=1}^\infty \int_{t_n}^{\bar{t}_n} \left[\frac{N}{2} - M(t - t_n)\right]^m dt$$
$$= \sum_{n=1}^\infty \frac{1}{M(m+1)} \left(\frac{N}{2}\right)^{m+1} = \infty,$$

contradicting (3.9.28).

If $N = 0$, then $N_1 < 0$, and a similar argument will again yield a contradiction. □

For convenience, we recall the definitions of strongly nonoscillatory and weakly oscillatory solutions.

Definition 3.9.1. A solution y of (3.9.1) is called strongly nonoscillatory if $y(t)y'(t) \neq 0$ for large t. It is called weakly oscillatory if $y(t) \neq 0$ for large t but y' oscillates.

The following result gives information about the asymptotic behavior of solutions in the super-half-linear case.

Theorem 3.9.2. *Let $\lambda \ge p$, assume that* (3.5.3) *and* (3.9.7) *hold, and*

$$\int_0^\infty a^{-\frac{1}{p}}(\sigma) \, d\sigma < \infty.$$

Then every nonoscillatory solution of (3.9.1) *tends to a finite limit as* $t \to \infty$, *and any oscillatory solution y of* (3.9.1) *satisfies* $\lim_{t\to\infty} y(t) = 0$.

Proof. Let y be a proper solution of (3.9.1). Then, parts (i) and (iii) of Theorem 3.9.2 imply $F(t) \le C < \infty$ on $\mathbb{R}_+$, for some positive constant C. From this and Lemma 3.9.1, we obtain

$$|y(t)| \le \gamma^{-\frac{1}{\lambda+1}} R^{-\frac{\beta}{\lambda+1}}(t) F^{\frac{1}{\lambda+1}}(t) \le C_1 R^{-\frac{\beta}{\lambda+1}}$$

and

$$|y'(t)| \leq a^{-\frac{1}{p}}(t)R^{\frac{\alpha}{p+1}}(t)F^{\frac{1}{p+1}}(t) \leq C^{\frac{1}{p+1}}a^{-\frac{1}{p}}(t)R^{\frac{\alpha}{p+1}}(t)$$

for $t \in \mathbb{R}_+$, where $C_1 = \left(\frac{C}{\gamma}\right)^{\frac{1}{\lambda+1}}$. Thus,

$$\left|y^{\frac{1}{p}}(t)y'(t)\right| \leq C_2 a^{-\frac{1}{p}}(t) \tag{3.9.31}$$

for $t \in R_+$, where $C_2 = C^{\frac{1}{p+1}}C_1^{\frac{1}{p}}$.

Let y be a strongly nonoscillatory solution. Then there exists $t_0 \in \mathbb{R}_+$ such that $y(t)y'(t) \neq 0$ on $[t_0, \infty)$. An integration of (3.9.31) yields

$$\left|y^{\delta}(t) - y^{\delta}(t_0)\right| \leq C_2\delta \int_{t_0}^{\infty} a^{-\frac{1}{p}}(\sigma)\,d\sigma < \infty.$$

Hence, y is bounded on $\mathbb{R}_+$, and the boundedness and monotonicity of y imply that y has a finite limit as $t \to \infty$.

Suppose y is oscillatory. Let $\{t_k\}_{k=1}^{\infty}$ and $\{\tau_k\}_{k=1}^{\infty}$ be sequences such that $0 \leq t_k < \tau_k < t_{k+1}$, $k = 1, 2, \ldots$, $\lim_{k\to\infty} t_k = \infty$, and

$$y(t_k) = 0, \quad y'(\tau_k) = 0, \quad y(t) > 0 \quad \text{and} \quad y'(t) > 0 \quad \text{on } (t_k, \tau_k).$$

(The case $y(t) < 0$ and $y'(t) < 0$ for $t \in (t_k, \tau_k)$ can be handled similarly.) Integrating (3.9.31), we have

$$y^{\delta}(\tau_k) = \delta \int_{t_k}^{\tau_k} y^{\frac{1}{p}}(\sigma)y'(\sigma)\,d\sigma \leq C_2\delta \int_{t_k}^{\tau_k} a^{-\frac{1}{p}}(\sigma)\,d\sigma \to 0$$

as $k \to \infty$, so $\lim_{t\to\infty} y(t) = 0$.

Let y be a weakly oscillatory solution. If either $y' \geq 0$ or $y' \leq 0$ for large t, then the proof is the same as for strongly nonoscillatory solutions, so suppose y' changes its sign in every neighborhood of ∞. Let $y > 0$ on $[t_0, \infty)$; the case $y < 0$ is similar. If $\limsup_{t\to\infty} y(t) = \infty$, then we define sequences $\{t_k\}_1^{\infty}$ and $\{\sigma_k\}_1^{\infty}$ in the following way. First, let $\bar{y}(t) = y^{\delta}(t)$ for $t \in \mathbb{R}_+$. Then $\bar{y}'$ has a zero at $t \in \mathbb{R}_+$ if and only if either y or y' has a zero at t. Hence, $\bar{y} > 0$ for $t \geq t_0$ and $\bar{y}'$ changes sign for large t. Let $t_0 < t_k < \sigma_k < t_{k+1}$, $\lim_{k\to\infty} t_k = \infty$, and t_1 and σ_1 be fixed consecutive zeros of $\bar{y}'$ such that $\bar{y}(t) > 0$ on (t_1, σ_1). If t_k and σ_k are defined, then t_{k+1}

and σ_{k+1} are chosen so that

$$\bar{y}(t_{k+1}) = \bar{y}(\sigma_k), \quad \bar{y}'(\sigma_{k+1}) = 0, \quad \bar{y}(t) > 0 \quad \text{on } (t_{k+1}, \sigma_{k+1}),$$

$k = 1, 2, \ldots$. This choice is possible and it is clear that

$$y^\delta(\sigma_k) = y^\delta(t_1) + \sum_{i=1}^{k} \left(y^\delta(\sigma_i) - y^\delta(t_i)\right), \quad k = 1, 2, \ldots. \tag{3.9.32}$$

Furthermore, it follows from an integration of (3.9.31)

$$y^\delta(\sigma_k) - y^\delta(t_k) \le C_2 \delta \int_{t_k}^{\sigma_k} a^{-\frac{1}{p}}(t)\, dt,$$

and hence, (3.9.32) yields

$$y^\delta(\sigma_k) \le y^\delta(t_1) + C_2\delta \int_{t_1}^{\sigma_k} a^{-\frac{1}{p}}(t)\, dt \le y^\delta(t_1) + C_2\delta \int_{t_1}^{\infty} a^{-\frac{1}{p}}(t)\, dt = B. \tag{3.9.33}$$

Since y has a local extrema at σ_k, (3.9.33) implies that y is bounded by B at σ_k for $k = 1, 2, \ldots$. This contradicts the assumption that $\limsup_{t\to\infty} y(t) = \infty$ and proves that y is bounded on $\mathbb{R}_+$.

Now suppose that

$$\limsup_{t\to\infty} y(t) = C_2, \quad \liminf_{t\to\infty} y(t) = C_1,$$

and $0 \le C_1 < C_2 < \infty$. Then, there are sequences $\{\bar{t}_k\}_1^\infty$ and $\{\bar{\sigma}_k\}_1^\infty$ such that $t_0 \le \bar{t}_k < \bar{\sigma}_k < \bar{t}_{k+1}$, $k = 1, 2, \ldots$, $\lim_{k\to\infty} \bar{t}_k = \infty$, $y(\bar{t}_k) = C_1 + \frac{1}{3}(C_2 - C_1)$, and $y(\bar{\sigma}_k) = C_1 + \frac{2}{3}(C_2 - C_1)$, $k = 1, 2, \ldots$. Then (3.9.31) yields

$$\begin{aligned} 0 &< \frac{1}{\delta}\left[\left(C_1 + \frac{2}{3}(C_2 - C_1)\right)^\delta - \left(C_1 + \frac{1}{3}(C_2 - C_1)\right)^\delta\right] \\ &= \frac{1}{\delta}\left[y^\delta(\bar{\sigma}_k) - y^\delta(\bar{t}_k)\right] \\ &= \int_{\bar{t}_k}^{\bar{\sigma}_k} y^{\frac{1}{p}}(t) y'(t)\, dt \le C_2 \int_{\bar{t}_k}^{\bar{\sigma}_k} a^{-\frac{1}{p}}(t)\, dt \to 0 \end{aligned}$$

as $k \to \infty$. This contradiction proves that y has a finite limit as $t \to \infty$. □

Theorem 3.9.3. *Let* (3.5.3) *and* (3.9.8) *hold and let* y *be a solution of* (3.9.1). *If either*

$$\lambda \ge p, \quad (3.9.13) \ \textit{holds}, \quad \textit{and} \quad R(t) \le R_0 < \infty \quad \textit{on } \mathbb{R}_+, \tag{3.9.34}$$

or

$$\lambda < p, \quad (3.9.14) \ \textit{holds}, \quad \textit{and} \quad R(t) \le R_0 < \infty \quad \textit{on } \mathbb{R}_+, \tag{3.9.35}$$

or

$$\lambda \ge p \quad \textit{and} \quad \lim_{t\to\infty} R(t) = 0, \tag{3.9.36}$$

then $\lim_{t\to\infty} y^{[1]}(t) = 0$.

Proof. By Lemma 3.9.1,

$$\left| y^{[1]}(t) \right| \le R^{\beta_2}(t) F^{\frac{p}{p+1}}(t) \tag{3.9.37}$$

for $t \in \mathbb{R}_+$. If (3.9.34) (respectively, (3.9.35)) holds, then (3.9.13) (respectively, (3.9.14)) implies $\lim_{t\to\infty} F(t) = 0$, and the conclusion follows from (3.9.37). If (3.9.36) holds, then parts (i) and (iii) of Theorem 3.9.2 imply that F is bounded on $\mathbb{R}_+$. The conclusion again follows from (3.9.37). □

The next two propositions are in a sense only partial results due to the fact that they only pertain to solutions satisfying an additional qualitative property, viz., condition (3.9.13). However, we believe that these results will prove to be useful in future investigations.

Proposition 3.9.1. *Let* $\lambda > p$, (3.5.3) *and* (3.9.8) *hold, and assume that there is a constant* $K_2 > 0$ *such that*

$$R^{-\beta}(t) \left(\int_t^\infty |g'(\sigma)|\, d\sigma \right)^{\frac{\lambda+1}{\lambda}} \le K_2 \quad \textit{on } \mathbb{R}_+. \tag{3.9.38}$$

Then every solution of (3.9.1) *satisfying* (3.9.13) *is bounded.*

Proof. Let y be a proper solution of (3.9.1) satisfying (3.9.13). Then Lemma 3.9.1 implies

$$\begin{aligned} |y(t)|^{\lambda+1} &\le \gamma^{-1} R^{-\beta}(t) F(t) \\ &\le \gamma^{-1} R^{-\beta}(t) M \left(\int_t^\infty |g'(\sigma)|\, d\sigma \right)^{\frac{\lambda+1}{\lambda}} \le M \gamma^{-1} K_2 \end{aligned}$$

for large t. Hence, y is bounded on $\mathbb{R}_+$. □

Proposition 3.9.2. *Let $\lambda > p$, (3.5.3) and (3.9.8) hold, and assume that there is a constant $K_3 > 0$ such that*

$$\int_0^\infty a^{-\frac{1}{p}}(s)\left(\int_s^\infty |g'(\sigma)|\,d\sigma\right)^{\frac{p(\lambda+2)+1}{(\lambda+1)p\lambda}} ds = K_3 < \infty. \tag{3.9.39}$$

Then every solution of (3.9.1) satisfying (3.9.13) is bounded.

Proof. Let y be a solution of (3.9.1) satisfying (3.9.13). Then Lemma 3.9.1 yields

$$\left|y^{\frac{1}{p}}(t)y'(t)\right| \le Ca^{-\frac{1}{p}}(t)F^{\frac{\beta_1}{p}}(t), \quad t \in \mathbb{R}_+,$$

where $C = \gamma^{-1/[(\lambda+1)p]}$. From this and from (3.9.13), we have

$$\left|y^{\frac{1}{p}}(t)y'(t)\right| \le C_1 a^{-\frac{1}{p}}(t)\left(\int_t^\infty |g'(\sigma)|\,d\sigma\right)^{\frac{p(\lambda+2)+1}{(p+1)p\lambda}}, \tag{3.9.40}$$

for $t \ge t_0$, where t_0 is sufficiently large and $C_1 = CM^{\frac{\beta_1}{p}}$.

Let y be an oscillatory solution and choose sequences $\{t_k\}_{k=1}^\infty$ and $\{\tau_k\}_{k=1}^\infty$ such that $t_0 \leqq t_k < \tau_k < t_{k+1}$, $k = 1, 2, \ldots$, $\lim_{k\to\infty} t_k = \infty$, $y(t_k) = 0$, $y'(\tau_k) = 0$, and $y(t) > 0$ and $y'(t) > 0$ on (t_k, τ_k). (The case $y < 0$ and $y' < 0$ on (t_k, τ_k) is similar.) Then, (3.9.40) yields

$$\begin{aligned} y^\delta(\tau_k) &= \delta \int_{t_k}^{\tau_k} y^{\frac{1}{p}}(\sigma)y'(\sigma)\,d\sigma \\ &\le C_1\delta \int_{t_k}^\infty a^{-\frac{1}{p}}(\sigma)\left(\int_\sigma^\infty |g'(s)|\,ds\right)^{\frac{p(\lambda+2)+1}{(p+1)p\lambda}} d\sigma \\ &\le C_1\delta K_3, \quad k = 1, 2, \ldots. \end{aligned} \tag{3.9.41}$$

Hence, y is bounded on $\mathbb{R}_+$.

Let y be strongly nonoscillatory and t_0 be such that $y(t)y'(t) \ne 0$ on $[t_0, \infty)$. Then, an inequality similar to (3.9.41) holds with $\tau_k = t$ and $t_k = t_0$, and again we see that y is bounded on $\mathbb{R}_+$. If y is weakly oscillatory and $y > 0$ on $[t_0, \infty)$, then the proof is similar to that of Theorem 3.9.2 for weakly oscillatory solutions. □

Conditions (3.9.38) and (3.9.39) are difficult to compare under the assumption that (3.5.3) holds, but we do have examples where (3.9.38) holds and (3.9.39) does not. We believe that this may be true in general, i.e., that (3.9.38) is the better condition, but we have been unable to prove it.

Our next theorem gives sufficient conditions for equation (3.9.1) to be of the strong nonlinear limit-circle type.

Theorem 3.9.4. *Assume that* (3.5.3) *and* (3.9.8) *hold and either* (i) $\lambda = p$, *or* (ii) $\lambda < p$,

$$\int_0^\infty \frac{|e(\sigma)|}{a(\sigma)} d\sigma < \infty, \tag{3.9.42}$$

$$\int_0^\infty \frac{|e(\sigma)|}{r(\sigma)} d\sigma < \infty, \tag{3.9.43}$$

and

$$\liminf_{t\to\infty} R^\beta(t) \left(\int_t^\infty |g'(s)|\, ds \right)^{-\omega} \exp\left\{ \int_0^t \left(R^{-1}(\sigma)\right)'_+ R(\sigma)\, d\sigma \right\} = 0. \tag{3.9.44}$$

If

$$\int_0^\infty R^{-\beta}(\sigma) d\sigma < \infty,$$

then equation (3.9.1) *is of the strong nonlinear limit-circle type.*

Proof. Let $\lambda < p$ and $Z(t) = R^{-\beta}(t)F(t)$ for $t \in \mathbb{R}_+$; then (3.9.1) yields

$$Z'(t) = \left(\frac{1}{R(t)}\right)' |y^{[1]}(t)|^\delta + \delta \frac{e(t)}{r(t)} y'(t).$$

Since $p\delta > 1$, the inequality $|u| \le |u|^{p\delta} + 1$ holds for all u, so

$$\begin{aligned}
\left| \frac{e(t)}{r(t)} |y'(t)| \operatorname{sgn} y'(t) \right| &\le \frac{|e(t)|}{r(t)} [|y'(t)|^{p\delta} + 1] \\
&\le \frac{|e(t)|}{a^\delta(t) r(t)} |y^{[1]}(t)|^\delta + \frac{|e(t)|}{r(t)} = \frac{|e(t)|}{a(t)} \frac{|y^{[1]}(t)|^\delta}{R(t)} + \frac{|e(t)|}{r(t)} \\
&\le \frac{|e(t)|}{a(t)} Z(t) + \frac{|e(t)|}{r(t)}.
\end{aligned}$$

Hence,

$$Z'(t) \le \left[\frac{R'_-(t)}{R(t)} + \delta \frac{|e(t)|}{a(t)} \right] Z(t) + \delta \frac{|e(t)|}{r(t)}.$$

Integrating and then applying Gronwall's inequality and conditions (3.9.42) and (3.9.43), we obtain

$$Z(t) \leq Z_1 \exp\left\{\int_0^t \frac{R'_-(\sigma)}{R(\sigma)} d\sigma\right\}$$

for some constant $Z_1 > 0$. If (3.9.15) holds, then

$$R^{-\beta}(t) M_2 \left(\int_t^\infty |g'(\sigma)|\, d\sigma\right)^\omega \leq R^{-\beta}(t) F(t)$$
$$= Z(t) \leq Z_1 \exp\left\{\int_0^t (R^{-1}(\sigma))'_+ R(\sigma) d\sigma\right\},$$

or

$$M_2 \leq Z_1 R^\beta(t) \left(\int_t^\infty |g'(\sigma)|\, d\sigma\right)^{-\omega} \exp\left\{\int_0^t (R^{-1}(\sigma))'_+ R(\sigma) d\sigma\right\}.$$

This contradicts (3.9.44) so the case $\lambda < p$ and (3.9.15) holding is impossible.

Thus, in view of Theorem 3.9.2, in either case (i) or (ii), we have that $F(t)$ is bounded from above, say, $F(t) \leq B < \infty$ for $t \in \mathbb{R}_+$. We then have

$$\int_0^\infty \left[\frac{|y^{[1]}(\sigma)|^\delta}{R(\sigma)} + \gamma |y(\sigma)|^{\lambda+1}\right] d\sigma \leq B \int_0^\infty R^{-\beta}(\sigma) d\sigma < \infty,$$

that is, equation (3.9.1) is of the strong nonlinear limit-circle type. □

Remark 3.9.1. It is possible to obtain an alternate form of Theorem 3.9.4 (ii) by dropping condition (3.9.42) and replacing condition (3.9.44) with

$$\liminf_{t\to\infty} R^\beta(t) \left(\int_t^\infty |g'(\sigma)|\, ds\right)^{-\omega}$$
$$\times \exp\left\{\int_0^t \left[(R^{-1}(\sigma))'_+ R(\sigma) + \delta \frac{e(\sigma)}{a(\sigma)}\right] d\sigma\right\} = 0.$$

The modifications needed to the above proof should be clear.

Our next theorem gives sufficient conditions for equation (3.9.1) to be of the nonlinear limit-point type as well as conditions for it to have a strong nonlinear limit-point type solution.

Theorem 3.9.5. *Assume that* (3.5.3) *and* (3.9.8) *hold,*

$$\lim_{t\to\infty} \frac{a'(t)}{a^{1-\frac{\beta}{p}}(t) r^{\alpha}(t)} = 0, \tag{3.9.45}$$

and

$$\lim_{t\to\infty} \frac{e(t)}{r(t)} R^{\lambda\beta_2}(t) = 0. \tag{3.9.46}$$

If

$$\int_0^\infty R^{-\beta}(\sigma)\, d\sigma = \infty, \tag{3.9.47}$$

then equation (3.9.1) *is of the nonlinear limit-point type. If, in addition,*

$$r(t) \ge r_0 > 0 \quad \textit{for } t \in \mathbb{R}_+, \tag{3.9.48}$$

then equation (3.9.1) *has a strong nonlinear limit-point type solution.*

Proof. By Lemma 3.9.3, equation (3.9.1) has a solution y for which there exist a constant $C > 0$ and $t_0 \in \mathbb{R}_+$ such that

$$\frac{C}{2} \le F(t) \le 2C \tag{3.9.49}$$

for $t \ge t_0$. From this and (3.5.2), we have

$$\int_{t_0}^t \frac{|y^{[1]}(\sigma)|^\delta}{R(\sigma)}\, d\sigma + \gamma \int_{t_0}^t |y(\sigma)|^{\lambda+1}\, d\sigma = \int_{t_0}^t F(\sigma) R^{-\beta}(\sigma)\, d\sigma$$
$$\ge \frac{C}{2}\int_{t_0}^t R^{-\beta}(\sigma)\, d\sigma \tag{3.9.50}$$

for $t \ge t_0$. Now (3.9.47) and (3.9.50) imply that either

$$\int_0^\infty \frac{|y^{[1]}(\sigma)|^\delta}{R(\sigma)}\, d\sigma = \infty \quad \text{or} \quad \int_0^\infty |y(\sigma)|^{\lambda+1}\, d\sigma = \infty. \tag{3.9.51}$$

Furthermore,

$$\begin{aligned}\int_{t_0}^t |y(\sigma)|^{\lambda+1}\, d\sigma &= -\int_{t_0}^t \frac{y(\sigma)(y^{[1]}(\sigma))'}{r(\sigma)}\, d\sigma + \int_{t_0}^t \frac{e(\sigma)}{r(\sigma)} y(\sigma)\, d\sigma \\ &= -\frac{y(t)y^{[1]}(t)}{r(t)} + \frac{y(t_0)y^{[1]}(t_0)}{r(t_0)} + \int_{t_0}^t \frac{|y^{[1]}(\sigma)|^\delta}{R(\sigma)}\, d\sigma \\ &\quad + \int_{t_0}^t \left(\frac{1}{r(\sigma)}\right)' y(\sigma) y^{[1]}(\sigma)\, d\sigma + \int_{t_0}^t \frac{e(\sigma)}{r(\sigma)} y(\sigma)\, d\sigma,\end{aligned} \tag{3.9.52}$$

and from (3.9.9) and (3.9.49), we have

$$\begin{aligned} |y(t)y^{[1]}(t)| &\leq \gamma^{-\frac{1}{\lambda+1}} F^{\beta_1}(t) \leq \gamma^{-\frac{1}{\lambda+1}} (2C)^{\beta_1} := C_1, \\ |y(t)| &\leq \gamma^{-\frac{1}{\lambda+1}} R^{-\beta_2}(t) F^{\frac{1}{\lambda+1}}(t) \end{aligned} \tag{3.9.53}$$

for $t \geq t_0$. Moreover,

$$\begin{aligned} \int_{t_0}^{t} \left| \left(\frac{1}{r(\sigma)} \right)' \right| d\sigma &= \int_{t_0}^{t} \left| \left(\frac{a^{\frac{1}{p}}(\sigma)}{R(\sigma)} \right)' \right| d\sigma \\ &\leq \int_{t_0}^{t} \frac{|a'(\sigma)|}{p\, a(\sigma) r(\sigma)} \, d\sigma + \int_{t_0}^{t} |g(\sigma)| R^{-\beta}(\sigma) \, d\sigma, \end{aligned} \tag{3.9.54}$$

and using l'Hôspital rule, (3.9.45), and (3.9.47), we obtain

$$\begin{aligned} &\lim_{t\to\infty} \int_{t_0}^{t} \frac{|a'(\sigma)|}{a(\sigma) r(\sigma)} \, d\sigma \left(\int_{t_0}^{t} R^{-\beta}(\sigma) \, d\sigma \right)^{-1} \\ &= \lim_{t\to\infty} \frac{|a'(t)|}{a(t) r(t)} R^{\beta}(t) = \lim_{t\to\infty} \frac{|a'(t)|}{a^{1-\frac{\beta}{p}}(t) r^{\alpha}(t)} = 0. \end{aligned} \tag{3.9.55}$$

We will show that $\int_0^\infty |y(s)|^{\lambda+1} \, ds = \infty$, so suppose

$$\int_0^\infty |y(\sigma)|^{\lambda+1} \, d\sigma < \infty. \tag{3.9.56}$$

Then (3.9.51) implies

$$\int_0^\infty \frac{|y^{[1]}(\sigma)|^{\delta}}{R(\sigma)} \, d\sigma = \infty. \tag{3.9.57}$$

Now,

$$\lim_{t\to\infty} \int_{t_0}^{t} \frac{|e(s)|}{r(s)} R^{-\beta_2}(s) \, ds \left[\int_{t_0}^{t} R^{-\beta}(s) \, d\sigma \right]^{-1} = \lim_{t\to\infty} \frac{|e(t)|}{r(t)} R^{\frac{p\lambda}{(\lambda+2)p+1}}(t) = 0,$$

and in view of (3.9.46), (3.9.50), (3.9.55), (3.9.56), (3.9.57), and the fact that $\lim_{t\to\infty} g(t) = 0$, there exists $T \geq t_0$ such that

$$|g(t)| \leq \frac{C}{4C_1}, \quad \int_{t_0}^{t} \frac{|y^{[1]}(\sigma)|^{\delta}}{R(\sigma)} \, d\sigma \geq \frac{C}{3} \int_{t_0}^{t} R^{-\beta}(\sigma) \, d\sigma,$$

$$\frac{C_1}{p} \int_{t_0}^{t} \frac{|a'(\sigma)|}{a(\sigma) r(\sigma)} \, d\sigma \leq \frac{C}{24} \int_{t_0}^{t} R^{-\beta}(\sigma) \, d\sigma,$$

and

$$\int_{t_0}^{t} \frac{|e(\sigma)|}{r(\sigma)} R^{-\beta_2}(\sigma)\, d\sigma \le (48 \cdot 2^{\frac{1}{\lambda+1}} \gamma^{-\frac{1}{\lambda+1}})^{-1} C^{\frac{\lambda}{\lambda+1}} \int_{t_0}^{t} R^{-\beta}(\sigma)\, d\sigma$$

for $t \ge T$. Hence, (3.9.52), (3.9.53), and (3.9.54) yield

$$\begin{aligned}
\infty > \int_{t_0}^{\infty} |y(\sigma)|^{\lambda+1}\, d\sigma &\ge \int_{t_0}^{t} |y(\sigma)|^{\lambda+1}\, d\sigma \\
&\ge -\frac{y(t)y^{[1]}(t)}{r(t)} + C_2 + \int_{t_0}^{t} \frac{|y^{[1]}(\sigma)|^{\delta}}{R(\sigma)}\, d\sigma - \frac{C_1}{p} \int_{t_0}^{t} \frac{|a'(\sigma)|}{a(\sigma)r(\sigma)}\, d\sigma \\
&\quad - \frac{C}{4} \int_{t_0}^{t} R^{-\beta}(\sigma)\, d\sigma - \int_{t_0}^{t} \frac{|e(\sigma)|}{r(\sigma)} |y(\sigma)|\, d\sigma \\
&\ge -\frac{y(t)y^{[1]}(t)}{r(t)} + C_2 + \frac{C}{24} \int_{t_0}^{t} R^{-\beta}(\sigma)\, d\sigma \\
&\quad - \gamma^{-\frac{1}{\lambda+1}} (2C)^{\frac{1}{\lambda+1}} \int_{t_0}^{t} \frac{|e(\sigma)|}{r(\sigma)} R^{-\beta_2}(\sigma)\, d\sigma \\
&\ge C_2 - \frac{y(t)y^{[1]}(t)}{r(t)} + \frac{C}{48} \int_{t_0}^{t} R^{-\beta}(\sigma)\, d\sigma
\end{aligned} \tag{3.9.58}$$

for some constant C_2. If y is either oscillatory or weakly oscillatory and we take an increasing sequence $\{t_n\}_1^{\infty}$ of zeros of y' with $t_1 \ge T$, then (3.9.58) gives us a contradiction. If y is strongly nonoscillatory with $y(t)\, y'(t) < 0$ for large t, then (3.9.58) again yields a contradiction. If $y(t)\, y'(t) > 0$ for large t, then (3.9.56) cannot hold. Therefore,

$$\int_{0}^{\infty} |y(\sigma)|^{\lambda+1} d\sigma = \infty, \tag{3.9.59}$$

and so equation (3.9.1) is of the nonlinear limit-point type.

Now assume that (3.9.48) holds. In order to show that y is a strong nonlinear limit-point type solution, suppose that (3.9.57) does not hold. Since (3.9.59) holds, we can proceed in a manner similar to what we did above and choose $T_1 \ge t_0$ such that

$$|g(t)| \le \frac{C}{12C_1}, \quad \int_{t_0}^{t} |y(\sigma)|^{\lambda+1}\, d\sigma \ge \frac{C}{3} \int_{t_0}^{t} R^{-\beta}(\sigma)\, d\sigma,$$

$$\frac{C_1}{p} \int_{t_0}^{t} \frac{|a'(\sigma)|}{a(\sigma)r(\sigma)}\, d\sigma \le \frac{C}{12} \int_{t_0}^{t} R^{-\beta}(\sigma)\, d\sigma,$$

and

$$\int_{t_0}^{t} \frac{|e(\sigma)|}{r(\sigma)} R^{-\beta_2}(\sigma)\, d\sigma \leq \left(24 \cdot 2^{\frac{1}{\lambda+1}} \gamma^{-\frac{1}{\lambda+1}}\right)^{-1} C^{\frac{\lambda}{\lambda+1}} \int_{t_0}^{t} R^{-\beta}(\sigma)\, d\sigma$$

for $t \geq T_1$. Then, from (3.9.52), we obtain

$$\begin{aligned}\frac{C}{3} \int_{t_0}^{t} R^{-\beta}(\sigma)\, d\sigma &\leq \int_{t_0}^{t} |y(\sigma)|^{\lambda+1} d\sigma \\ &\leq -\frac{y(t)y^{[1]}(t)}{r(t)} + C_3 + \frac{5C}{24} \int_{t_0}^{t} R^{-\beta}(\sigma)\, d\sigma\end{aligned}$$

for some constant C_3. From (3.9.48) and (3.9.53), we have

$$\left|\frac{y(t)y^{[1]}(t)}{r(t)}\right| \leq \frac{C_1}{r_0}$$

for $t \geq t_0$, and again we have a contradiction. This completes the proof of the theorem. □

Remark 3.9.2. In view of (3.9.55), we could replace condition (3.9.45) with

$$\lim_{t\to\infty} \int_{t_0}^{t} \frac{|a'(\sigma)|}{a(\sigma)r(\sigma)}\, d\sigma \left(\int_{t_0}^{t} R^{-\beta}(\sigma)\, d\sigma\right)^{-1} = 0.$$

On the other hand, from (3.9.52) we see that we could also replace (3.9.45) with

$$\int_0^{\infty} \left|\left(\frac{1}{r(\sigma)}\right)'\right| d\sigma < \infty.$$

These would then yield alternate results to Theorem 3.9.5.

Theorem 3.9.6. *If $\lambda > p$ and*

$$\int_0^{\infty} R^{-\beta}(t) A^{q_1}(t)\, dt < \infty, \tag{3.9.60}$$

then equation (3.9.1) *is of the strong nonlinear limit-circle type.*

Proof. Let y be a solution of (3.9.1). By Lemma 3.9.4, the function G is bounded, say $0 \leq G(t) \leq G_0 < \infty$ for $t \in \mathbb{R}_+$. From this, (3.9.6), and

(3.9.60), we have

$$\begin{aligned}\int_0^\infty V(t)\,dt &= \gamma\int_0^\infty |y(t)|^{\lambda+1}\,dt + \int_0^\infty R^{-1}(t)|y^{[1]}(t)|^\delta\,dt\\ &= \int_0^\infty R^{-\beta}(t)F(t)\,dt = \int_0^\infty R^{-\beta}(t)A^{q_1}(t)G(t)\,dt < \infty,\end{aligned}$$

so y is of the strong nonlinear limit-circle type. □

Our next two theorems give conditions that guarantee that equation (3.9.1) is not of the strong nonlinear limit-circle type.

Theorem 3.9.7. *Assume that*

$$\int_0^\infty |e(t)|\exp\left\{\frac{p}{p+1}\int_0^t \frac{R'_-(s)}{R(s)}\,ds\right\}\,dt < \infty. \tag{3.9.61}$$

If

$$\int_0^\infty \exp\left\{-\int_0^t \frac{R'_+(s)}{R(s)}\,ds\right\}\,dt = \infty, \tag{3.9.62}$$

then (3.9.1) *is not of the strong nonlinear limit-circle type.*

Proof. For any solution y of (3.9.1), we see that

$$\begin{aligned}\frac{|e(t)|}{r(t)}|y'(t)| &= \frac{|e(t)|}{R(t)}|y^{[1]}(t)|^{1/p} = |e(t)|R^{-1/\delta}(t)\left(\frac{|y^{[1]}(t)|^\delta}{R(t)}\right)^{1/(p+1)}\\ &\le |e(t)|R^{-1/\delta}(t)V^{1/(p+1)}(t).\end{aligned} \tag{3.9.63}$$

A straightforward calculation gives

$$V'(t) = \left(\frac{1}{R(t)}\right)'|y^{[1]}(t)|^\delta + \delta\frac{e(t)}{r(t)}y'(t).$$

Hence, in view of (3.9.63),

$$V'(t) \ge -\frac{R'_+(t)}{R(t)}V(t) - |e(t)|R^{-1/\delta}(t)V^{1/(p+1)}(t).$$

Setting $Z = V^{1/\delta}$, we obtain

$$Z'(t) + \frac{R'_+(t)}{\delta R(t)}\, Z(t) \geq -\frac{|e(t)|}{\delta}\, R^{-1/\delta}(t),$$

and so

$$\begin{aligned}\left(Z(t)\exp\left\{\frac{1}{\delta}\int_0^t K(s)\,ds\right\}\right)' &\geq -\frac{1}{\delta}|e(t)|R^{-1/\delta}(t)\exp\left\{\frac{1}{\delta}\int_0^t K(s)\,ds\right\}\\ &= -\frac{1}{\delta}|e(t)|R^{-1/\delta}(0)\exp\left\{\frac{1}{\delta}\int_0^t \frac{R'_-(s)}{R(s)}\,ds\right\},\end{aligned}$$

where $K(t) = \frac{R'_+(t)}{R(t)}$. Integrating and applying (3.9.61), we have

$$\begin{aligned}Z(t)\exp\left\{\frac{1}{\delta}\int_0^t K(s)\,ds\right\} \geq Z(0) - \frac{1}{\delta}R^{-1/\delta}(0)\int_0^\infty |e(t)|\exp\\ \times\left\{\frac{1}{\delta}\int_0^t \frac{R'_-(s)}{R(s)}\,ds\right\}dt > -\infty \qquad (3.9.64)\end{aligned}$$

for $t \in \mathbb{R}_+$.

Now, let y be a solution of (3.9.1) such that

$$Z(0) \geq \frac{1}{\delta}R^{-1/\delta}(0)\int_0^\infty |e(t)|\exp\left\{\frac{1}{\delta}\int_0^t \frac{R'_-(s)}{R(s)}\,ds\right\}dt + 1.$$

From this and (3.9.64), we obtain

$$V(t) = Z^\delta(t) \geq \exp\left\{-\int_0^t K(s)\,ds\right\}, \quad t \in \mathbb{R}_+.$$

Integrating and applying condition (3.9.62), we see that

$$\int_0^t V(s)\,ds = \int_0^t \left[\frac{|y^{[1]}(s)|^\delta}{R(s)} + \gamma|y(s)|^{\lambda+1}\right]ds \to \infty \qquad (3.9.65)$$

as $t \to \infty$, and hence, y is not of the strong nonlinear limit-circle type. This completes the proof of the theorem. □

Now in the proof of Theorem 3.9.7, we showed that (3.9.65) holds. Notice that this does not prevent y from being a nonlinear limit-circle type solution. It is also possible that y is a strong nonlinear limit-point type solution.

Corollary 3.9.1. (i) *Let*

$$\int_0^\infty \frac{R'_-(t)}{R(t)}\,dt < \infty, \quad \int_0^\infty R^{-1}(t)\,dt = \infty, \quad \text{and} \quad \int_0^\infty |e(t)|\,dt < \infty. \tag{3.9.66}$$

Then (3.9.1) *is not of the strong nonlinear limit-circle type.*

(ii) *Let*

$$\int_0^\infty \frac{R'_+(t)}{R(t)}\,dt < \infty \quad \text{and} \quad \int_0^\infty R^{-1/\delta}(t)|e(t)|\,dt < \infty. \tag{3.9.67}$$

Then (3.9.1) *is not of the strong nonlinear limit-circle type.*

Proof. (i) The first and the third inequalities in (3.9.66) imply that (3.9.61) holds. Also,

$$\begin{aligned}\int_0^\infty \exp\left\{-\int_0^t \frac{R'_+(s)}{R(s)}\,ds\right\}dt &= \int_0^\infty \exp\left\{-\int_0^t \frac{R'(s)}{R(s)}\,ds - \int_0^t \frac{R'_-(s)}{R(s)}ds\right\}dt\\ &= \int_0^\infty \exp\left\{-\int_0^t \frac{R'_-(s)}{R(s)}ds\right\}\frac{R(0)}{R(t)}dt\\ &\geq \exp\left\{-\int_0^1 \frac{R'_-(s)}{R(s)}ds\right\}R(0)\int_0^\infty \frac{dt}{R(t)} = \infty.\end{aligned}$$

Hence, (3.9.62) holds and the conclusion follows from Theorem 3.9.7.

(ii) Now (3.9.62) follows from the first inequality in (3.9.67). Since

$$\begin{aligned}&\int_0^\infty |e(t)|\exp\left\{\frac{p}{p+1}\int_0^t \frac{R'_-(s)}{R(s)}\,ds\right\}dt\\ &= \int_0^\infty |e(t)|\exp\left\{-\frac{p}{p+1}\int_0^t \frac{R'(s)}{R(s)}\,ds + \frac{p}{p+1}\int_0^t \frac{R'_+(s)}{R(s)}\,ds\right\}dt\\ &\leq \exp\left\{\frac{p}{p+1}\int_0^\infty \frac{R'_+(s)}{R(s)}\,ds\right\}\int_0^\infty |e(t)|\left(\frac{R(0)}{R(t)}\right)^{1/\delta}dt < \infty,\end{aligned}$$

we see that (3.9.61) holds, and the conclusion again follows from Theorem 3.9.7. □

Theorem 3.9.8. *Assume that there is a positive constant K such that*

$$\int_0^\infty r^{\lambda+1}(t)\,dt < \infty \quad \text{and} \quad \frac{r(t)}{a(t)} \le K \tag{3.9.68}$$

for $t \in \mathbb{R}_+$. In addition, assume that one of the following conditions holds:

(i)

$$\int_0^\infty a^{-1/p}(t)\,dt = \infty \quad \text{and} \quad \int_0^\infty e(t)\,dt = \pm\infty;$$

or

(ii) *for $t \in \mathbb{R}_+$,*

$$R(t) \le R_0 < \infty, \quad r(t) \le r_0 < \infty, \quad |e(t)| \le M < \infty, \tag{3.9.69}$$

and

$$-\infty \le \liminf_{t\to\infty} \int_0^t e(s)\,ds < \limsup_{t\to\infty} \int_0^t e(s)\,ds \le \infty. \tag{3.9.70}$$

Then no nontrivial solution of (3.9.1) is of the strong nonlinear limit-circle type, and so equation (3.9.1) is not of the strong nonlinear limit-circle type.

Proof. Let y be a solution of (3.9.1) and suppose, to the contrary, that it is of the strong nonlinear limit-circle type. Then the second inequality in (3.9.68) implies

$$\int_0^\infty |y(t)|^{\lambda+1}\,dt < \infty \quad \text{and} \quad \int_0^\infty |y'(t)|^{p+1}\,dt < \infty. \tag{3.9.71}$$

From this and Sz.-Nagy's inequality (see [49, Chapter V, Theorem 1]), y is bounded on $\mathbb{R}_+$, say

$$|y(t)| \le M_1, \quad t \in \mathbb{R}_+. \tag{3.9.72}$$

Hence, (3.9.68), (3.9.71), and Hölder's inequality imply

$$\int_0^\infty r(t)|y(t)|^\lambda\,dt \le \left(\int_0^\infty |y(t)|^{\lambda+1}\,dt\right)^{\frac{\lambda}{\lambda+1}} \left(\int_0^\infty r^{\lambda+1}(t)\,dt\right)^{\frac{1}{\lambda+1}} < \infty,$$

and we see that

$$\int_0^\infty r(t)|y(t)|^\lambda \operatorname{sgn} y(t)\,dt = M_2 \in \mathbb{R}. \tag{3.9.73}$$

An integration of (3.9.1) on $[0,t]$ gives

$$y^{[1]}(t) = y^{[1]}(0) - \int_0^t r(s)|y(s)|^\lambda \operatorname{sgn} y(s)\,ds + \int_0^t e(s)\,ds. \tag{3.9.74}$$

(i) Suppose $\int_0^\infty e(t)\,dt = \infty$; the case $\int_0^\infty e(t)\,dt = -\infty$ can be handled similarly. Then (3.9.73)–(3.9.74) imply

$$\lim_{t\to\infty} y^{[1]}(t) = \infty.$$

Hence, from this and the hypotheses in part (i),

$$y(t) - y(0) = \int_0^t y'(s)\,ds = \int_0^t a^{-1/p}(s)|y^{[1]}(s)|^{1/p} \operatorname{sgn} y^{[1]}(s)\,ds \to \infty$$

as $t \to \infty$. This contradicts (3.9.72) and proves part (i).

(ii) Since $R(t) \le R_0$,

$$\int_0^\infty |y^{[1]}(t)|^\delta\,dt < \infty. \tag{3.9.75}$$

From (3.9.1), (3.9.69), and (3.9.72), we see that

$$\left(y^{[1]}(t)\right)' \text{ is bounded on } \mathbb{R}_+.$$

Applying Lemma 3.9.5 to (3.9.75) with $f = y^{[1]}$ and $m = \delta$ gives

$$\lim_{t\to\infty} y^{[1]}(t) = 0.$$

Then (3.9.73) and (3.9.74) imply $\lim_{t\to\infty} \int_{t_0}^t e(s)\,ds$ exists, which contradicts (3.9.70). This proves part (ii) and completes the proof of the theorem. □

The results above can be reformulated as giving necessary conditions for equation (3.9.1) to have a strong nonlinear limit-circle solution. For example, Theorem 3.9.7 could be presented as follows.

Theorem 3.9.9. *Assume* (3.9.61) *holds. If equation* (3.9.1) *has a strong nonlinear limit-circle solution, then*

$$\int_0^\infty \exp\left\{-\int_0^t \frac{R'_+(s)}{R(s)}\,ds\right\} dt < \infty.$$

3.10 Further Results for the Unforced Equation

For the sake of reference, our equation once again is

$$\left(a(t)|y'|^{p-1}y'\right)' + r(t)|y|^\lambda \operatorname{sgn} y = 0. \tag{3.10.1}$$

Recall that every solution of (3.10.1) is either proper or trivial. Our first theorem is concerned with the existence of oscillatory and nonoscillatory solutions.

In this section, as in the previous one, we do not require that condition (3.9.2) holds everywhere.

Theorem 3.10.1.

(i) *If*

$$\int_0^\infty \left(a^{-\frac{1}{p}}(\sigma) + r(\sigma)\right) d\sigma = \infty, \tag{3.10.2}$$

then, for every nonoscillatory solution y *of* (3.10.1), $\lim_{t\to\infty} F(t) \in \{0, \infty\}$. *Moreover, equation* (3.10.1) *has at least one oscillatory solution.*

(ii) *Let*

$$\int_0^\infty \left(a^{-\frac{1}{p}}(\sigma) + r(\sigma)\right) d\sigma < \infty \tag{3.10.3}$$

and let y *be a nontrivial solution of* (3.10.1). *Then* $\lim_{t\to\infty} F(t) \in (0, \infty)$ *and* y *is nonoscillatory.*

Proof. (i) See Lemma 3.6.5 for the case $\lambda \le p$. For $\lambda > p$, the proof is similar (see Remark 3.6.2).

(ii) According to Theorem 2 in [15], condition (3.10.3) is a necessary and sufficient condition for all nontrivial solutions of (3.10.1) to be nonoscillatory. Moreover, Lemma 5 in [15] and (3.10.3) holding imply

$$\int_0^\infty a^{-\frac{1}{p}}(\sigma) R^\alpha(\sigma)\, d\sigma = C < \infty. \tag{3.10.4}$$

Suppose, to the contrary, that y is a nontrivial solution of (3.10.1) such that

$$\lim_{t\to\infty} F(t) \in \{0, \infty\}. \tag{3.10.5}$$

Now,

$$\begin{aligned} F'(t) &= \beta R^{-\alpha}(t) R'(t) \left(\frac{|y^{[1]}(t)|^\delta}{R(t)} + \gamma |y(t)|^{\lambda+1} \right) + R^\beta(t) \left(R^{-1}(t)\right)' \left| y^{[1]}(t) \right|^\delta \\ &= -\alpha \frac{R'(t)}{R^\alpha(t)} \left(\frac{|y^{[1]}(t)|^\delta}{R(t)} - |y(t)|^{\lambda+1} \right), \end{aligned}$$

so

$$-C_1 a^{-\frac{1}{p}}(t)R^\alpha(t) \le \frac{F'(t)}{F(t)} = \alpha g(t) a^{-\frac{1}{p}}(t) R^\alpha(t) \left[\frac{|y^{[1]}(t)|^\delta}{R(t)} - |y(t)|^{\lambda+1}\right] \times \left[\frac{|y^{[1]}(t)|^\delta}{R(t)} + \gamma |y(t)|^{\lambda+1}\right]^{-1} \le C_1 a^{-\frac{1}{p}}(t) R^\alpha(t)$$

for $t \in \mathbb{R}_+$ and $C_1 = 1 + \left(1 + \frac{1}{\gamma}\right) \sup_{0 \le s < \infty} |g(s)|$. From this and from (3.10.4), we have

$$-C_1 C \le \ln \frac{F(t)}{F(0)} \le C_1 C,$$

so F is bounded from above and below by positive constants on $\mathbb{R}_+$. This contradicts (3.10.5) and completes the proof of the theorem. □

Example 3.10.1. Consider Eq. (3.10.1) with $a(t) = t^{2p}$, $r(t) = t^{-2}$, $t \ge 1$. Then $R(t) \equiv 1$ and conditions (3.5.3) and (3.10.3) hold. Thus, this equation has nonoscillatory solutions and so Theorem 3.10.1(ii) is not empty.

Theorem 3.10.2. *Assume that*

$$\int_0^\infty a^{-\frac{1}{p}}(\sigma)\, d\sigma = \infty \tag{3.10.6}$$

and $R(t) \ge R_0 > 0$ *for* $t \in \mathbb{R}_+$. *Then every nontrivial solution of* (3.10.1) *is oscillatory.*

Proof. From the definition of the function R, $r(t) \ge R_0 a^{-\frac{1}{p}}(t)$, $t \in \mathbb{R}_+$. Hence, $\int_0^\infty r(\sigma)\, d\sigma \ge R_0 \int_0^\infty a^{-\frac{1}{p}}(\sigma)\, d\sigma = \infty$. The statement follows from [138, Theorem 12.8]. □

Remark 3.10.1. It is known (see Theorem 2.4.3) that if (3.10.6) holds, then every limit-circle solution of (3.10.1) is oscillatory. A discussion of the relationship between the linear and nonlinear limit-point/limit-circle properties and the boundedness, oscillation, and convergence to zero of solutions can be found in [25].

The roles of the following propositions are similar to those of Propositions 3.9.1 and 3.9.2.

Proposition 3.10.1. *Let $\lambda > p$ and (3.5.3) hold and assume that there is a constant $K_4 > 0$ such that*

$$R^{-\beta}(t)\left(\int_t^\infty |g'(\sigma)|\, d\sigma\right)^\omega \le K_4 \quad on\ \mathbb{R}_+, \tag{3.10.7}$$

then every solution of (3.10.1) satisfying (3.6.4) is bounded.

Proof. The proof is the same as that of Proposition 3.9.1 except that here we use (3.6.4) instead of (3.9.13). □

Proposition 3.10.2. *Let $\lambda > p$ and (3.5.3) hold and assume that there is a constant $K_5 > 0$ such that*

$$\int_0^\infty a^{-\frac{1}{p}}(s)\Big(\int_s^\infty |g'(\sigma)|\, d\sigma\Big)^{\frac{p(\lambda+2)+1}{p(\lambda-p)}} ds \le K_5 \quad on\ \mathbb{R}_+.$$

Then every solution of (3.10.1) satisfying (3.6.4) is bounded on $\mathbb{R}_+$.

Proof. The proof is the same as that of Proposition 3.9.2 except that we must use (3.6.4) instead of (3.9.13). □

It should be pointed out that for the case $e(t) \equiv 0$, Proposition 3.10.1 (respectively, Proposition 3.10.2) gives a better result than Proposition 3.9.1 (respectively, Proposition 3.9.2).

Our next result gives sufficient conditions for equation (3.10.1) to be of the strong nonlinear limit-circle type, as well as a necessary and sufficient condition for this to happen.

Theorem 3.10.3. (i) *If $\lambda > p$ and*

$$\int_0^\infty R^{-\beta}(t)\Big(\int_0^t |g'(s)|\, ds + 1\Big)^\omega dt < \infty, \tag{3.10.8}$$

then equation (3.10.1) is of the strong nonlinear limit-circle type.

(ii) *Assume that*

$$\int_0^\infty |g'(s)|\, ds < \infty. \tag{3.10.9}$$

Then equation (3.10.1) is of the strong nonlinear limit-circle type if and only if

$$\int_0^\infty R^{-\beta}(t)\, dt < \infty. \tag{3.10.10}$$

Proof. (i) Let y be a solution of (3.10.1). If g is not identically constant on $\mathbb{R}_+$, then by Lemma 2.3.3, there is a constant $K > 0$ such that

$$G_1(t) \stackrel{\text{def}}{=} F(t) \left(\int_0^t |g'(s)|\, ds + 1 \right)^{-\omega} \le K \quad \text{for } t \in \mathbb{R}_+. \tag{3.10.11}$$

If g is identically constant on $\mathbb{R}_+$, the boundedness of $G_1 = F$ follows from Lemma 2.3.7. From this, (3.9.3), and (3.10.11), we have

$$\begin{aligned}\int_0^\infty V(t)\, dt &= \int_0^\infty R^{-\beta} F(t)\, dt \\ &= \int_0^\infty R^{-\beta}(t) \left(\int_0^t |g'(s)|\, ds + 1 \right)^{\omega} G_1(t)\, dt \\ &\le K \int_0^\infty R^{-\beta}(t) \left(\int_0^t |g'(s)|\, ds + 1 \right)^{\omega} dt < \infty\end{aligned}$$

by (3.10.8). Hence, y is of the strong nonlinear limit-circle type and part (i) is proved.

(ii) If (3.10.10) holds, then part (i) implies that equation (3.10.1) is of the strong nonlinear limit-circle type. Let

$$\int_0^\infty R^{-\beta}(t)\, dt = \infty.$$

Then, by Lemma 2.3.2, there exist a solution y of (3.10.1), a constant $C_0 > 0$, and a $t_0 \in \mathbb{R}_+$ such that

$$0 < \frac{3}{4} C_0 \le F(t) \quad \text{for } t \ge t_0. \tag{3.10.12}$$

From this and (3.9.3),

$$\begin{aligned}&\gamma \int_{t_0}^\infty |y(t)|^{\lambda+1}\, dt + \int_{t_0}^\infty R^{-1}(t) |y^{[1]}(t)|^{\delta}\, dt \\ &\quad = \int_{t_0}^\infty R^{-\beta}(t) F(t)\, dt \ge \frac{3}{4} C_0 \int_{t_0}^\infty R^{-\beta}(t)\, dt = \infty.\end{aligned}$$

Hence, either $\int_0^\infty |y(t)|^{\lambda+1}\, dt = \infty$ or $\int_0^\infty R^{-1}(t)|y^{[1]}(t)|^{\delta}\, dt = \infty$ and so y is not of the strong nonlinear limit-circle type. This proves (ii). □

Remark 3.10.2. A result similar to Theorem 3.10.3(i), but for equation (3.9.1), is proved in Theorem 2.6.1 and Remark 2.6.1 above if

$$\int_0^\infty R^{-\beta_2}(t)|e(t)|\,dt < \infty \tag{3.10.13}$$

holds and ω in (3.10.8) is replaced by q_1.

Remark 3.10.3. (i) For equation (3.10.1), note that $q = \beta_1$ if and only if $p \ge \frac{\lambda}{\lambda+2}$; in this case $q_1 = \omega$. Hence, if $p \ge \frac{\lambda}{\lambda+2}$, then Theorem 3.10.3(i) is a special case of Theorem 3.9.6. If $p < \frac{\lambda}{\lambda+2}$, then $q_1 > \omega$ and the conditions in Theorem 3.10.3 are weaker than those in Theorem 3.9.6. In this case, (3.9.6) and (3.10.11) imply $\lim_{t\to\infty} G(t) = 0$.

(ii) If $p < \frac{\lambda}{\lambda+2}$, so that $q_1 > \omega$, then condition (3.10.8) is better than the condition used in Theorem 2.6.1 and Remark 2.6.1 where we have a result similar to Theorem 3.10.3(i) for the forced equation (3.9.1).

(iii) Consider equation (3.10.1) with (3.10.9) holding. Then $G(t) \ge F(t)\left(\int_0^\infty |g'(s)|\,ds + 1\right)^{-q_1}$, and by Lemma 2.3.2, there exist a solution y, a constant $C > 0$, and $t_0 \in \mathbb{R}_+$ such that (3.10.12) holds. Hence, $\liminf_{t\to\infty} G(t) > 0$ and so in general the boundedness of G cannot be replaced by $\lim_{t\to\infty} G(t) = 0$ in Lemma 3.9.4.

3.11 Equations with $\int_0^\infty (a^{-\frac{1}{p}}(t) + r(t))\,dt < \infty$

With rare exception, in previous sections in this chapter we assumed that

$$\int_0^\infty (a^{-\frac{1}{p}}(t) + r(t))\,dt = \infty$$

(see (3.5.4), (3.9.2), and (3.10.2)). In this section, we will be explicitly requiring that this condition does not hold and again examining the nonlinear limit-point/limit-circle properties of solutions.

Due to the nature of the results here, we will actually be considering a more general form of the equations in this chapter, namely the equation

$$\left(a(t)|y'|^{p-1}y'\right)' + r(t)\,f(y) = 0, \tag{3.11.1}$$

where $p > 0$, $a \in C^1(\mathbb{R}_+)$, $r \in C^1(\mathbb{R}_+)$, $a > 0$, $r > 0$, and $f : \mathbb{R} \to \mathbb{R}$ is continuous, nondecreasing, and satisfies $xf(x) > 0$ for $x \ne 0$. Clearly, if

$$f(y) = |y|^\lambda \operatorname{sgn} y, \tag{3.11.2}$$

then equation (3.11.1) identically becomes the equation studied in earlier sections in this chapter. By [13, Theorem 3] every nontrivial solution of (3.11.1) is proper.

For equation (3.11.1), the nonlinear limit-point and limit-circle definitions take the following form.

Definition 3.11.1. A solution y of equation (3.11.1) defined on $\mathbb{R}_+$ is said to be of the nonlinear limit-circle type if

$$\int_0^\infty y(\sigma)f(y(\sigma))\,d\sigma < \infty, \tag{3.11.3}$$

and it is said to be of the nonlinear limit-point type otherwise, i.e., if

$$\int_0^\infty y(\sigma)f(y(\sigma))\,d\sigma = \infty. \tag{3.11.4}$$

Equation (3.11.1) will be said to be of the nonlinear limit-circle type if every solution y of (3.11.1) defined on $\mathbb{R}_+$ satisfies (3.11.3) and to be of the *nonlinear limit-point* type if there is at least one solution y defined on $\mathbb{R}_+$ for which (3.11.4) holds.

Definition 3.11.2. A solution y of (3.11.1) defined on $\mathbb{R}_+$ is said to be of the strong nonlinear limit-point type if

$$\int_0^\infty y(\sigma)f(y(\sigma))\,d\sigma = \infty$$

and

$$\int_0^\infty \frac{|y^{[1]}(\sigma)|^\delta}{R(\sigma)}\,d\sigma = \infty.$$

Equation (3.11.1) is said to be of the strong nonlinear limit-point type if every nontrivial solution is of the strong nonlinear limit-point type.

Definition 3.11.3. A solution y of (3.11.1) defined on $\mathbb{R}_+$ is said to be of the strong nonlinear limit-circle type if

$$\int_0^\infty y(\sigma)f(y(\sigma))\,d\sigma < \infty$$

and

$$\int_0^\infty \frac{|y^{[1]}(\sigma)|^\delta}{R(\sigma)}\, d\sigma < \infty.$$

Equation (3.11.1) is said to be of the strong nonlinear limit-circle type if every solution is of the strong nonlinear limit-circle type.

We will use the condition

$$\int_0^\infty (a^{-\frac{1}{p}}(t) + r(t))\, dt < \infty. \tag{3.11.5}$$

For any solution $y : \mathbb{R}_+ \to \mathbb{R}$ of (3.11.1), we set

$$\begin{aligned} Z(t) &= \frac{a(t)}{r(t)} |y'(t)|^{p+1} + \delta \int_0^{y(t)} f(s)\, ds \\ &= \frac{|y^{[1]}(t)|^\delta}{R(t)} + \delta \int_0^{y(t)} f(s)\, ds. \end{aligned} \tag{3.11.6}$$

Theorem 3 in [13] also ensures that $Z > 0$ on $\mathbb{R}_+$ for every nontrivial solution y of (3.11.1). Notice that if (3.11.2) holds, then $Z(t)$ is same as $V(t)$ used in Section 3.9 above and agrees with similar expressions throughout this chapter.

We begin with a lemma giving conditions under which the function Z is bounded from below away from zero. We need the condition

$$\int_0^\infty \frac{R'_+(t)}{R(t)}\, dt < \infty. \tag{3.11.7}$$

Notice that this implies that the function R is bounded from above, i.e., there exists a constant R_1 such that

$$0 \le R(t) \le R_1 < \infty \quad \text{for } t \in \mathbb{R}_+. \tag{3.11.8}$$

Lemma 3.11.1. *Assume that condition* (3.11.7) *holds. For any nontrivial solution y of equation* (3.11.1), *there exists a positive constant Z_0 such that*

$$0 < Z_0 \le Z(t) \quad \textit{for } t \in \mathbb{R}_+. \tag{3.11.9}$$

Proof. Let y be a nontrivial solution of (3.11.1). Then,

$$Z'(t) = -\frac{R'(t)}{R^2(t)} \left|y^{[1]}(t)\right|^\delta$$

and so

$$-\frac{R'_+(t)}{R(t)} \leq -\frac{R'(t)}{R(t)} \frac{\left|y^{[1]}(t)^\delta\right|}{R(t)\,Z(t)} = \frac{Z'(t)}{Z(t)}$$

for $t \in \mathbb{R}_+$. Integrating, we obtain

$$\exp\left\{-\int_0^t \frac{R'_+(s)}{R(s)}\,ds\right\} \leq \frac{Z(t)}{Z(0)},$$

and so (3.11.7) implies (3.11.9) holds. □

Our next two lemmas are modeled after results in [15].

Lemma 3.11.2. *The equation*

$$\left(|\omega'|^{p-1}\omega'\right)' + f(\omega) = 0 \tag{3.11.10}$$

has a periodic solution satisfying $\omega(0) = d$ *and* $\omega'(0) = 0$ *with* $d > 0$. *If* $2T$ *is the smallest positive period of* ω, *then* $\omega(0) = -\omega(T)$. *Moreover,*

$$\omega \textit{ is decreasing on } [0, T] \textit{ and increasing on } [T, 2T].$$

Proof. In view of [138, Corollary 6.1], the solution ω is oscillatory. Then the proof is same as the proof of [15, Lemma 1] with no changes. □

Lemma 3.11.3. *For every nontrivial solution* y *of* (3.11.1), *there exists a continuous positive function* φ *such that*

$$\int_0^{y(t)} f(s)\,ds = Z(t)\int_0^{\omega(\varphi(t))} f(s)\,ds \tag{3.11.11}$$

and

$$y^{[1]}(t) = (R(t)\,Z(t))^{1/\delta}|\omega'(\varphi(t))|^{p-1}\omega'(\varphi(t)) \tag{3.11.12}$$

for $t \geq 0$, *where* ω *is the periodic solution of* (3.11.10) *with* $d > 0$ *satisfying*

$$\int_0^d f(s)\,ds = \frac{1}{\delta}. \tag{3.11.13}$$

Moreover, y *is nonoscillatory if and only if* φ *is bounded on* $\mathbb{R}_+$, *and it is oscillatory if and only if* $\lim_{t\to\infty}\varphi(t) = \infty$.

Proof. The proof is similar to that of [15, Lemma 2] (where $f(x) = |x|^\lambda \operatorname{sgn} x$). Define

$$\omega^{[1]}(t) = |\omega'(t)|^{p-1}\omega'(t) \quad \text{for } t \in \mathbb{R}_+.$$

Let y be a nontrivial solution of (3.11.1) and let $\{\tau_n\}_{n=1}^{\infty}$ be the sequence (finite or infinite) of all zeros of y' on $\mathbb{R}$ if such zeros of y' exist. By [13, Corollary 1], y has only a finite number of zeros on any finite interval and it has no double zeros. Hence, from (3.11.1) we see that the sequence $\{\tau_n\}$ exists and the zeros of y and $y^{[1]}$ separate each other.

Set

$$h(t) = \operatorname{sgn} y(t)\, Z^{-1}(t) \int_0^{y(t)} f(s)\, ds \quad \text{for } t \in \mathbb{R}_+.$$

From (3.11.6), we have

$$-\frac{1}{\delta} < h(t) < \frac{1}{\delta} \quad \text{for } t \neq \tau_k,\ k = 1, 2, \ldots, \tag{3.11.14}$$

and

$$h(\tau_k) = \frac{1}{\delta} \operatorname{sgn} y(\tau_k) \quad \text{for } k = 1, 2, \ldots. \tag{3.11.15}$$

Now define the function $h_1(t)$ by

$$\operatorname{sgn} h_1(t) \int_0^{h_1(t)} f(s)\, ds = h(t) \quad \text{for } t \in \mathbb{R}.$$

From this and (3.11.13)–(3.11.15), we have

$$-d < h_1(t) < d \quad \text{for } t \neq \tau_k \tag{3.11.16}$$

and

$$h_1(\tau_k) = d \operatorname{sgn} y(\tau_k) \quad \text{for } k = 1, 2, \ldots. \tag{3.11.17}$$

Next, we construct the function φ satisfying (3.11.11).

Consider three cases:

(i) y' does not have zeros on $\mathbb{R}_+$;
(ii) y' has a finite number N of zeros on $\mathbb{R}_+$;
(iii) y' has infinitely many zeros on $\mathbb{R}_+$.

In case (i), using (3.11.16), we define

$$\varphi(t) = \omega_0(h_1(t)) \quad \text{for } t \in \mathbb{R}_+,$$

where ω_0 is the inverse function of ω on $[0, T]$ (respectively, on $[T, 2T]$) if $y'(0) < 0$ (respectively, if $y'(0) > 0$).

Let either case (ii) or (iii) hold and suppose $y(\tau_1) < 0$; the proof if $y(\tau_1) > 0$ is similar. If $\tau_1 > 0$, define

$$\varphi(t) = \omega_0\left(h_1(t)\right) \quad \text{for } t \in [0, \tau_1).$$

Then, for $t \geq \tau_1 \geq 0$, define

$$\varphi(t) = \omega_n\left(h_1(t)\right) \quad \text{on } [\tau_n, \tau_{n+1}),$$

where ω_n is the inverse function of ω on the interval $[nT, (n+1)T)$; it is possible to do this in view of (3.11.16) and (3.11.17). In case (ii), define

$$\varphi(t) = \omega_{N+1}\left(h_1(t)\right) \quad \text{on } [\tau_N, \infty),$$

where ω_{N+1} is the inverse function of ω on the interval $[(N+1)T, (N+2)T)$. Hence, φ is defined on $\mathbb{R}_+$ and it is continuous there since (3.11.17) implies

$$\lim_{t \to \tau_n -} \varphi(t) = \lim_{t \to \tau_n -} \omega_n\left(h_1(t)\right) = nT = \varphi(\tau_n).$$

Note that we also have

$$nT < \varphi(t) < (n+1)T \quad \text{on } (\tau_n, \tau_{n+1}).$$

Thus, (3.11.11) holds, and y is nonoscillatory (oscillatory) if and only if φ is bounded ($\lim_{t \to \infty} \varphi(t) = \infty$). Moreover,

$$\operatorname{sgn} y^{[1]}(t) = \operatorname{sgn} \omega'(\varphi(t)). \tag{3.11.18}$$

In order to prove (3.11.12), note that

$$\left(|\omega^{[1]}(t)| + \delta \int_0^{\omega(t)} f(s)\, ds\right)' = 0,$$

and so integrating and using (3.11.13) and the initial conditions for ω at $t = 0$, we have

$$\left|\omega^{[1]}(t)\right|^{\delta} + \delta \int_0^{\omega(t)} f(s)\, ds = 1 \quad \text{for } t \in \mathbb{R}_+. \tag{3.11.19}$$

From this, (3.11.6), and (3.11.11), we obtain

$$\begin{aligned} Z(t) - \frac{|y^{[1]}(t)|^{\delta}}{R(t)} &= \delta \int_0^{y(t)} f(s)\, ds \\ &= \delta Z(t) \int_0^{\omega(\varphi(t))} f(s)\, ds = Z(t)\left(1 - \left|\omega^{[1]}(\varphi(t))\right|^{\delta}\right), \end{aligned}$$

or

$$\frac{|y^{[1]}(t)|^\delta}{R(t)} = Z(t)\left|\omega^{[1]}(\varphi(t))\right|^\delta \quad \text{for } t \in \mathbb{R}_+.$$

From this and (3.11.18) we see that (3.11.12) holds. □

Remark 3.11.1. Notice that Lemmas 3.11.2 and 3.11.3 hold without assuming that f is nondecreasing or that (3.11.5) and (3.11.7) hold.

For purposes of our discussion, it will be convenient to divide the set N of all nonoscillatory solutions of (3.11.1) into the subsets:

$$N_0 = \left\{y \in N : \lim_{t\to\infty} y(t) = 0 \quad \text{and} \quad \lim_{t\to\infty} |y^{[1]}(t)| \in (0,\infty)\right\};$$

$$N_1 = \left\{y \in N : \lim_{t\to\infty} |y(t)| \in (0,\infty) \quad \text{and} \quad \lim_{t\to\infty} |y^{[1]}(t)| \in (0,\infty)\right\};$$

and

$$N_2 = \left\{y \in N : \lim_{t\to\infty} |y(t)| \in (0,\infty) \quad \text{and} \quad \lim_{t\to\infty} y^{[1]}(t) = 0\right\}.$$

Our next lemma shows that nonoscillatory solutions of each of these types exist.

Lemma 3.11.4. *Assume that conditions* (3.11.5) *and* (3.11.7) *hold. Then* $N = N_0 \cup N_1 \cup N_2$ *and* $N_i \neq \emptyset$ *for* $i = 0, 1, 2$.

Proof. Since f is nondecreasing on $\mathbb{R}_+$ and (3.11.5) holds, Lemma 1.1 in [136] implies that any nonoscillatory solution of equation (3.11.1) satisfies $y^{[1]}(t) \neq 0$ for all large t. Moreover, Theorems 1 and 4 in [89] imply $N = N_0 \cup N_1 \cup N_2 \cup N_3$ where

$$N_3 = \left\{y \in N : \lim_{t\to\infty} y(t) = \lim_{t\to\infty} y^{[1]}(t) = 0\right\}.$$

If $y \in N_3$, then, by Lemma 3.11.1, (3.11.9) holds. Since y is nonoscillatory, Lemma 3.11.3 implies φ is bounded, and so (3.11.9) and (3.11.11) imply $\lim_{t\to\infty} \omega(\varphi(t)) = 0$. Hence, $\lim_{t\to\infty} \varphi(t) = T \in (0,\infty)$ exists. Moreover, $\omega(T) = 0$ and it follows from (3.11.19) that $\lim_{t\to\infty} |\omega^{[1]}(\varphi(t))| = 1$. From

(3.11.12), we have

$$\lim_{t\to\infty} R(t)Z(t) = 0.$$

Since y is nonoscillatory, assume that $y(t) > 0$ for large t; the case $y(t) < 0$ is similar. Then, $y'(t) < 0$ eventually and

$$-y'(t) = a^{-\frac{1}{p}}(t)(R(t)\,Z(t))^{\frac{1}{p+1}}|\omega'(\varphi(t))|.$$

Since $\lim_{t\to\infty} \omega^{[1]}(\varphi(t)) = -1$, an integration yields

$$y(t) \le \int_t^\infty a^{-\frac{1}{p}}(\sigma)(R(\sigma)\,Z(\sigma))^{\frac{1}{p+1}}\omega'(\varphi(\sigma))\,d\sigma \tag{3.11.20}$$

with

$$\lim_{t\to\infty} (R(t)\,Z(t))^{\frac{1}{p+1}}\omega'(\varphi(t)) = 0.$$

On the other hand, Lemma 3 in [89] ensures the existence of a constant $B > 0$ such that

$$B\int_t^\infty a^{-\frac{1}{p}}(\sigma)\,d\sigma \le y(t) \quad \text{for } t \ge t_2.$$

This contradicts (3.11.20) and proves that $N_3 = \emptyset$.

The conclusion $N_0 \ne 0$ follows from [136, Theorem 2.2], while Theorem 3 in [89] implies $N_1 \ne \emptyset$ and $N_2 \ne \emptyset$. □

For our next lemma, we will assume that there exist $\varepsilon > 0$, $M > 0$, and $q > 0$ such that

$$|f(x)| \le |x|^q \quad \text{either for } |x| \le \varepsilon \quad \text{or for } |x| \ge M. \tag{3.11.21}$$

Lemma 3.11.5. *Let* (3.11.5) *and* (3.11.21) *hold. Then any nontrivial solution of* (3.11.1) *is nonoscillatory.*

Proof. If the first part of (3.11.21) holds, then it follows from [138, Corollary 7.1] (with $k = 1$ and 2 and $\varphi_2(t,u) = r(t)\max_{0\le s\le u}(f(s) + |f(-s)|)$ there) that every nontrivial solution y of (3.11.1) is either nonoscillatory or is oscillatory and satisfies $\lim_{t\to\infty} y(t) = \lim_{t\to\infty} y^{[1]}(t) = 0$. If the second part of (3.11.21) holds, this is also true by [138, Corollary 7.3] (with $k = 1$ and $k = 2$).

We will show that there are no oscillatory solutions. Suppose, for the sake of obtaining a contradiction, that y is an oscillatory solution of (3.11.1). Then Lemma 3.11.1 implies that (3.11.9) holds, and by Lemma 3.11.3, $\lim_{t\to\infty}\varphi(t) = \infty$. Now,

$$\lim_{t\to\infty}\int_0^{y(t)} f(s)\,ds = 0,$$

but

$$\lim_{t\to\infty} Z(t)\int_0^{\omega(\varphi(t))} f(s)\,ds$$

does not exist. In view of (3.11.11), this is a contradiction, and so all nontrivial solutions of (3.11.1) are nonoscillatory. □

Our final lemma is more technical in nature.

Lemma 3.11.6. *If* (3.11.5) *holds, then* $\int_0^\infty a^{-\frac{1}{p}}(s)R^\alpha(s)\,ds < \infty$.

Proof. By Hölder's inequality and (3.11.5), we have

$$\begin{aligned}\int_0^\infty a^{-\frac{1}{p}}(s)R^\alpha(s)\,ds &= \int_0^\infty a^{-\frac{\beta}{p}}(s)r^\alpha(s)\,ds\\ &\le \left[\int_0^\infty \left(a^{-\frac{\beta}{p}}(t)\right)^{\frac{1}{\beta}} dt\right]^\beta \left[\int_0^\infty (r^\alpha(t))^{\frac{1}{\alpha}}\,dt\right]^\alpha\\ &= \left[\int_0^\infty a^{-\frac{1}{p}}(t)\,dt\right]^\beta \left[\int_0^\infty r(t)\,dt\right]^\alpha < \infty.\end{aligned}$$

□

Our first theorem gives information about the behavior of solutions of equation (3.11.1) especially as it relates to the limit-point/limit-circle properties of solutions.

Theorem 3.11.1. *Assume that* (3.11.5) *and* (3.11.7) *hold.*

(i) $N = N_0 \cup N_1 \cup N_2$ *and* $N_i \ne \emptyset$, $i = 0, 1, 2$.

(ii) *Let* $y \in N_0$, *and assume that for* $\operatorname{sgn} c_1 = \operatorname{sgn} c \ne 0$, $\lim_{u\to 0} f(c_1u)/f(cu)$ *exists as a finite positive number. Then*

$$\int_0^\infty \frac{|y^{[1]}(\sigma)|^\delta}{R(\sigma)}\,d\sigma = \infty.$$

Moreover,

$$\int_0^\infty y(\sigma)\,f\left(y(\sigma)\right)\,d\sigma < \infty$$

if and only if

$$\int_0^\infty \left| f\left(\int_s^\infty a^{-\frac{1}{p}}(\tau)\, d\tau\right)\right| \int_s^\infty a^{-\frac{1}{p}}(\sigma)\, d\sigma\, ds < \infty.$$

(iii) *If $y \in N_1$, then y is of the strong nonlinear limit-point type, i.e.,*

$$\int_0^\infty y(\sigma)\, f\left(y(\sigma)\right)\, d\sigma = \infty \quad \text{and} \quad \int_0^\infty \frac{|y^{[1]}(\sigma)|^\delta}{R(\sigma)}\, d\sigma = \infty.$$

(iv) *If $y \in N_2$, then*

$$\int_0^\infty y(\sigma)\, f\left(y(\sigma)\right)\, d\sigma = \infty.$$

Moreover,

$$\int_0^\infty \frac{|y^{[1]}(\sigma)|^\delta}{R(\sigma)}\, d\sigma < \infty$$

if and only if

$$\int_0^\infty \frac{1}{R(s)}\left(\int_s^\infty r(\sigma)\, d\sigma\right)^\delta ds < \infty.$$

(v) *If* (3.11.21) *holds, then every nontrivial solution of* (3.11.1) *is nonoscillatory.*

Proof.

Part (i) follows from Lemma 3.11.4 and part (v) follows from Lemma 3.11.5.

If $y \in N_1$, then (3.11.8), (3.11.9), and the fact that f is nondecreasing clearly show that y is of the strong nonlinear limit-point type, so (iii) is proved.

If $y \in N_2$, then $\int_0^\infty y(t)\, f\left(y(t)\right)\, dt = \infty$ and there exist $t_0 \in \mathbb{R}_+$ and $B > 0$ such that

$$\frac{B}{2} \le y(t) \le B \quad \text{for } t \in [t_0, \infty).$$

(The case $y(t) < 0$ is similar.) By an integration of (3.11.1) on $[t, \infty)$ and the fact that $\lim_{t\to\infty} y^{[1]}(t) = 0$, we have

$$\begin{aligned} f\left(\frac{B}{2}\right) \int_t^\infty r(s)\, ds &\le y^{[1]}(t) \\ &= \int_t^\infty r(s) f\left(y(s)\right)\, ds \le f(B) \int_t^\infty r(s)\, ds \end{aligned}$$

for $t \geq t_0$. Raising this inequality to the power δ, dividing by $R(t)$, and integrating again, we obtain

$$f^\delta\left(\frac{B}{2}\right)\int_{t_0}^{\infty} R^{-1}(\sigma)\left(\int_\sigma^\infty r(s)\,ds\right)^\delta d\sigma \leq \int_{t_0}^{\infty} \frac{|y^{[1]}(\sigma)|^\delta}{R(\sigma)}\,d\sigma$$

$$\leq f^\delta(B)\int_{t_0}^{\infty} R^{-1}(\sigma)\left(\int_\sigma^\infty r(s)\,ds\right)^\delta d\sigma,$$

and so (iv) is proved.

If $y \in N_0$, then (3.11.7) implies $\int_0^\infty \frac{|y^{[1]}(\sigma)|^\delta}{R(\sigma)}\,d\sigma = \infty$. Since $y^{[1]}$ does not oscillate, y' does not oscillate as well. Suppose that $y(t) > 0$ for large t; the case $y(t) < 0$ is similar. Then there exists $t_1 \in \mathbb{R}_+$ such that

$$y'(t) < 0 \quad \text{and} \quad y^{[1]}(t) < 0 \quad \text{on } [t_1, \infty).$$

Since y is nonoscillatory, the function φ defined by Lemma 3.11.3 is bounded, so (3.11.9) and (3.11.11) imply

$$\lim_{t\to\infty} \omega(\varphi(t)) = 0.$$

Thus, there exists $T \in (0, \infty)$ such that $\lim_{t\to\infty} \varphi(t) = T$ and $\omega(T) = 0$. Hence, $\omega^{[1]}(T) = 1$, so (3.11.19) and (3.11.13) imply that there exists $t_2 \geq t_1$ such that

$$\frac{1}{2} \leq \left|\omega^{[1]}(\varphi(t))\right| \leq 1 \quad \text{for } t \geq t_2. \tag{3.11.22}$$

Since $y \in N_0$, (3.11.12) and (3.11.22) imply that there exist $t_3 \geq t_2$, $M_1 > 0$, and $M_2 > 0$ such that

$$0 < M_1 \leq R(t)Z(t) \leq M_2 \quad \text{for } t \geq t_3.$$

From (3.11.12), we have

$$-y'(t) = a^{-\frac{1}{p}}(t)\,(R(t)\,Z(t))^{\frac{1}{p+1}}\,|\omega'(\varphi(t))|,$$

or

$$C_0\, a^{-\frac{1}{p}}(t) \leq -y'(t) \leq C_1\, a^{-\frac{1}{p}}(t) \quad \text{for } t \geq t_2$$

with $C_0 = \frac{1}{2}(M_1)^{\frac{1}{p+1}}$ and $C_1 = (M_2)^{\frac{1}{p+1}}$. Integrating and using the fact that $\lim_{t\to\infty} y(t) = 0$, we obtain

$$C_0\int_t^\infty a^{-\frac{1}{p}}(t)\,dt \leq y(t) \leq C_1\int_t^\infty a^{-\frac{1}{p}}(t)\,dt,$$

or

$$\begin{aligned}
&C_0 \int_t^\infty f\left(C_0 \int_s^\infty a^{-\frac{1}{p}}(\tau)\, d\tau\right) \int_s^\infty a^{-\frac{1}{p}}(\sigma)\, d\sigma\, ds \\
&\leq \int_t^\infty y(s)\, f(y(s))\, ds \\
&\leq C_1 \int_t^\infty f\left(C_1 \int_s^\infty a^{-\frac{1}{p}}(\tau)\, d\tau\right) \int_s^\infty a^{-\frac{1}{p}}(\sigma)\, d\sigma\, ds \qquad (3.11.23)
\end{aligned}$$

for $t \geq t_2$. Using l'Hôpital's rule and the assumptions in (ii), we have

$$\begin{aligned}
1 &\leq \lim_{t\to\infty} \frac{\int_t^\infty f(C_3 \int_s^\infty a^{-\frac{1}{p}}(\tau)\, d\tau) \int_s^\infty a^{-\frac{1}{p}}(\sigma)\, d\sigma\, ds}{\int_t^\infty f(C_2 \int_s^\infty a^{-\frac{1}{p}}(\tau)\, d\tau) \int_s^\infty a^{-\frac{1}{p}}(\sigma)\, d\sigma\, ds} \\
&= \lim_{t\to\infty} \frac{\int_t^\infty f(C_3 \int_s^\infty a^{-\frac{1}{p}}(\sigma)\, d\sigma)\, ds}{\int_t^\infty f(C_2 \int_s^\infty a^{-\frac{1}{p}}(\sigma)\, d\sigma)\, ds} = M < \infty
\end{aligned}$$

for $C_3 > C_2 > 0$. Thus, part (ii) is proved, and this completes the proof of the theorem. □

Remark 3.11.2. Condition (3.11.7) is not needed in part (iv) of Theorem 3.11.1. In part (iii), condition (3.11.7) can be replaced by

$$\int_0^\infty \frac{1}{R(\sigma)} d\sigma = \infty. \qquad (3.11.24)$$

Among other things, Theorem 3.11.1 gives conditions under which equation (3.11.1) always has a strong nonlinear limit-point type solution, and with some additional conditions holding, gives conditions under which all solutions of (3.11.1) are of the strong nonlinear limit-point type.

The following corollary gives necessary and sufficient conditions for any solution $y \in N_0$ to be of the nonlinear limit-circle type. Notice that in view of Theorem 3.11.1(ii), such a solution could not be of the strong nonlinear limit-circle type, but there could be strong nonlinear limit-point type solutions.

Corollary 3.11.1. *Let* (3.11.5) *and* (3.11.7) *hold and let* $y \in N_0$. *Then*

$$\int_0^\infty y(t)\, f(y(t))\, dt < \infty$$

if and only if

$$\int_0^\infty \left| f\left(C\int_s^\infty a^{-\frac{1}{p}}(\tau)\,d\tau\right)\right| \int_s^\infty a^{-\frac{1}{p}}(\sigma)\,d\sigma\,ds < \infty$$

for every $0 < |C| < \infty$.

Proof. The result follows from (3.11.23) and the fact that f is nondecreasing. □

We can drop condition (3.11.7) and part of condition (3.11.5) and obtain a somewhat better nonlinear limit-point type result for the solutions in N_0.

Corollary 3.11.2. *Assume that the first integral in* (3.11.5) *converges and let* $y \in N_0$. *If*

$$\int_0^\infty \left| f\left(C\int_s^\infty a^{-\frac{1}{p}}(\tau)\,d\tau\right)\right| \int_s^\infty a^{-\frac{1}{p}}(\sigma)\,d\sigma\,ds = \infty \tag{3.11.25}$$

for all $C \neq 0$, *then*

$$\int_0^\infty y(t)\,f(y(t))\,dt = \infty.$$

Proof. Let $y \in N_0$, say $y(t) > 0$ for $t \geq T$. Then Lemma 3 in [89] implies that there exists $C > 0$ such that

$$C\int_t^\infty a^{-\frac{1}{p}}(s)\,ds \leq y(t) \quad \text{for } t \geq T.$$

Now f is nondecreasing, so

$$C\int_t^\infty f\left(C\int_s^\infty a^{-\frac{1}{p}}(\tau)\,d\tau\right)\int_s^\infty a^{-\frac{1}{p}}(\sigma)\,d\sigma\,ds \leq \int_t^\infty y(s)\,f(y(s))\,ds. \tag{3.11.26}$$

The conclusion then follows from (3.11.25) and (3.11.26). □

Remark 3.11.3. If $a^{\frac{1}{p}}(t) = t^s$ and (3.11.2) holds, then (3.11.25) holds if and only if

$$1 < s \leq 1 + \frac{1}{\lambda + 1}.$$

Observe here that we do not obtain the conclusion that y is of the strong nonlinear limit-point type since we are not requiring that (3.11.7) or (3.11.24) holds.

The next theorem formulates our results in case (3.11.2) holds.

Theorem 3.11.2. *Assume that* (3.11.2), (3.11.5) *and* (3.11.7) *hold. Then every nontrivial solution of* (3.11.1) *belongs to* $N_0 \cup N_1 \cup N_2$ *and* $N_i \neq \emptyset$ *for* $i = 0, 1, 2$.

(i) *If* $y \in N_0$, *then* $\int_0^\infty \frac{|y^{[1]}(\sigma)|^\delta}{R(\sigma)}\, d\sigma = \infty$. *Moreover,*

$$\int_0^\infty |y(\sigma)|^{\lambda+1}\, d\sigma < \infty$$

if and only if

$$\int_0^\infty \left(\int_s^\infty a^{-\frac{1}{p}}(\sigma)\, d\sigma \right)^{\lambda+1} ds < \infty.$$

(ii) *If* $y \in N_1$, *then*

$$\int_0^\infty |y(\sigma)|^{\lambda+1}\, d\sigma = \infty \quad \text{and} \quad \int_0^\infty \frac{|y^{[1]}(\sigma)|^\delta}{R(\sigma)}\, d\sigma = \infty.$$

(iii) *If* $y \in N_2$, *then* $\int_0^\infty |y(\sigma)|^{\lambda+1}\, d\sigma = \infty$. *Moreover,*

$$\int_0^\infty \frac{|y^{[1]}(\sigma)|^\delta}{R(\sigma)}\, d\sigma < \infty$$

if and only if

$$\int_0^\infty \frac{1}{R(s)} \left(\int_s^\infty r(\sigma)\, d\sigma \right)^\delta ds < \infty.$$

Proof. The conclusions follow immediately from Lemma 3.11.5 and Theorem 3.11.1 since

$$\lim_{u\to\infty} \frac{f(C_1 u)}{f(C_2 u)} = \left(\frac{C_1}{C_2} \right)^\lambda < \infty \quad \text{for } \operatorname{sgn} C_1 = \operatorname{sgn} C_2 \neq 0.$$

□

Remark 3.11.4. For the case $\lambda > p$ in Theorem 3.11.2, the fact that all solutions are nonoscillatory can be obtained from [138, Theorem 14.2], but his result does not apply if $\lambda < p$ without additional hypotheses.

It is worth pointing out that condition (3.11.7) may be replaced by

$$g_+(t) = \left(\frac{a^{\frac{1}{p}}(t) R'(t)}{R^{1+\alpha}(t)} \right)_+ \quad \text{is bounded on } \mathbb{R}. \tag{3.11.27}$$

This can easily be seen from the fact that

$$\frac{R'_+(t)}{R(t)} = g_+(t)a^{-\frac{1}{p}}(t)R^{\alpha}(t)$$

and then applying (3.11.27) and Lemma 3.11.6. The condition

$$R \in AC^1_{\text{loc}}(\mathbb{R}_+), \quad \lim_{t\to\infty} g(t) = 0, \quad \text{and} \quad \int_0^\infty |g'(\sigma)|\, d\sigma < \infty, \qquad (3.11.28)$$

is used earlier in this chapter often without conditions (3.11.5) and (3.11.7) holding.

Remark 3.11.5. It follows from [15, Theorem 2] that for (3.11.2), $\lambda \neq p$, and (3.11.28) holding, all solutions are nonoscillatory if and only if (3.11.5) holds. Our result that all solutions are nonoscillatory for all p and λ if (3.11.5) and (3.11.7) hold is somewhat better.

We do wish to point out that the results in this section do not explicitly depend on a relationship between p and λ. This can be seen from Example 3.15.11 below where it does not matter if $\lambda = 2 > 1 = p$ or $\lambda = \frac{1}{2} < 1 = p$.

3.12 Thomas–Fermi Equations

Thus far in this chapter we discussed the notion of strong nonlinear limit-point and limit-circle properties of solution of equation beginning with Emden–Fowler equations in Section 3.2, moving on to equations with p-Laplacians, which we might refer to as generalized Emden–Fowler equations, and eventually to such equations with a forcing term. In this section, we will be considering Thomas–Fermi type equations, i.e., equations similar to (3.5.1), (3.8.1), and (3.9.1) but with $r(t) < 0$.

We consider the equation

$$\left(a(t)|y'|^{p-1}y'\right)' = r(t)\,|y|^{\lambda}\operatorname{sgn} y \qquad (3.12.1)$$

as well as the forced version of this equation

$$\left(a(t)|y'|^{p-1}y'\right)' = r(t)\,|y|^{\lambda}\operatorname{sgn} y + e(t). \qquad (3.12.2)$$

Here we assume that a, r, and e are continuous functions defined on $\mathbb{R}_+ = [0,\infty)$, $a(t) > 0$, $r(t) > 0$, and p and λ are positive constants.

Equation (3.12.1) is equivalent to the system of two differential equations

$$\begin{cases} y_1' = a^{-\frac{1}{p}}(t)|y_2|^{\frac{1}{p}} \operatorname{sgn} y_2, \\ y_2' = r(t)|y_1|^{\lambda} \operatorname{sgn} y_1, \end{cases} \tag{3.12.3}$$

and (3.12.2) is equivalent to

$$\begin{cases} y_1' = a^{-\frac{1}{p}}(t)|y_2|^{\frac{1}{p}} \operatorname{sgn} y_2, \\ y_2' = r(t)|y_1|^{\lambda} \operatorname{sgn} y_1 + e(t). \end{cases} \tag{3.12.4}$$

Then the relationship between a solution y of (3.12.1) or (3.12.2) and a solution (y_1, y_2) of the corresponding system (3.12.3) or (3.12.4) is

$$y_1(t) = y(t) \quad \text{and} \quad y_2(t) = a(t)|y'(t)|^{p-1}y'(t). \tag{3.12.5}$$

Definition 3.12.1. A solution y of (3.12.2) is continuable if it is defined on $\mathbb{R}_+$. It is said to be noncontinuable if it is defined on $[0, \tau)$ with $\tau < \infty$, and it cannot be defined at $t = \tau$. A continuable solution y is called proper if it is nontrivial in any neighborhood of ∞. A proper solution y of (3.12.2) is said to be oscillatory if it has a sequence of zeros tending to ∞, and it is nonoscillatory otherwise. A solution y is said to be weakly oscillatory if y is nonoscillatory but y' is oscillatory.

The definition of nonlinear limit-point and nonlinear limit-circle types of solutions as well as the strong nonlinear limit-point and nonlinear limit-circle types are exactly the same as for the Emden–Fowler equation and its extensions.

As for equation (3.12.1), both types of singular solutions may exist (i.e., noncontinuable solutions or nontrivial solutions that are trivial in a neighborhood of infinity) and so we say the following. Equation (3.12.1) is said to be of the strong nonlinear limit-circle type if every solution defined on $\mathbb{R}_+$ is of the strong nonlinear limit-circle type. Equation (3.12.1) is said to be of the strong nonlinear limit-point type if every proper solution is of the strong nonlinear limit-point type and proper solutions exist. We also define the function $R : \mathbb{R}_+ \to \mathbb{R}_+$ and the constant δ as we did earlier in this chapter.

3.13 Classification of Solutions

We let $\mathcal{S}$ denote the set of all continuable solutions of (3.12.2). These solutions can be divided into the following classes. Let

$$\mathcal{N}_1 = \{y \in S : \text{for some } t_y \in \mathbb{R}_+,\ y(t)y'(t) > 0 \text{ for } t \in [t_y, \infty)\},$$

$$\mathcal{N}_{11} = \Big\{y \in \mathcal{N}_1 : \lim_{t\to\infty} |y_1(t)| = C < \infty \quad \text{and} \quad \lim_{t\to\infty} |y_2(t)| = C_1 < \infty, \quad \text{with } CC_1 > 0\Big\},$$

$$\mathcal{N}_{12} = \Big\{y \in \mathcal{N}_1 : \lim_{t\to\infty} |y_1(t)| = C < \infty \text{ and } \lim_{t\to\infty} |y_2(t)| = \infty,\ C > 0\Big\},$$

$$\mathcal{N}_{13} = \Big\{y \in \mathcal{N}_1 : \lim_{t\to\infty} |y_1(t)| = \infty \text{ and } \lim_{t\to\infty} |y_2(t)| = C_1 < \infty,\ C_1 > 0\Big\},$$

$$\mathcal{N}_{14} = \Big\{y \in \mathcal{N}_1 : \lim_{t\to\infty} |y_1(t)| = \infty \text{ and } \lim_{t\to\infty} |y_2(t)| = \infty\Big\},$$

$$\mathcal{N}_2 = \{y \in S : \text{for some } t_y \in \mathbb{R}_+,\ y(t)\,y'(t) < 0 \text{ for } t \in [t_y, \infty)\},$$

$$\mathcal{N}_{21} = \Big\{y \in \mathcal{N}_2 : \lim_{t\to\infty} y_1(t) = 0 \text{ and } \lim_{t\to\infty} y_2(t) = 0\Big\},$$

$$N_{22} = \Big\{y \in \mathcal{N}_2 : \lim_{t\to\infty} y_1(t) = 0 \text{ and } \lim_{t\to\infty} y_2(t) = C_1 \neq 0\Big\},$$

$$\mathcal{N}_{23} = \Big\{y \in \mathcal{N}_2 : \lim_{t\to\infty} y_1(t) = C \neq 0 \text{ and } \lim_{t\to\infty} y_2(t) = C_1,\ CC_1 \le 0\Big\},$$

$$\mathcal{N}_3 = \{y \in S : \text{for some } t_y \in \mathbb{R}_+,\ y(t) \neq 0 \text{ and } y' \text{ oscillates for } t \in [t_y, \infty)\},$$

$$\mathcal{N}_4 = \{y \in S : y \equiv 0 \text{ for large } t\},$$

$$\mathcal{O} = \{y \in S : y \text{ is proper and oscillatory}\}.$$

Our first lemma gives some basic information about the set $\mathcal{S}$.

Lemma 3.13.1. *The following results hold.*

(i) *We have* $\mathcal{S} = \mathcal{N}_1 \cup \mathcal{N}_2 \cup \mathcal{N}_3 \cup \mathcal{N}_4 \cup \mathcal{O}$.
(ii) *If* $e \equiv 0$ *on* $\mathbb{R}_+$, *then* $\mathcal{S} = \mathcal{N}_1 \cup \mathcal{N}_2 \cup \mathcal{N}_4$ *and* $\mathcal{N}_3 = \mathcal{O} = \emptyset$. *Moreover, if* $\lambda \ge p$, *then* $\mathcal{N}_4 = \{y \in \mathcal{S} : y \equiv 0 \text{ on } \mathbb{R}_+\}$.

Proof. Part (i) is obvious. Part (ii) follows from (3.12.4), and if $\lambda \ge p$, then $\mathcal{N}_4 = \{y \in \mathcal{S} : y \equiv 0 \text{ on } \mathbb{R}_+\}$ follows from [138, Theorem 9.2]. □

The next lemma shows that in some cases the class $\mathcal{N}_1$ is empty.

Lemma 3.13.2.

(i) *If* $\lambda \le p$, *then all solutions of* (3.12.2) *are continuable.*

(ii) *Let* $\lambda > p$ *and assume that there are constants* $K_1 > 0$ *and* $\sigma \le 0$ *such that*

$$\int_0^\infty s^{\frac{\sigma}{p+1}} a^{-\frac{1}{p}}(s)\, ds = \infty, \quad R(t) \ge K_1 t^\sigma \quad \text{for large } t, \tag{3.13.1}$$

and

$$\lim_{t\to\infty} \frac{|e(t)|}{r(t)} = 0. \tag{3.13.2}$$

Then every solution y *of* (3.12.2) *satisfying*

$$y_1(t) > 0 \quad \text{and} \quad y_2(t) > 0 \quad \text{for } t \ge t_y \in \mathbb{R}_+$$

is noncontinuable, i.e., $y \notin \mathcal{S}$ *and* $\mathcal{N}_1 = \emptyset$.

Proof. Part (i) follows from [138, Theorem 1.1]. To prove (ii), let y be a continuable solution of (3.12.2) and assume for simplicity, that there exists $T \ge t_y$ such that $T \ge 1$, $2|e(t)|/r(t) \le y_1^\lambda(t_y)$ for $t \ge T$,

$$y_1(t) > 0 \text{ and } y_2(t) > 0 \quad \text{for } t \ge T,$$

and (3.13.1) holds for $t \ge T$. Then, y_1 is increasing on $[T, \infty)$, and using (3.13.1) and (3.13.2), we have

$$\begin{aligned}
\delta^{-1} y_2^\delta(t) &\ge \int_T^t y_2^{\frac{1}{p}}(s) y_2'(s)\, ds = \int_T^t a^{\frac{1}{p}}(s) y_1'(s) \left[r(s)\, y_1^\lambda(s) + e(s)\right] ds \\
&\ge \int_T^t a^{\frac{1}{p}}(s) y_1'(s) y_1^\lambda(s)\, r(s) \left[1 - \frac{|e(s)|}{r(s)\, y_1^\lambda(T)}\right] ds \\
&\ge \frac{1}{2} K_1 t^\sigma \int_T^t y_1^\lambda(s)\, y_1'(s)\, ds = \frac{K_1}{2(\lambda+1)} t^\sigma \left(y_1^{\lambda+1}(t) - y_1^{\lambda+1}(T)\right) \\
&\ge C_1 t^\sigma y_1^{\lambda+1}(t)
\end{aligned}$$

for $t \ge t_1 = T + 1$, where

$$C_1 = \frac{K_1}{2(\lambda+1)} \left[1 - \left(\frac{y_1(T)}{y_1(t_1)}\right)^{\lambda+1}\right]^{-1}.$$

From this, we obtain

$$(y_1'(t))^{p+1} \ge \delta C_1 a^{-\delta}(t) t^\sigma\, y_1^{\lambda+1}(t),$$

so

$$y_1'(t)/y_1^{\frac{\lambda+1}{p+1}}(t) \geq C_2 t^{\frac{\sigma}{p+1}} a^{-\frac{1}{p}}(t)$$

for $t \geq t_1$, where $C_2 = (\delta C_1)^{\frac{1}{p+1}}$. In view of condition (3.13.1), an integration yields a contradiction for large t since $\frac{\lambda+1}{p+1} > 1$. □

Our next lemma shows that the strong limit-circle property holds for solutions in certain classes.

Lemma 3.13.3. *Let $y \in \mathcal{N}_2 \cup \mathcal{N}_3 \cup \mathcal{N}_4 \cup \mathcal{O}$,*

$$\int_0^\infty |e(t)|\, dt < \infty, \quad \textit{and} \quad r(t) - |e(t)| \geq r_0 > 0 \quad \textit{on } \mathbb{R}_+. \tag{3.13.3}$$

Then

$$\int_0^\infty |y(s)|^{\lambda+1}\, ds < \infty \quad \textit{and} \quad \int_0^\infty \frac{a(s)}{r(s)} |y'(s)|^{p+1}\, ds < \infty. \tag{3.13.4}$$

Proof. Let y be a solution of (3.12.2) defined on $\mathbb{R}_+$. Then

$$\begin{aligned}(y_1(t)y_2(t))' &= y_1'(t)y_2(t) + y_1(t)y_2'(t)\\ &= y_1'(t)\, a(t)\, |y_1'(t)|^{p-1}\, y_1'(t)\\ &\quad + r(t)\, |y_1(t)|^\lambda \operatorname{sgn} y_1(t)\, y_1(t) + e(t)\, y_1(t)\end{aligned}$$

on $\mathbb{R}_+$. Integrating, we have

$$\begin{aligned}y_1(t)\, y_2(t) - y_1(0)\, y_2(0) &= \int_0^t r(s) \left[\frac{a(s)}{r(s)} |y_1'(s)|^{p+1} + |y_1(s)|^{\lambda+1}\right] ds\\ &\quad + \int_0^t e(s)\, y_1(s)\, ds.\end{aligned}$$

If $|y_1| \geq 1$, then $|y_1|^{\lambda+1} \geq |y_1|$, so we have

$$\begin{aligned}y_1(t)\, y_2(t) - y_1(0)\, y_2(0) \geq \int_0^t (r(s) - |e(s)|) &\left[\frac{a(s)}{r(s)} |y_1'(s)|^{p+1}\right.\\ &\left. + |y_1(s)|^{\lambda+1}\right] ds - \int_0^t |e(s)|\, ds. \end{aligned} \tag{3.13.5}$$

Since $y \in \mathcal{S} \setminus \mathcal{N}_1$, there exists a sequence $\{t_k\}_{k=1}^{\infty}$, such that $\lim_{k\to\infty} t_k = \infty$ and $y_1(t_k)\, y_2(t_k) \le 0$, $k = 1, 2, \ldots$. From this and (3.13.5), we have

$$\infty > -y_1(0)\, y_2(0) \ge -\int_0^{\infty} |e(s)|\, ds + r_0 \int_0^{t_k} \frac{a(s)}{r(s)} |y_1'(s)|^{p+1}\, ds + r_0 \int_0^{t_k} |y_1(s)|^{\lambda+1}\, ds$$

and the conclusion follows as $k \to \infty$. □

Our next several lemmas deal with the unforced equation (3.12.1). We need to introduce the following notation. Let

$$J_1 = \int_0^{\infty} a^{-\frac{1}{p}}(\sigma) \left(\int_0^{\sigma} r(s)\, ds \right)^{\frac{1}{p}} d\sigma,$$

$$J_2 = \lim_{t\to\infty} \int_0^{t} a^{-\frac{1}{p}}(\sigma) \left(\int_{\sigma}^{t} r(s)\, ds \right)^{\frac{1}{p}} d\sigma,$$

$$J_3 = \int_0^{\infty} r(\sigma) \left(\int_0^{\sigma} a^{-\frac{1}{p}}(s)\, ds \right)^{\lambda} d\sigma,$$

and

$$J_4 = \lim_{t\to\infty} \int_0^{t} r(\sigma) \left(\int_{\sigma}^{t} a^{-\frac{1}{p}}(s)\, ds \right)^{\lambda} d\sigma.$$

Our next lemma gives necessary and sufficient conditions for some of our classes of solutions to be nonempty.

Lemma 3.13.4. *Let* $e(t) \equiv 0$.

(a) $\mathcal{N}_{11} \ne \emptyset$ *if and only if* $J_1 < \infty$ *and* $J_3 < \infty$;
(b) $\mathcal{N}_{12} \ne \emptyset$ *if and only if* $J_1 < \infty$ *and* $J_3 = \infty$;
(c) $\mathcal{N}_{13} \ne \emptyset$ *if and only if* $J_1 = \infty$ *and* $J_3 < \infty$;
(d) $\mathcal{N}_{22} \ne \emptyset$ *if and only if* $J_4 < \infty$;
(e) $\mathcal{N}_{23} \ne \emptyset$ *if and only if* $J_2 < \infty$;
(f) *if* $\lambda \le p$ $(\lambda \ge p)$, *then* $\mathcal{N}_1 \ne \emptyset$ $(\mathcal{N}_2 \ne \emptyset)$;
(g) $\mathcal{N}_1 = \mathcal{N}_{11} \cup \mathcal{N}_{12} \cup \mathcal{N}_{13} \cup \mathcal{N}_{14}$ *and* $\mathcal{N}_2 = \mathcal{N}_{21} \cup \mathcal{N}_{22} \cup \mathcal{N}_{23}$.

Proof. Parts (a)–(e) and (g) follow from [56, Theorem 10] and part (f) is proved in [138, Theorem 9.1 (Theorem 9.2)] (also see [56, Theorems 6 and 7]). □

Next, we give conditions under which the class $\mathcal{N}_1$ is empty if $\lambda > p$.

Lemma 3.13.5. *Let $\lambda > p$ and assume that at least one of the following conditions holds.*

(i) *There exist constants $K_1 > 0$ and $\sigma \leq 0$ such that (3.13.1) holds.*
(ii) *There exists $K_2 > 0$ such that $\int_0^\infty a^{-\frac{1}{p}}(\sigma)\,d\sigma = \infty$ and*

$$r(t)\left(\int_0^t a^{-\frac{1}{p}}(\sigma)\,d\sigma\right)^{1+\lambda} \geq K_2 a^{-\frac{1}{p}}(t)$$

for large t.

Then for every solution y of (3.12.1) satisfying $y(T)\,y'(T) > 0$ at some $T \in \mathbb{R}_+$ is noncontinuable; in particular, $\mathcal{N}_1 = \emptyset$.

Proof. (i) Let y be a continuable solution of (3.12.1) and let $T \in \mathbb{R}_+$ be such that $y(T)\,y'(T) > 0$. Then, $y_1(T)y_2(T) > 0$ and so (3.12.3) yields $y \in \mathcal{N}_1$ with $t_y \leq T$. All conditions of Lemma 3.13.2 are satisfied, and so $y \notin S$. This contradiction completes the proof of this case.

(ii) This follows from [138, Theorem 17.3]. □

The following example shows that part (ii) in Lemma 3.13.5 is better than part (i) at least in the case where all the coefficient functions are powers of t.

Example 3.13.1. Let $p = 1$, $\lambda \geq 4/3$, and let $a(t) = t^{-2}$ and $r(t) = t^{-5}$ for large t. Then Lemma 3.13.5(ii) holds but part (i) fails for any $\sigma \leq 0$.

Clearly, solutions in the class $\mathcal{N}_1$ are of the nonlinear limit-point type. The next lemma gives conditions under which they (i) are or (ii) are not of the strong nonlinear limit-point type.

Lemma 3.13.6. *Let $y \in \mathcal{N}_1$. Then*

$$\int_0^\infty |y(t)|^{\lambda+1}\,dt = \infty.$$

(i) *If*

$$\int_0^\infty R^{-1}(t)\left[\int_0^t r(\tau)\left(\int_0^\tau a^{-\frac{1}{p}}(\sigma)\left(\int_0^\sigma r(s)\,ds\right)^{\frac{1}{p}}d\sigma\right)^\lambda d\tau\right]^\delta dt=\infty, \tag{3.13.6}$$

then

$$\int_0^\infty \frac{a(t)}{r(t)}\,|y'(t)|^{p+1}\,dt=\infty.$$

(ii) *If* $\lambda < p$ *and*

$$\int_0^\infty R^{-1}(t)\left(\int_0^t r(\sigma)\left(\int_0^\sigma a^{-\frac{1}{p}}(s)\,ds\right)^\lambda d\sigma\right)^{\frac{p+1}{p-\lambda}} dt<\infty, \tag{3.13.7}$$

then

$$\int_0^\infty \frac{a(t)}{r(t)}\,|y'(t)|^{p+1}\,dt<\infty.$$

Proof. (i) Let $y \in \mathcal{N}_1$ and for simplicity, assume $y(t) > 0$ on $[t_y, \infty)$; the case $y < 0$ on $[t_y, \infty)$ can be handled similarly. Then, y_1 is increasing, and from (3.12.3), we have

$$y_2'(t) \ge Cr(t) \quad \text{and} \quad y_2(t) \ge C\int_T^t r(\sigma)\,d\sigma \quad \text{for } t\in[T,\infty),$$

with $T = t_y$ and $C = y^\lambda(T)$. Hence, (3.12.3) yields

$$y_1'(t)=a^{-\frac{1}{p}}(t)\,y_2^{\frac{1}{p}}(t)\ge C^{\frac{1}{p}}a^{-\frac{1}{p}}(t)\left(\int_T^t r(\sigma)\,d\sigma\right)^{\frac{1}{p}},$$

or

$$y_1(t)\ge C^{\frac{1}{p}}\int_T^t a^{-\frac{1}{p}}(\sigma)\left(\int_T^\sigma r(s)\,ds\right)^{\frac{1}{p}}d\sigma.$$

From this and the second equality in (3.12.5), we obtain

$$y_2'(t)\ge C_1 r(t)\left(\int_T^t a^{-\frac{1}{p}}(\sigma)\left(\int_T^\sigma r(s)\,ds\right)^{\frac{1}{p}}d\sigma\right)^\lambda,$$

or

$$y_2(t)\ge C_1\int_T^t r(\tau)\left(\int_T^\tau a^{-\frac{1}{p}}(\sigma)\left(\int_T^\sigma r(s)\,ds\right)^{\frac{1}{p}}d\sigma\right)^\lambda d\tau$$

for $C_1 = C^{\frac{\lambda}{p}}$. From this and (3.13.6), we have

$$\int_0^\infty \frac{a(t)}{r(t)} |y'(t)|^{p+1}\, dt = \int_0^\infty \frac{|y_2(t)|^\delta}{R(t)}\, dt \geq C_1^\delta \int_0^\infty R^{-1}(t)$$
$$\times \left[\int_0^t r(\tau)\left(\int_0^\tau a^{-\frac{1}{p}}(\sigma)\left(\int_0^\sigma r(s)\,ds\right)^{\frac{1}{p}} d\sigma\right)^\lambda d\tau\right]^\delta dt = \infty.$$

(ii) Let $y \in \mathcal{N}_1$. Then there exists $T \in \mathbb{R}_+$ such that $y_1 > 0$ and $y_2 > 0$ are increasing on $[T, \infty)$ (the case where $y_1 < 0$ is similar). Let $C = y_1(T)$. Then there exist $T_1 \geq T$ and $C_1 > 0$ such that

$$y_1(t) = C + \int_T^t a^{-\frac{1}{p}}(\sigma) y_2^{\frac{1}{p}}(\sigma)\, d\sigma$$

and

$$\begin{aligned} y_2'(t) = r(t) y_1^\lambda(t) &= r(t)\left[C + \int_T^t a^{-\frac{1}{p}}(\sigma) y_2^{\frac{1}{p}}(\sigma)\, d\sigma\right]^\lambda \\ &\leq 2^\lambda r(t)\left[C^\lambda + y_2^{\frac{\lambda}{p}}(t)\left(\int_T^t a^{-\frac{1}{p}}(\sigma)\, d\sigma\right)^\lambda\right] \\ &\leq C_1 r(t) y_2^{\frac{\lambda}{p}}(t)\left(\int_T^t a^{-\frac{1}{p}}(\sigma)\, d\sigma\right)^\lambda \end{aligned}$$

for $t \geq T_1$. Integrating, we obtain

$$y_2^{1-\frac{\lambda}{p}}(t) \leq y_2^{1-\frac{\lambda}{p}}(T_1) + C_1\left(1 - \frac{\lambda}{p}\right)\int_{T_1}^t r(\sigma)\left(\int_T^\sigma a^{-\frac{1}{p}}(s)\, ds\right)^\lambda d\sigma$$

for $t \geq T_1$. Since there exist $t_0 \in [T_1, \infty)$ and $M > 0$ such that

$$y_2^{1-\frac{\lambda}{p}}(t) \leq M\, C_1\left(1 - \frac{\lambda}{p}\right)\int_T^t r(\sigma)\left(\int_T^\sigma a^{-\frac{1}{p}}(s)\, ds\right)^\lambda d\sigma \tag{3.13.8}$$

for $t \geq t_0$, (3.13.8) yields

$$y_2(t) \leq C_2\left(\int_T^t r(\sigma)\left(\int_T^\sigma a^{-\frac{1}{p}}(s)\, ds\right)^\lambda d\sigma\right)^{\frac{p}{p-\lambda}}$$

for $t \geq t_0$ and $C_2 = (M\, C_1(1-\frac{\lambda}{p}))^{\frac{p}{p-\lambda}}$. From this, we obtain

$$\int_{t_0}^{\infty} \frac{a(t)}{r(t)} |y'(t)|^{p+1}\, dt = \int_{t_0}^{\infty} \frac{|y_2(t)|^{\delta}}{R(t)}\, dt$$

$$\leq C_2^{\delta} \int_{t_0}^{\infty} \frac{1}{R(t)} \left(\int_T^t r(\sigma) \left(\int_T^{\sigma} a^{-\frac{1}{p}}(s)\, ds \right)^{\lambda} d\sigma \right)^{\frac{p+1}{p-\lambda}} dt < \infty,$$

and the statement is proved. □

Remark 3.13.1. Notice that condition (3.13.6) is satisfied if, for example, $\int_0^{\infty} \frac{dt}{R(t)} = \infty$.

The next two lemmas concern solutions in the class $\mathcal{N}_2$.

Lemma 3.13.7. *Let $y \in \mathcal{N}_2$.*

(i) *If*

$$\int_0^{\infty} R^{-1}(\sigma)\, d\sigma < \infty,$$

then

$$\int_0^{\infty} \frac{a(t)}{r(t)} |y'(t)|^{p+1}\, dt < \infty.$$

(ii) *If $r(t) \geq r_0 > 0$ on $\mathbb{R}_+$, then*

$$\int_0^{\infty} |y(\sigma)|^{\lambda+1}\, d\sigma < \infty \quad \text{and} \quad \int_0^{\infty} \frac{a(t)}{r(t)} |y'(t)|^{p+1}\, dt < \infty.$$

Proof. Let $y \in \mathcal{N}_2$. Suppose that $y_1(t) > 0$ and $y_2(t) < 0$ for $t \geq \tau \geq 0$; the case $y_1(t) < 0$ and $y_2(t) > 0$ is similar.

(i) It follows from Lemma 3.13.4(g) and the definitions of $\mathcal{N}_{21}$, $\mathcal{N}_{22}$, and $\mathcal{N}_{23}$ that $|y_2(t)| \leq M < \infty$ for $t \geq T$ and some $M > 0$. Hence,

$$\int_T^{\infty} \frac{a(\sigma)}{r(\sigma)} |y'(\sigma)|^{p+1}\, d\sigma = \int_T^{\infty} \frac{|y_2(\sigma)|^{\delta}}{R(\sigma)}\, d\sigma \leq M^{\delta} \int_T^{\infty} R^{-1}(\sigma)\, d\sigma < \infty,$$

and so (i) holds.

(ii) This result is a special case of Lemma 3.13.3 for $e \equiv 0$ on $\mathbb{R}_+$. □

Lemma 3.13.8. *Let $\lambda < p$ and assume that there are constants $R_0 > 0$ and $a_0 > 0$ such that*

$$R(t) \geq R_0 \quad \text{and} \quad a(t) \leq a_0 \quad \text{for } t \in \mathbb{R}_+.$$

Then $\mathcal{N}_{21} = \emptyset$.

Proof. By Lemma 3.13.2(i), all solutions are continuable. Let y be a proper solution with $y_1(t) > 0$ and $y_2(t) < 0$ for $t \geq T$; the case $y_1 < 0$ and $y_2 > 0$ is similar. Suppose $y \in \mathcal{N}_{21}$; then,

$$\lim_{t\to\infty} y_1(t) = \lim_{t\to\infty} y_2(t) = 0. \tag{3.13.9}$$

From this, we have

$$\begin{aligned} \delta^{-1} y_2^{\delta}(t) &= -\int_t^\infty y_2^{\frac{1}{p}}(s)\, y_2'(s)\, ds = -\int_t^\infty R(s)\, y_1'(s)\, y_1^{\lambda}(s)\, ds \\ &\geq \frac{R_0}{\lambda+1}\, y_1^{\lambda+1}(t) \end{aligned}$$

for $t \geq T$, or

$$(y_1'(t))^{p+1} \geq \frac{\delta R_0}{\lambda+1}\, a^{-\delta}(t)\, y_1^{\lambda+1}(t) \geq C\, y_1^{\lambda+1}(t)$$

with $C = \delta R_0\, a_0^{-\delta} (\lambda+1)^{-1}$. This implies

$$y_1' \geq C^{\frac{1}{p+1}}\, y_1^{\frac{\lambda+1}{p+1}}(t)$$

for $t \geq T$, and an integration yields

$$(y_1(t))^{\frac{p-\lambda}{p+1}} \geq (y_1(T))^{\frac{p-\lambda}{p+1}} + C^{\frac{1}{p+1}} \frac{p-\lambda}{p+1} (t-T)$$

for $t \geq T$. This contradicts (3.13.9) for large t and completes the proof of the lemma. □

3.14 Limit-Point/Limit-Circle Solutions of Thomas–Fermi Type Equations

Our first theorem gives sufficient conditions for the existence of limit-point type solutions to the forced equation.

Theorem 3.14.1. *Let $\lambda \le p$ and assume that there exists $M \in (0,\infty)$ such that*

$$|e(t)| \le M\, r(t) \quad \text{for } t \in \mathbb{R}_+. \tag{3.14.1}$$

Then equation (3.12.2) *is of the nonlinear limit-point type.*

Proof. Let y be a solution of (3.12.2) given by the Cauchy initial conditions $y_1(0) = M^{\frac{1}{\lambda}}$, $y_2(0) = C > 0$. Then, according to Lemma 3.13.2(i), y is defined on $\mathbb{R}_+$, so $y \in \mathcal{S}$. We will prove that $y_2(t) > 0$ on $\mathbb{R}_+$. Suppose, to the contrary, that there is a $t_0 > 0$ such that

$$y_2(t_0) = 0 \quad \text{and} \quad y_2(t) > 0 \quad \text{on } [0, t_0).$$

Then, $y_2'(t_0) \le 0$ and y_1 is increasing on $[0, t_0)$. Hence, (3.12.4) and (3.14.1) yield

$$\begin{aligned} y_2'(t_0) &= r(t_0)\, y_1^{\lambda}(t_0) + e(t_0) \ge r(t_0)\, y_1^{\lambda}(t_0) - M\, r(t_0) \\ &> r(t_0)\left[y_1^{\lambda}(0) - M\right] = 0. \end{aligned}$$

This contradiction proves that $y_2(t) > 0$ and, hence, $y_1'(t) > 0$ on $\mathbb{R}_+$. Since y_1 is increasing and positive, $\int_0^\infty |y_1(t)|^{\lambda+1}\, dt = \infty$ and the theorem is proved. □

The next result is for the case $\lambda > p$; we are able to obtain that all solutions are of the strong nonlinear limit-circle type.

Theorem 3.14.2. *Let $\lambda > p$ and assume that there are constants $K_1 > 0$, $r_0 > 0$, and $\sigma \le 0$ such that* (3.13.1) *holds, $r(t) \ge r_0$ for large t,*

$$\lim_{t\to\infty} \frac{|e(t)|}{r(t)} = 0, \quad \text{and} \quad \int_0^\infty |e(t)|\, dt < \infty.$$

Then equation (3.12.2) *is of the strong nonlinear limit-circle type.*

Proof. Since the conditions of Lemmas 3.13.2 and 3.13.3 are satisfied, $\mathcal{N}_1 = \emptyset$ and $y \in \mathcal{S} \setminus \mathcal{N}_1$ is of the strong nonlinear limit-circle type. □

Our first result for the unforced equation is a limit-point type result.

Theorem 3.14.3. *Assume that either*

(i) $\lambda \leq p$, *or*
(ii) $\lambda > p$ *and one of the following conditions holds:*

$$\int_0^\infty a^{-\frac{1}{p}}(\sigma)\left(\int_\sigma^\infty r(s)\,ds\right)^{\frac{1}{p}} d\sigma < \infty, \tag{3.14.2}$$

$$\int_0^\infty a^{-\frac{1}{p}}(\sigma)\left(\int_0^\sigma r(s)\,ds\right)^{\frac{1}{p}} d\sigma < \infty, \tag{3.14.3}$$

or

$$\int_0^\infty r(\sigma)\left(\int_0^\sigma a^{-\frac{1}{p}}(s)\,ds\right)^{\lambda} d\sigma < \infty. \tag{3.14.4}$$

Then equation (3.12.1) *is of the nonlinear limit-point type.*

Proof. Part (i) follows from Theorem 3.14.1.

Let (3.14.2) hold. Then Lemma 3.13.4(e) implies $\mathcal{N}_{23} \neq \emptyset$ and there is a $y \in \mathcal{N}_{23}$ such that $\lim_{t\to\infty} y(t) \neq 0$; hence $\int_0^\infty |y(s)|^{\lambda+1}\,ds = \infty$.

Finally, let (3.14.3) or (3.14.4) hold. Then parts (a)–(c) of Lemma 3.13.4 guarantee that $\mathcal{N}_{11} \cup \mathcal{N}_{12} \cup \mathcal{N}_{13} \neq \emptyset$ and there exists $y \in \mathcal{N}_{11} \cup \mathcal{N}_{12} \cup \mathcal{N}_{13}$ such that $|y(t)| \geq C > 0$ on $[t_y, \infty)$. The conclusion follows immediately. □

Notice that part (i) of Theorem 3.14.3 says that if $\lambda \leq p$, then equation (3.12.1) always has a nonlinear limit-point type solution. The following theorem is a strong nonlinear limit-circle result.

Theorem 3.14.4. *Let* $\lambda > p$ *and* $r(t) \geq r_0 > 0$ *on* $\mathbb{R}_+$. *Assume that one of the following two conditions holds.*

(i) *There exist constants* $K_1 > 0$ *and* $\sigma \leq 0$ *such that* (3.13.1) *holds.*
(ii) *There exists* $K_2 > 0$ *such that*

$$\int_0^\infty a^{-\frac{1}{p}}(\sigma)\,d\sigma = \infty \quad \textit{and} \quad r(t)\left(\int_0^t a^{-\frac{1}{p}}(\sigma)\,d\sigma\right)^{1+\lambda} \geq K_2 a^{-\frac{1}{p}}(t)$$

for large t.

Then (3.12.1) *is of the strong nonlinear limit-circle type.*

Proof. By Lemma 3.13.1(ii), $\mathcal{N}_4$ contains only the trivial solution. Furthermore, by [56, Theorem 6], $\mathcal{N}_2 \neq \emptyset$. Lemma 3.13.7(ii) implies $y \in \mathcal{N}_2$ is of the strong nonlinear limit-circle type. Since $\mathcal{N}_1 = \emptyset$ (see Lemma 3.13.5), the conclusion follows from Lemma 3.13.2(ii). □

The next two results give conditions under which the second integral condition in the definition of a strong limit-circle type solution is satisfied. One is for the case $\lambda > p$ and the other is for $\lambda < p$.

Theorem 3.14.5. *Let $\lambda > p$, either parts* (i) *or* (ii) *of Lemma* 3.13.5 *hold, and*

$$\int_0^\infty \frac{d\sigma}{R(\sigma)} < \infty.$$

Then $\mathcal{S} \neq \emptyset$, and for any solution $y \in \mathcal{S}$,

$$\int_0^\infty \frac{a(t)}{r(t)} |y'(t)|^{p+1}\, dt < \infty. \tag{3.14.5}$$

Proof. By Lemma 3.13.1(ii), $\mathcal{S} = \mathcal{N}_1 \cup \mathcal{N}_2 \cup \mathcal{N}_4$ and Lemma 3.13.5 implies $\mathcal{N}_2 = \emptyset$. Lemma 3.13.4(f) implies $\mathcal{N}_1 \neq \emptyset$, and so the statement follows from Lemma 3.13.7(i). □

Theorem 3.14.6. *Let $\lambda < p$ and*

$$\int_0^\infty R^{-1}(t) \left(\int_0^t r(\sigma) \left(\int_0^\sigma a^{-\frac{1}{p}}(s)\, ds \right)^\lambda d\sigma \right)^{\frac{p+1}{p-\lambda}} dt < \infty. \tag{3.14.6}$$

Then (3.14.5) *is satisfied for any $y \in \mathcal{S}$ and, moreover, $\mathcal{S} \neq \emptyset$.*

Proof. It follows from Lemma 3.13.4(f) that $\mathcal{S} \neq \emptyset$. Now condition (3.14.6) implies $\int_0^\infty R^{-1}(\sigma)\, d\sigma < \infty$, so the conclusion follows from Lemma 3.13.6(ii), Lemma 3.13.7(i), and the definition of $\mathcal{N}_4$. □

Our next two theorems are strong nonlinear limit-point results.

Theorem 3.14.7. *Let $\lambda < p$ and let $R_0 > 0$ and $a_0 > 0$ be such that*

$$R(t) \geq R_0 \quad \text{and} \quad a(t) \leq a_0 \quad \text{for } t \in \mathbb{R}_+, \tag{3.14.7}$$

and

$$\int_0^\infty R^{-1}(t) \left[\int_0^t r(\tau) \left(\int_0^\tau a^{-\frac{1}{p}}(\sigma) \left(\int_0^\sigma r(s)\, ds \right)^{\frac{1}{p}} d\sigma \right)^\lambda d\tau \right]^\delta dt = \infty.$$

Then (3.12.1) *is of the strong nonlinear limit-point type.*

Proof. According to Lemmas 3.13.4(f) and Lemma 3.13.8, $\mathcal{N}_1 \neq \emptyset$ and $\mathcal{N}_{21} = \emptyset$. From (3.14.7), $r(t) \geq R_0 / a_0^{\frac{1}{p}}$ on $\mathbb{R}_+$, $J_2 = \infty$, and $J_4 = \infty$, so parts (d) and (e) of Lemma 3.13.4 yield $\mathcal{N}_{22} = \mathcal{N}_{23} = \emptyset$; hence $\mathcal{N}_2 = \emptyset$. The conclusion then follows from Lemma 3.13.6. □

Theorem 3.14.8. *Let $\lambda < p$ and assume that* (3.13.6) *holds,*

$$\int_0^\infty a^{-\frac{1}{p}}(\sigma)\,d\sigma = \infty, \tag{3.14.8}$$

and there is a constant $M > 0$ such that

$$r(t)\left(\int_0^t a^{-\frac{1}{p}}(\sigma)\,d\sigma\right)^{p+1} \geq Ma^{-\frac{1}{p}}(t) \tag{3.14.9}$$

for large t. Then equation (3.12.1) *is of the strong nonlinear limit-point type.*

Proof. By Lemma 3.13.1(ii), $\mathcal{S} = \mathcal{N}_1 \cup \mathcal{N}_2 \cup \mathcal{N}_4$. Since (3.14.8) and (3.14.9) hold, Theorems 16.2 and 17.3 in [138] imply $\mathcal{N}_2 = \emptyset$. Thus, proper solutions can only belong to the class $\mathcal{N}_1$, and since Lemma 3.13.4(f) implies $\mathcal{N}_1 \neq \emptyset$, the conclusion follows from Lemma 3.13.6(i). □

3.15 Examples and Further Discussion

To illustrate our results in this chapter, we give the following examples.

Example 3.15.1. Consider the equation

$$y'' + \alpha(\alpha+1)\frac{t^{\alpha(\lambda-1)-2}}{(t^\alpha - 1)^\lambda}|y|^\lambda \operatorname{sgn} y = 0, \quad t \geq 2, \tag{3.15.1}$$

with $\alpha > 1$ and $\lambda > 1$. It is easy to see that the hypotheses of Theorem 3.4.4 are satisfied, and $y(t) = (t^\alpha - 1)/t^\alpha \to 1$ is a solution of this equation with

$$\int_2^\infty \frac{(y'(t))^2}{r(t)}\,dt < \infty.$$

Example 3.15.2. Consider the equation

$$y'' + t^b|y|^\lambda \operatorname{sgn} y = 0, \quad t \geq 1. \tag{3.15.2}$$

First, we see that (3.2.2) holds if and only if $b \geq -(\lambda+3)/2$, and $g(t) \to 0$ as $t \to \infty$ if $b > -(\lambda+3)/2$. It is also not hard to see that (3.2.4) holds if $b > -(\lambda+1)$ and (3.3.4) holds if $b > -2$. Moreover, (3.3.8) holds if $b < 0$, (3.3.9) holds if $-2 < b < 0$, and (3.3.10) holds if $b < -2$. By Theorem 3.4.1, we have that for (i) $\lambda \geq 1$ and $b > -2$ or (ii) $\lambda < 1$ and

$b > -(\lambda+1)$, equation (3.15.2) is of the nonlinear limit-circle type if and only if $b > (\lambda+3)/(\lambda+1)$. If either (i) $\lambda \geq 1$ and $b > -2$ or (ii) $\lambda < 1$ and $b > -(\lambda+1)$, then Theorem 3.4.2 implies that (3.15.2) is of the strong nonlinear limit-point type if and only if $b \leq (\lambda+3)/(\lambda+1)$.

From Theorem 3.4.3 we conclude that (3.15.2) is of the strong nonlinear limit-point type if (i) $\lambda < 1$ and $-(\lambda+3)/2 < b < 0$ or (ii) $\lambda = 1$ and $-2 < b < 0$. By Theorems 3.4.3 and 3.4.4, if $\lambda > 1$ and $-(\lambda+3)/2 \leq b < -2$, then (3.15.2) is of the strong nonlinear limit-point type. From part (i) of Theorem 3.4.5, we have that (3.15.2) is of the strong nonlinear limit-point type if $\lambda \leq 1$ and $b < 0$. Finally, by Corollary 3.4.1, if $b > 0$, then equation (3.15.2) is of the strong nonlinear limit-point type if and only if $0 < b \leq (\lambda+3)/(\lambda+1)$.

If $\lambda < 1$, then (3.5.3) holds if $b > -(\lambda+3)/2$ and (3.6.6) holds if $b > -(\lambda+1)$. (Since $\lambda < 1$, these reduce to $b > -(\lambda+1)$.) Condition (3.6.12) holds if $b \leq (\lambda+3)/(\lambda+1)$ while (3.5.4) and (3.6.11) are automatic. Part (i) of Lemma 3.6.7 also applies to equation (3.15.2).

Example 3.15.3. Consider the equation

$$\left(|y'|^3 y'\right)' + t^b y^3 = 0, \quad t \geq 1, \tag{3.15.3}$$

that is, in equation (3.5.1) we have $a(t) \equiv 1$, $r(t) = t^b$, $p = 4$, and $\lambda = 3$. Condition (3.5.3) holds if $b\alpha > -1$, that is, if $b > -21/5$, and condition (3.6.6) holds if $b > -4$. By Theorem 3.7.1, we then have that equation (3.15.3) is of the strong nonlinear limit-circle (limit-point) type if and only if $b > 21/16$ $(-4 < b \leq 21/16)$.

Example 3.15.4. Consider the equation

$$\left(e^{at}|y'|^{p-1}y'\right)' + e^{bt}|y|^\lambda \operatorname{sgn} y = 0, \quad t \geq 0. \tag{3.15.4}$$

Assume that $\lambda \leq p$,

$$\frac{a}{p} + b > 0, \quad b\left(\frac{p+1}{\lambda+1}\right) > a,$$

and either $a \leq 0$ or $b \geq 0$. If either (i) $\lambda = p$, or (ii) $\lambda < p$ and $a < b$, then Theorem 3.7.1(a) implies that equation (3.15.4) is of the strong nonlinear limit-circle type.

On the other hand, if $\lambda > p$ and

$$\frac{a}{p} + b \le 0, \quad a \le 0, \quad \text{and} \quad b > a,$$

then Theorem 3.8.1(b) implies that equation (3.15.4) is of the strong nonlinear limit-point type.

Finally, if $\lambda > p$,

$$\frac{a}{p} + b > 0, \quad b \ge 0,$$

and either (i) $a \ge 0$ and $b > a\lambda/p$, or (ii) if $a \le 0$, then equation (3.15.4) is of the strong nonlinear limit-circle type by Theorem 3.8.1(a).

Example 3.15.5. Consider the equation

$$\left(|y'|^{p-1}y'\right)' + t^b|y|^\lambda \operatorname{sgn} y = 0, \quad t \ge 1. \tag{3.15.5}$$

If $p \ge \lambda > 0$ and $0 \ge b > -[(\lambda+2)p+1]/(p+1)$, then the hypotheses of Theorem 3.7.2 are satisfied, so equation (3.15.5) is of the strong nonlinear limit-point type. Moreover, if $p = \lambda$, Theorem 3.7.1 implies that (3.15.5) is of the strong nonlinear limit-circle type if and only if $b > (\lambda+1)/\lambda$, and it is of the strong nonlinear limit-point type if and only if $(\lambda+1)/\lambda \ge b > -(\lambda+1)$.

Now if $b > [(\lambda+2)p+1]/(\lambda+1)p$, then Theorem 3.8.1(a) implies that equation (3.15.5) is of the strong nonlinear limit-circle type, and if $-(p+1) < b \le [(\lambda+2)p+1]/(\lambda+1)p$, then Theorem 3.8.1(b) implies it is of the strong nonlinear limit-point type. If $\lambda = p = 1$, then $[(\lambda+2)p+1]/(\lambda+1)p = 2$, which is the dividing line for the linear limit-circle/limit-point results of Dunford and Schwartz [72] (also see the discussion in [25]).

If, instead, we have $-[(\lambda+2)p+1]/(p+1) < b < -(p+1)$, then Theorem 3.8.3 shows that equation (3.15.5) is of the nonlinear limit-point type. Notice that Theorem 3.8.1 does not yield any information about this range of values on b. In addition, for the case $\lambda > 2p+1$, we have $-[(\lambda+2)p+1]/(p+1) < -(2p+1)$, so we can conclude the following:

- if $-(2p+1) \le b < -(p+1)$, then (3.15.5) is of the strong nonlinear limit-point type;
- if $-\frac{(\lambda+2)p+1}{p+1} < b < -(2p+1)$, then (3.15.5) is of the nonlinear limit-point type but not of the strong nonlinear limit-point type.

If $\lambda \le 2p+1$, then $-(2p+1) < -\frac{(\lambda+2)p+1}{p+1}$, and so (3.15.5) is of the strong nonlinear limit-point type.

We also see that if $b \geq -\frac{1}{\alpha}$, then (3.10.9) holds, so Theorem 3.10.3(ii) implies equation (3.15.5) is of the strong nonlinear limit circle type if and only if $b > \frac{(\lambda+2)p+1}{(\lambda+1)p} = \frac{1}{\beta}$. Or if $b < -\frac{1}{\alpha}$, then Theorem 3.9.7 implies (3.15.5) is not of the nonlinear limit-circle type. Thus, (3.15.5) is of the strong nonlinear limit-circle type if and only if $b > \frac{1}{\beta}$.

Example 3.15.6. Consider the equation

$$\left(t^{2p}|y'|^{p-1}y'\right)' + t^b|y|^\lambda \operatorname{sgn} y = t^s, \quad \lambda > p. \tag{3.15.6}$$

Conditions (3.5.3) and (3.9.38) hold if $b > -2 + 1/\alpha$, and (3.9.8) holds if $s < -(b+2)/[(\lambda+2)p+1]$. Hence, by Proposition 3.9.1, every solution of (3.15.6) satisfying (3.9.13) is bounded.

Example 3.15.7. Consider the unforced version of the previous example, namely,

$$\left(t^{2p}|y'|^{p-1}y'\right)' + t^b|y|^\lambda \operatorname{sgn} y = 0, \quad \lambda > p. \tag{3.15.7}$$

We know that condition (3.5.3) holds if $b > -2 + 1/\alpha$. Now condition (3.10.7) holds if $b \geq p - 1$, and moreover we have $-2 + 1/\alpha > p - 1$ since $\lambda > p$. Thus, by Proposition 3.10.1, every solution of (3.15.7) satisfying (3.6.4) is bounded provided $b > -2+1/\alpha$. Now $b \geq -1$ since $-2+1/\alpha > -1$, so condition (3.10.2) holds, and by Theorem 3.10.1(i), every nonoscillatory solution of (3.15.7) satisfies $\lim_{t\to\infty} F(t) \in \{0, \infty\}$. But Lemma 3.9.2(i) also applies to (3.15.7), so we then know that solutions must satisfy (3.9.13). That is, if $b > -2 + 1/\alpha$, then every nonoscillatory solution of (3.15.7) satisfies (3.9.13). We further know from [15, Lemma 7], that this equation has nonoscillatory solutions if and only if one of the following conditions holds:

$$\int_0^\infty a^{-\frac{1}{p}}(\sigma) = \infty \quad \text{and} \quad \int_0^\infty a^{-\frac{1}{p}}(s)\left(\int_s^\infty r(\sigma)d\sigma\right)^{\frac{1}{p}} ds < \infty,$$

or

$$\int_0^\infty r(\sigma)d\sigma = \infty \quad \text{and} \quad \int_0^\infty r(s)\left(\int_s^\infty a^{-\frac{1}{p}}(\sigma)d\sigma\right)^{\lambda} ds < \infty. \tag{3.15.8}$$

Since (3.15.8) holds if $-1 \leq b < -1 + \lambda$, we know that solutions satisfying (3.9.13) exist and Proposition 3.9.1 is satisfied provided $-2 + 1/\alpha < b < -1 + \lambda$.

Example 3.15.8. Consider the forced equation

$$\left(|y'|^{p-1}y'\right)' + t^b|y|^\lambda \operatorname{sgn} y = t^s, \quad t \geq 1. \tag{3.15.9}$$

If $\lambda = p$, $b > \frac{\lambda+1}{\lambda}$, and $s \le -2 - \frac{b}{(\lambda+1)^2}$, then by Theorem 3.9.4(i), equation (3.15.9) is of the strong nonlinear limit-circle type. If $\lambda < p$, $b > \frac{(\lambda+2)p+1}{(\lambda+1)p}$ and $s \le -2 - \frac{b}{(\lambda+2)p+1}$, then by Theorem 3.9.4(ii), equation (3.15.9) is of the strong nonlinear limit-circle type.

If, instead, $-\frac{(\lambda+2)p+1}{p+1} < b \le \frac{(\lambda+2)p+1}{(\lambda+1)p}$ and again $s \le -2 - \frac{b}{(\lambda+2)p+1}$, then Theorem 3.9.5 implies equation (3.15.9) is of the nonlinear limit-point type.

Example 3.15.9. In order to see that Theorems 3.9.7 and 3.9.8 are independent of each other even in case $R(t) \le R_0 < \infty$, consider the equation

$$y'' + t^{-7}(2+\sin t)y^3 = \sin t, \quad t \ge 1. \tag{3.15.10}$$

The hypotheses of Theorem 3.9.8(ii) are satisfied with $p = 1$, $\lambda = 3$, and $a \equiv 1$, so equation (3.15.10) is not of the strong nonlinear limit-circle type. Theorem 3.9.7 cannot be applied since (3.9.62) does not hold. To see this, let $k_0 \in \{1, 2, \dots\}$ and let $\{c_k\}_{k=k_0}^{\infty}$, and $\{d_k\}_{k=k_0}^{\infty}$ be sequences such that $c_k \in \left(\frac{3\pi}{2} + 2k\pi, \frac{5}{3}\pi + 2k\pi\right)$, $d_k \in \left(\frac{7}{3}\pi + 2k\pi, \frac{5\pi}{2} + 2k\pi\right)$, $\frac{\cos t}{3} - \frac{7}{c_{k_0}} \ge 0$ on $[c_k, d_k]$, $k = k_0, k_0+1, \dots$, and let $K = \frac{1}{3} - \frac{7\pi}{c_{k_0}} > 0$. Then,

$$\begin{aligned}\int_{c_k}^{d_k} \frac{R'_+(t)}{R(t)}\,dt &\ge \int_{c_k}^{d_k} \left(\frac{\cos t}{3} - \frac{7}{c_{k_0}}\right) dt \\ &\ge \frac{1}{3} - \frac{7}{c_{k_0}}(d_k - c_k) \ge \frac{1}{3} - \frac{7\pi}{c_{k_0}} = K\end{aligned}$$

and

$$\begin{aligned}\int_0^{\infty} \exp\left\{-\int_0^t \frac{R'_+(s)}{R(s)}\,ds\right\} dt &\le C + \sum_{k=k_0}^{\infty} \exp\left\{-\sum_{i=k_0}^{k} \int_{c_i}^{d_i} \frac{R'_+(s)}{R(s)}\,ds\right\} \\ &\le C + \sum_{k=k_0}^{\infty} \exp\{-K(k-k_0)\} < \infty,\end{aligned}$$

where $C = \int_0^{c_{k_0}} \exp\{-\int_0^t \frac{R'_+(s)}{R(s)}\,ds\}\,dt$. Thus, (3.9.62) does not hold.

Example 3.15.10. Consider the equation

$$y'' + t^{-1/4}y^3 = -t^{-3/4}\sin t - \frac{3}{2}t^{-7/4}\cos t + \frac{21}{16}t^{-11/4}\sin t + t^{-5/2}\sin^3 t.$$

We have $p = 1$, $\lambda = 3$, $a \equiv 1$, and $r(t) = t^{-1/4}$. This equation has the strong nonlinear limit-circle solution $y(t) = t^{-3/4} \sin t$, $t \geq 1$. Note that the condition $\int_0^\infty r^{\lambda+1}(t)\, dt < \infty$ in Theorem 3.9.8 is not satisfied. In addition, we see that $\int_0^\infty e(t)\, dt$ is convergent by [107, Paragraph 3.761]. Thus, we see that strong nonlinear limit-circle solutions may exist even if R is positive and small and $\int_0^\infty e(t)\, dt$ exists.

Example 3.15.11. To see that it is possible for an equation to have a nonlinear limit-circle type solution that is not of the strong nonlinear limit-circle type, and at the same time, have a nonlinear limit-point type solution that is not of the strong nonlinear limit-point type, consider

$$\left((t+1)^2 y'\right)' + (t+1)^{-2}|y|^\lambda \operatorname{sgn} y = 0, \quad t \geq 0. \tag{3.15.11}$$

The hypotheses of Theorem 3.11.2 are satisfied,

$$\int_0^\infty \left(\int_t^\infty a^{-\frac{1}{p}}(\sigma)\, d\sigma\right)^{\lambda+1} dt = \int_0^\infty (t+1)^{-1-\lambda} dt < \infty,$$

and

$$\int_0^\infty \frac{1}{R(t)} \left(\int_t^\infty r(\sigma)\, d\sigma\right)^\delta dt = \int_0^\infty (t+1)^{-2}\, dt < \infty.$$

Thus, there is a solution y_1 of (3.15.11) such that

$$\int_0^\infty |y_1(\sigma)|^{\lambda+1}\, d\sigma < \infty \quad \text{and} \quad \int_0^\infty \frac{|y_1^{[1]}(\sigma)|^\delta}{R(\sigma)}\, d\sigma = \infty$$

and there is a solution y_2 such that

$$\int_0^\infty |y_2(\sigma)|^{\lambda+1}\, d\sigma = \infty \quad \text{and} \quad \int_0^\infty \frac{|y_2^{[1]}(\sigma)|^\delta}{R(\sigma)}\, d\sigma < \infty.$$

Moreover, $y_1 \in N_0$ and $y_2 \in N_2$. Theorem 3.11.2(ii) also shows that this equation has a solution in N_1 that is of the strong nonlinear limit-point type. We want to emphasize here that Theorem 3.11.2 ensures that all nontrivial solutions of (3.15.11) are nonoscillatory.

Example 3.15.12. By Theorem 3.11.2, both of the equations

$$(t^{\frac{3}{2}} y')' + t^{-\frac{3}{2}} y = 0, \quad t \geq 0,$$

and

$$\left(t^4 (y')^3\right)' + t^{-\frac{4}{3}} |y|^2 \operatorname{sgn} y = 0, \quad t \geq 0,$$

are of the strong nonlinear limit-point type. This follows from the fact that all solutions are nonoscillatory and

$$\int_0^\infty \left(\int_t^\infty a^{-\frac{1}{p}}(s)\, ds \right)^{\lambda+1} dt = \int_0^\infty \frac{1}{R(t)} \left(\int_t^\infty r(s)\, ds \right)^{\delta} dt = \infty$$

for both of these equations.

Example 3.15.13. Consider the equation

$$\left(t^a |y'|^{p-1} y'\right)' + t^b |y|^\lambda \operatorname{sgn} y = 0, \quad t \geq 1, \tag{3.15.12}$$

where $p > 0$, $\lambda > 0$, and $a/p \leq -b$. If

$$p < a \leq \frac{\lambda+2}{\lambda+1} p \quad \text{and} \quad a - 2p - 1 \leq b < -1,$$

then, by Theorem 3.11.2, equation (3.15.12) is of the strong nonlinear limit-point type.

Example 3.15.14. Consider the equation

$$\left(t^\alpha |y'|^{p-1} y'\right)' = t^\beta\, |y|^\lambda \operatorname{sgn} y. \tag{3.15.13}$$

By taking $\sigma = 0$ in part (i) of Theorem 3.14.4, if

$$\lambda > p \geq \alpha \quad \text{and} \quad \beta \geq \max\left\{0, -\frac{\alpha}{p}\right\},$$

then equation (3.15.13) is of the strong nonlinear limit-circle type. On the other hand, if we choose $\sigma = \beta + \alpha/p \leq 0$, then (3.15.13) is of the strong nonlinear limit-circle type provided

$$\lambda > p \quad \text{and} \quad \beta \geq \max\{0, \alpha - p - 1\}.$$

If

$$\lambda > p \geq \alpha \quad \text{and} \quad \beta \geq 0,$$

then the hypotheses of Theorem 3.14.4(ii) are satisfied, and equation (3.15.13) is of the strong nonlinear limit-circle type.

If

$$\lambda > p \geq \alpha \quad \text{and} \quad \beta > \frac{p-\alpha}{p},$$

then by Theorem 3.14.5, any continuable solution y of (3.15.13) satisfies (3.14.5).

If $\lambda < p < \alpha$ and either

$$\frac{p-\alpha}{p} < \beta < -1$$

or

$$-1 < \beta < -1 + \frac{p-\lambda}{\lambda}\left(\frac{\alpha}{p} - 2\right),$$

then, by Theorem 3.14.6, any continuable solution y of (3.15.13) satisfies (3.14.5).

If

$$\alpha < 0 \quad \text{and} \quad \beta \geq -\alpha/p,$$

then equation (3.12.1) is of the strong nonlinear limit-point type by Theorem 3.14.7.

If

$$\lambda < p, \quad \alpha \leq p, \quad \text{and} \quad \alpha - p - 1 \leq \beta \leq (p-\alpha)/p,$$

then the hypotheses of Theorem 3.14.8 are satisfied, and so equation (3.15.13) is of the strong nonlinear limit-point type.

Chapter 4

Damped Equations

4.1 Introduction

In this chapter, we consider Emden–Fowler type equations with a p-Laplacian as considered in Chapter 2 or Chapter 3 but with a damping term. That is, we consider equations of the form

$$\left(a(t)|y'|^{p-1}y'\right)' + b(t)|y'|^{q-1}y' + r(t)|y|^{\lambda-1}y = 0 \tag{4.1.1}$$

and its special case

$$\left(a(t)|y'|^{p-1}y'\right)' + b(t)|y'|^{p-1}y' + r(t)|y|^{\lambda-1}y = 0. \tag{4.1.2}$$

We assume throughout this chapter that $a \in C^1(\mathbb{R}_+)$, $b \in C^0(\mathbb{R}_+)$, $a^{\frac{1}{p}}r \in C^1(\mathbb{R}_+)$, $a(t) > 0$, $b(t) \not\equiv 0$, and $r(t) > 0$. We will examine the sub-half-linear case $(0 < \lambda \leq p \leq q)$ in Sections 4.3 and 4.4 and the super-half-linear case $(0 < q \leq p \leq \lambda)$ in Sections 4.5 and 4.6. The case $p = q$, i.e., equation (4.1.2), is considered separately in Section 4.7. Examples of the results and some additional discussion appears in the last section of the chapter.

Definition 4.1.1. A solution y of (4.1.1) is continuable if it is defined on $\mathbb{R}_+$. It is said to be noncontinuable if it is defined on $[0, \tau)$ with $\tau < \infty$, and it cannot be defined at $t = \tau$. A continuable solution y is called proper if it is nontrivial in any neighborhood of ∞. A proper solution y of (4.1.1) is said to be oscillatory if it has a sequence of zeros tending to ∞, and it is nonoscillatory otherwise.

The definitions of a limit-point and limit-circle solution are not affected by the presence of the damping terms in the above equations. That is, the definition of a nonlinear limit-point or limit-circle solution of equation (4.1.1) is the same as it is in Definition 2.2.3. As a consequence of the fact that singular solutions may exist, the definitions of a strong nonlinear limit-point and a strong nonlinear limit-circle solution take the following form.

Definition 4.1.2. A solution y of (4.1.1) defined on $\mathbb{R}_+$ is said to be of the strong nonlinear limit-point type if

$$\int_0^\infty |y(\sigma)|^{\lambda+1}\,d\sigma = \infty$$

and

$$\int_0^\infty \frac{|y^{[1]}(\sigma)|^\delta}{R(\sigma)}\,d\sigma = \infty.$$

Equation (4.1.1) is said to be of the strong nonlinear limit-point type if every proper solution is of the strong nonlinear limit-point type and such solutions exist.

Definition 4.1.3. A solution y of (4.1.1) defined on $\mathbb{R}_+$ is said to be of the strong nonlinear limit-circle type if

$$\int_0^\infty \left|y(\sigma)\right|^{\lambda+1}\,d\sigma < \infty$$

and

$$\int_0^\infty \frac{|y^{[1]}(\sigma)|^\delta}{R(\sigma)}\,d\sigma < \infty.$$

Equation (4.1.1) is said to be of the strong nonlinear limit-circle type if every solution defined on $\mathbb{R}_+$ is of the strong nonlinear limit-circle type.

4.2 Notation

Because of what we did in earlier chapters, it should come as no surprise that we introduce the following notation. We let

$$y^{[1]}(t) = a(t)|y'(t)|^{p-1}y'(t),$$

define the function $R : \mathbb{R}_+ \to \mathbb{R}$ by

$$R(t) = a^{1/p}(t)r(t),$$

and let δ denote the constant

$$\delta = \frac{p+1}{p}.$$

We will write equation (4.1.1) as the equivalent system

$$\begin{aligned} y_1' &= a^{-\frac{1}{p}}(t)|y_2|^{\frac{1}{p}}\operatorname{sgn} y_2, \\ y_2' &= -b(t)a^{-\frac{q}{p}}(t)|y_2|^{\frac{q}{p}}\operatorname{sgn} y_2 - r(t)|y_1|^{\lambda}\operatorname{sgn} y_1, \end{aligned} \tag{4.2.1}$$

where the relationship between a solution y of (4.1.1) and a solution (y_1, y_2) of the system (4.2.1) is given by

$$y_1(t) = y(t) \quad \text{and} \quad y_2(t) = a(t)|y'(t)|^{p-1}y'(t).$$

It will also be useful to define the following constants some of which agree with those in Chapter 2 or Chapter 3:

$$\alpha = \frac{p+1}{(\lambda+2)p+1}, \quad \beta = \frac{(\lambda+1)p}{(\lambda+2)p+1}, \quad \gamma = \frac{p+1}{p(\lambda+1)},$$

$$\beta_1 = \frac{p}{(\lambda+2)p+1}, \quad \omega_2 = \frac{1}{\lambda+1} + \frac{q}{p+1}, \quad \alpha_1 = \alpha\,\gamma^{-\frac{1}{\lambda+1}},$$

$$\beta_2 = \frac{(\lambda+1)(p+1)}{p-\lambda+(v-1)p(\lambda+1)} \quad \text{for either } p > \lambda \text{ or } v > 1,$$

$$\omega_1 = \frac{vp+1}{p+1}, \quad v_1 = \beta_1(v-1), \quad \omega = \frac{1}{\lambda+1} + \frac{p}{p+1}, \quad v = \frac{q}{p} \geq 1.$$

Notice that $\alpha = 1-\beta$, $\omega_2 - 1 = 1/\beta_2$, $\omega_2 \geq \omega_1$, and $\omega \geq 1$.

As before, we define $g\colon \mathbb{R}_+ \to \mathbb{R}$ by

$$g(t) = -\frac{a^{\frac{1}{p}}(t)R'(t)}{R^{\alpha+1}(t)},$$

and we will often make use of the condition

$$\lim_{t\to\infty} g(t) = 0 \quad \text{and} \quad \int_0^{\infty} |g'(s)|\,ds < \infty. \tag{4.2.2}$$

If (4.2.2) holds, we define the constants

$$\gamma_1 = \alpha\,\gamma^{-\frac{1}{\lambda+1}} \sup_{s\in\mathbb{R}_+} |g(s)| \quad \text{and} \quad \gamma_2 = \delta + \gamma_1.$$

For any solution $y\colon \mathbb{R}_+ \to \mathbb{R}$ of (4.1.1), we let

$$F(t) = R^\beta(t)\left[\frac{a(t)}{r(t)}|y'(t)|^{p+1} + \gamma|y(t)|^{\lambda+1}\right]$$
$$= R^\beta(t)\left(\frac{|y_2(t)|^\delta}{R(t)} + \gamma|y(t)|^{\lambda+1}\right) \tag{4.2.3}$$

and observe that $F \geq 0$ on $\mathbb{R}_+$ for every solution y of (4.1.1).

4.3 The Sub-Half-Linear Case: Preliminary Lemmas

Since we are in the sub-half-linear case, we have the equation

$$\left(a(t)|y'|^{p-1}y'\right)' + b(t)|y'|^{q-1}y' + r(t)|y|^{\lambda-1}y = 0, \quad 0 < \lambda \leq p \leq q. \tag{4.3.1}$$

We begin with the following facts about the solutions (4.3.1).

Remark 4.3.1. The functions a, b, and r are smooth enough so that all nontrivial solutions of (4.3.1) defined on $\mathbb{R}_+$ are proper (see [48]). Moreover, if either $q = p$ or $b(t) \geq 0$ on $\mathbb{R}_+$, then all nontrivial solutions of (4.3.1) are defined on $\mathbb{R}_+$.

Lemma 4.3.1. *For every nontrivial solution y of* (4.3.1) *defined on $\mathbb{R}_+$, $F(t) > 0$ for $t \geq 0$.*

Proof. Suppose, to the contrary, that (4.3.1) has a nontrivial solution y such that $F(t_0) = 0$ for a number $t_0 \in \mathbb{R}_+$. Then (4.2.3) implies $y(t_0) = y'(t_0) = 0$ and so equation (4.3.1) has the solution $\bar{y}$ defined by

$$\bar{y}(t) = y(t) \quad \text{for } t \in [0, t_0] \quad \text{and} \quad \bar{y}(t) = 0 \quad \text{for } t \geq t_0.$$

But this contradicts Remark 4.3.1 and proves the lemma. □

Lemma 4.3.2. *Let y be a solution of* (4.3.1). *Then:*

(i) *for $t \in \mathbb{R}_+$, we have*

$$|y(t)| \leq \gamma^{-\frac{1}{\lambda+1}} R^{-\beta_1}(t) F^{\frac{1}{\lambda+1}}(t) \quad \text{and} \quad |y_2(t)| \leq R^{\beta_1}(t) F^{\frac{p}{p+1}}(t); \tag{4.3.2}$$

(ii) *for* $0 \le \tau < t$, *we have*

$$F(t) = F(\tau) - \alpha g(\tau)\, y(\tau)\, y_2(\tau) + \alpha g(t)\, y(t)\, y_2(t) - \alpha \int_\tau^t g'(s)\, y(s)\, y_2(s)\, ds - \int_\tau^t \Big[\delta R^{-\alpha}(s)|y_2(s)|^{\frac{1}{p}} \operatorname{sgn} y_2(s) - \alpha g(s)\, y(s)\Big] \frac{b(s)}{a^v(s)} |y_2(s)|^{v-1} y_2(s)\, ds \tag{4.3.3}$$

and

$$\left| \int_\tau^t [\delta R^{-\alpha}(s)|y_2(s)|^{\frac{1}{p}} \operatorname{sgn} y_2(s) - \alpha g(s)\, y(s)] \frac{b(s)}{a^v(s)} |y_2(s)|^{v-1} y_2(s)\, ds \right| \le \delta \int_\tau^t \frac{|b(s)|}{a^v(s)} R^{v_1}(s)\, F^{\omega_1}(s)\, ds + \gamma_1 \int_\tau^t \frac{|b(s)|}{a^v(s)} R^{v_1}(s)\, F^{\omega_2}(s)\, ds. \tag{4.3.4}$$

Proof. Let y be a solution of (4.3.1). Then it is a solution of the equation

$$\big(a(t)|z'|^{p-1}z'\big)' + r(t)|z|^{\lambda-1}z = e(t) \tag{4.3.5}$$

with $e(t) = -b(t)|y'(t)|^{q-1}y'(t) = -\frac{b(t)}{a^v(t)}|y_2(t)|^{v-1}y_2(t)$. Then (4.3.2) and (4.3.3) follow from Lemma 3.9.1 applied to (4.3.5). Relation (4.3.4) follows from (4.3.2). □

The following two lemmas give us sufficient conditions for the boundedness of F from above and from below by positive constants.

Lemma 4.3.3. *Let* (4.2.2) *hold and assume that*

$$\int_0^\infty \frac{|b(s)|}{a^v(s)} R^{v_1}(s)\, ds < \infty. \tag{4.3.6}$$

Then for any nontrivial solution y of equation (4.3.1) *defined on* $\mathbb{R}_+$, *the function F is bounded from below on* $\mathbb{R}_+$ *by a positive constant depending on y.*

Proof. Suppose, to the contrary, that there is a nontrivial solution of (4.3.1) such that

$$\liminf_{t\to\infty} F(t) = 0.$$

By Lemma 4.3.1, $F(t) > 0$ on $\mathbb{R}_+$. Let $\bar{t} \in \mathbb{R}_+$ be such that

$$2\alpha_1 \sup_{s\in[\bar{t},\infty)} |g(s)| + \alpha_1 \int_{\bar{t}}^{\infty} |g'(s)|\, ds + (\gamma_1 + \delta) \int_{\bar{t}}^{\infty} \frac{|b(s)|}{a^v(s)} R^{v_1}(s)\, ds \le \frac{1}{2}; \tag{4.3.7}$$

the existence of such a $\bar{t}$ follows from (4.2.2) and (4.3.6). Then, for any $t_0 \ge \bar{t}$ such that $F(t_0) \le 1$, there exist τ and σ such that $t_0 \le \sigma < \tau$ and

$$2F(\tau) = F(\sigma) = F(t_0) > 0 \quad \text{and} \quad F(\tau) \le F(t) \le F(\sigma) \tag{4.3.8}$$

for $\sigma \le t \le \tau$. Then (4.3.2) implies

$$|y(t)\, y_2(t)| \le \gamma^{-\frac{1}{\lambda+1}} F^{\omega}(t) \tag{4.3.9}$$

on $\mathbb{R}_+$. From this, (4.3.3) (with $\tau = \sigma$ and $t = \tau$), (4.3.4), (4.3.8), and the fact that $F(\sigma) \le 1$, we have

$$\frac{F(\sigma)}{2} = F(\sigma) - F(\tau) \le \left[\alpha_1 |g(\tau)| + \alpha_1 |g(\sigma)| + \alpha_1 \int_{\bar{t}}^{\infty} |g'(s)|\, ds\right] F^{\omega}(\sigma)$$

$$+ \gamma_1 \int_{\bar{t}}^{\infty} \frac{|b(s)|}{a^v(s)} R^{v_1}(s)\, ds F^{\omega_2}(\sigma) + \delta \int_{\bar{t}}^{\infty} \frac{|b(s)|}{a^v(s)} R^{v_1}(s)\, ds F^{\omega_1}(\sigma).$$

Hence, using (4.3.7) and the facts that $\omega_2 \ge \omega_1 \ge 1$, $\omega \ge 1$, and $F(\sigma) \le 1$, we obtain

$$F(\sigma) \le \frac{1}{2} F(\sigma).$$

This contradiction to $F(\sigma) > 0$ proves the lemma. □

Lemma 4.3.4. *Assume that $b \ge 0$ for large t, (4.2.2) holds,*

$$\int_0^{\infty} \frac{b(t)}{a^v(t)} R^{v_1}(t)\, dt < \infty, \tag{4.3.10}$$

and either (i) $\lambda = p = q$, *or* (ii) $q > \lambda$ *and*

$$\liminf_{t\to\infty} R^{\beta}(t) \left(\int_t^{\infty} \left[|g'(s)| + \frac{b(s)}{a^v(s)} R^{v_1}(s) \right] ds \right)^{\beta_2} \exp\left\{ \int_0^t \frac{R'_-(s)}{R(s)}\, ds \right\} = 0. \tag{4.3.11}$$

Then for every solution of (4.3.1) *the function F is bounded on $\mathbb{R}_+$.*

Proof. Let y be a nontrivial solution of (4.3.1). Then according to Remark 4.3.1 and Lemma 4.3.1, y is defined on $\mathbb{R}_+$ and $F(t) > 0$ on $\mathbb{R}_+$. In view of (4.2.2) and (4.3.10), we can choose $\bar{t} \in \mathbb{R}_+$ such that

$$\int_{\bar{t}}^{\infty} \left[|g'(s)| + \frac{b(s)}{a^v(s)} R^{v_1}(s) \right] ds \leq \frac{1}{2} \left[3\alpha_1 + 2^{\omega_1}\delta + 2^{\omega_2}\gamma_1 \right]^{-1}. \tag{4.3.12}$$

Suppose that F is not bounded, i.e.,

$$\limsup_{t \to \infty} F(t) = \infty.$$

Then, for any $t_0 \geq \bar{t}$ with $F(t_0) \geq 1$, there exist σ and τ such that $t_0 \leq \sigma < \tau$, $\frac{1}{2}F(\tau) = F(\sigma) = F(t_0)$, and

$$1 \leq F(\sigma) \leq F(t) \leq F(\tau) \quad \text{for} \quad \sigma \leq t \leq \tau.$$

Since g is of bounded variation and $\lim_{t \to \infty} g(t) = 0$, we see that

$$|g(\sigma)| = |g(\sigma) - g(\infty)| \leq \int_{\sigma}^{\infty} |g'(s)|\, ds. \tag{4.3.13}$$

Setting $\tau = \sigma$ and $t = \tau$ in (4.3.2)–(4.3.4), we have (4.3.9) and

$$\begin{aligned} F(\sigma) = F(\tau) - F(\sigma) \leq \Bigg[&\alpha_1 |g(\sigma)| + \alpha_1 |g(\tau)| + \alpha_1 \int_{\sigma}^{\tau} |g'(s)|\, ds \\ &+ \gamma_1 \int_{\sigma}^{\tau} \frac{b(s)}{a^v(s)} R^{v_1}(s)\, ds \Bigg] F^{\omega_2}(\tau) \\ &+ \delta \int_{\sigma}^{\infty} \frac{b(s)}{a^v(s)} R^{v_1}(s)\, ds\, F^{\omega_1}(\tau). \end{aligned}$$

From this, (4.3.7), (4.3.12), and (4.3.13), we obtain

$$\begin{aligned} F(\sigma) &\leq \left[3\alpha_1 \int_{\sigma}^{\infty} |g'(s)|\, ds + (2^{\omega_2}\gamma_1 + 2^{\omega_1}\delta) \int_{\sigma}^{\infty} \frac{b(s)}{a^v(s)} R^{v_1}(s)\, ds \right] F^{\omega_2}(\sigma) \\ &\leq K \int_{\sigma}^{\infty} \left[|g'(s)| + \frac{b(s)}{a^v(s)} R^{v_1}(s) \right] ds\, F^{\omega_2}(\sigma) \leq \frac{1}{2} F^{\omega_2}(\sigma), \end{aligned} \tag{4.3.14}$$

where $K = 3\alpha_1 + 2^{\omega_1}\delta + 2^{\omega_2}\gamma_1$.

If $\lambda = p = q$, then $\omega_2 = 1$ and (4.3.14) gives us a contradiction.

Now let $q > \lambda$ and (4.3.11) hold. Then $\omega_2 > 1$ and (4.3.14) implies

$$F(t_0) = F(\sigma) \geq K^{-\beta_2}\left(\int_\sigma^\infty \left[|g'(s)| + \frac{b(s)}{a^v(s)}R^{v_1}(s)\right] ds\right)^{-\beta_2}.$$

Hence,

$$F(t) \geq K_1\left(\int_t^\infty \left[|g'(s)| + \frac{b(s)}{a^v(s)}R^{v_1}(s)\right] ds\right)^{-\beta_2} \tag{4.3.15}$$

for all $t \geq \bar{t}$ such that $F(t) \geq 1$, where $K_1 = K^{-\beta_2}$. At the same time, (4.3.14) implies $F(t) \geq 2^{\beta_2} > 1$ for these values of t. Thus, (4.3.15) holds for all $t \geq \bar{t}$. On the other hand, if $z(t) = F(t)R^{-\beta}(t)$, then (4.2.3) implies

$$\begin{aligned} z'(t) &= (R^{-1}(t))'|y_2(t)|^\delta - \delta r^{-1}(t)\, y'(t)\, b(t)|y'(t)|^q \operatorname{sgn} y'(t) \\ &\leq (R^{-1}(t))'|y_2(t)|^\delta \leq \frac{R'_-(t)}{R(t)}R^{-\beta}(t)F(t) = \frac{R'_-(t)}{R(t)}\, z(t) \end{aligned}$$

for $t \geq \bar{t}$. So,

$$z(t) \leq z(\bar{t}) \exp \int_0^t \frac{R'_-(s)}{R(s)}\, ds.$$

From this and (4.3.15),

$$\begin{aligned} &K_1\left\{\int_t^\infty \left[|g'(s)| + \frac{b(s)}{a^v(s)}R^{v_1}(s)\right] ds\right\}^{-\beta_2} \\ &\quad \leq F(t) = R^\beta(t)z(t) \\ &\quad \leq z(\bar{t})R^\beta(t) \exp \int_0^t \frac{R'_-(s)}{R(s)}\, ds, \end{aligned}$$

which contradicts (4.3.11). Hence, F is bounded from above on $\mathbb{R}_+$. Since $F > 0$ on $\mathbb{R}_+$, the conclusion follows. □

Lemma 4.3.5. *Let* (4.2.2) *and* (4.3.6) *hold. Then there exist a solution* y *of* (4.3.1) *defined on* $\mathbb{R}_+$, *a constant* $c_0 > 0$, *and* $t_0 \in \mathbb{R}_+$ *such that*

$$0 < \frac{3}{4}c_0 \leq F(t) \leq \frac{3}{2}c_0 \quad \text{for } t \geq t_0.$$

Moreover, c_0 *can be chosen arbitrary small.*

Proof. Condition (4.2.2) implies that g is bounded, so we can choose $M > 0$, $t_0 \in \mathbb{R}_+$, and c_0 such that

$$|g(t)| \leq M \quad \text{for } t \geq t_0, \quad \int_{t_0}^{\infty} |g'(s)|\,ds \leq M, \quad \int_{t_0}^{\infty} \frac{|b(s)|}{a^v(s)} R^{v_1}(s)\,ds \leq M,$$

$$M \leq \frac{1}{4}\left(\frac{3}{2}\right)^{-\omega_2} [3\alpha\gamma^{-\frac{1}{\lambda+1}} + \delta + \gamma_1]^{-1}, \quad \text{and} \quad 0 < c_0 \leq \frac{2}{3}.$$

Consider a solution y of (4.3.1) such that $F(t_0) = c_0$. First, we will show that

$$F(t) \leq \frac{3}{2}c_0 \leq 1 \quad \text{for } t \geq t_0. \tag{4.3.16}$$

Suppose (4.3.16) does not hold. Then there exist $t_2 > t_1 \geq t_0$ such that

$$F(t_2) = \frac{3}{2}c_0, \quad F(t_1) = c_0, \quad \text{and} \quad c_0 < F(t) < \frac{3}{2}c_0$$

for $t \in (t_1, t_2)$. Lemma 4.3.2 (with $\tau = t_1$ and $t = t_2$), and the facts that $\omega \leq \omega_2$, $\omega_1 \leq \omega_2$, and $c_0 < 1$ imply

$$\begin{aligned}\frac{c_0}{2} = \frac{F(t_1)}{2} &= F(t_2) - F(t_1) \leq 3\alpha\gamma^{-\frac{1}{\lambda+1}} M \left(\frac{3}{2}c_0\right)^{\omega} \\ &\quad + M(\delta F^{\omega_1}(t_2)) + \gamma_1 M F^{\omega_2}(t_2) \\ &\leq M(3\alpha\gamma^{-\frac{1}{\lambda+1}} + \delta + \gamma_1)\left(\frac{3}{2}\right)^{\omega_2} c_0^{\omega} \leq \frac{1}{4}c_0^{\omega}.\end{aligned}$$

Hence, $c_0^{\omega-1} \geq 2$ which contradicts the choice of c_0, and so (4.3.16) holds.

Now, Lemma 4.3.2 (with $t = t$, $\tau = t_0$) similarly implies

$$\begin{aligned}|F(t) - c_0| &\leq 3\alpha\gamma^{-\frac{1}{\lambda+1}} M \left(\frac{3}{2}c_0\right)^{\omega} + M\left[\left(\frac{3}{2}\right)^{\omega_1} \delta c_0^{\omega_1} + \left(\frac{3}{2}\right)^{\omega_2} \gamma_1 c_0^{\omega_2}\right] \\ &\leq M c_0 \left[3\left(\frac{3}{2}\right)^{\omega} \alpha\gamma^{-\frac{1}{\lambda+1}} + \left(\frac{3}{2}\right)^{\omega_1} \delta + \left(\frac{3}{2}\right)^{\omega_2} \gamma_1\right] \leq \frac{c_0}{4},\end{aligned}$$

and the statement of the lemma is proved. □

Lemma 4.3.6. *Suppose that (4.2.2) and (4.3.6) hold and*

$$\int_0^\infty R^{-\beta}(t)\,dt = \infty. \tag{4.3.17}$$

In addition, assume that either

$$\int_0^\infty \left(\left| \left(\frac{1}{r(s)} \right)' \right| + \frac{|b(s)|}{a^v(s)r(s)} R^{\beta_1(v-1)}(s) \right) ds < \infty \tag{4.3.18}$$

or

$$\int_0^\infty \left(\frac{|a'(s)|}{a(s)r(s)} + \frac{|b(s)|}{a^v(s)r(s)} R^{v_1}(s) \right) ds < \infty \tag{4.3.19}$$

holds. If y is a solution of (4.3.1) with

$$c_1 \le F(t) \le c_2 \tag{4.3.20}$$

on $\mathbb{R}_+$ for some positive constants c_1 and c_2, then

$$\int_0^\infty |y(t)|^{\lambda+1}\,dt = \infty. \tag{4.3.21}$$

Moreover, if r does not tend to zero as $t \to \infty$, then

$$\int_0^\infty \frac{|y_2(t)|^\delta}{R(t)}\,dt = \infty. \tag{4.3.22}$$

Proof. Let y be a nontrivial solution of (4.3.1) satisfying (4.3.20). Then in view of (4.2.3), (4.3.17), and (4.3.20)

$$\gamma \int_0^t |y(s)|^{\lambda+1}\,ds + \int_0^t \frac{|y_2(s)|^\delta}{R(s)}\,ds = \int_0^t \frac{F(s)}{R^\beta(s)}\,ds \to \infty \tag{4.3.23}$$

as $t \to \infty$. Now, (4.3.2) and (4.3.20) imply

$$|y(t)\,y_2(t)| \le \gamma^{-\frac{1}{\lambda+1}} c_2^\omega \stackrel{\text{def}}{=} M_1 \tag{4.3.24}$$

for $t \ge 0$ so there exists $t_0 \ge 0$ such that

$$|g(t)| \le \frac{c_1}{2M_1 \max(1,\gamma)} \tag{4.3.25}$$

for $t \geq t_0$. It follows from (4.3.1) that

$$\int_0^t |y(s)|^{\lambda+1}\,ds = -\int_0^t \frac{y(s)y_2'(s)}{r(s)}\,ds - \int_0^t \frac{b(s)}{a^v(s)r(s)} y(s)|y_2(s)|^{v-1}y_2(s)\,ds$$

$$= -\frac{y(t)y_2(t)}{r(t)} + D + \int_0^t \frac{|y_2(s)|^\delta}{R(s)}\,ds + J(t) \tag{4.3.26}$$

where $D = y(0)y_2(0)r^{-1}(0)$ and

$$J(t) = \int_0^t \left[\left(\frac{1}{r(s)}\right)' - \frac{b(s)}{a^v(s)r(s)}|y_2(s)|^{v-1}\right] y(s)y_2(s)\,ds. \tag{4.3.27}$$

Hence, from (4.3.24),

$$|J(t)| \leq M_1 D_1 \int_0^t \left[\left|\left(\frac{1}{r(s)}\right)'\right| + \frac{|b(s)|}{a^v(s)r(s)} R^{\beta_1(v-1)}(s)\right] ds \tag{4.3.28}$$

with $D_1 = 1 + c_2^{\frac{q-p}{p+1}}$ for $t \geq t_0$. Moreover, (4.2.3), (4.3.20), and (4.3.23) imply

$$c_3 \int_{t_0}^t R^{-\beta}(s)\,ds \leq \int_{t_0}^t |y(s)|^{\lambda+1}\,ds + \int_{t_0}^t \frac{|y_2(s)|^\delta}{R(s)}\,ds \leq c_4 \int_{t_0}^t R^{-\beta}(s)\,ds \tag{4.3.29}$$

for $t \geq t_0$ with $c_3 = \frac{c_1}{\max(1,\gamma)}$ and $c_4 = \frac{c_2}{\min(1,\gamma)}$.

If (4.3.18) holds, then (4.3.28) implies J is bounded on $\mathbb{R}_+$, and in view of (4.3.26), we have

$$\left|\int_{t_0}^t |y(s)|^{\lambda+1}\,ds - \int_{t_0}^t \frac{|y_2(s)|^\delta}{R(s)}\,ds + \frac{y(t)y_2(t)}{r(t)}\right| \leq J_1(t) + m\int_{t_0}^t R^{-\beta}(s)\,ds \tag{4.3.30}$$

for $t \geq t_0$ with $J_1(t) = |J(t)| + \frac{|y(0)y_2(0)|}{r(0)}$ and $m = 0$. Note that J_1 is bounded on $\mathbb{R}_+$.

Now let (4.3.19) hold. Then, using (4.3.2), (4.3.20), and (4.3.25) and setting $c_5 = \max(1, c_2^{\frac{q-p}{p+1}})$, we have

$$\int_{t_0}^{t} \left[\left| \left(\frac{1}{r(s)} \right)' \right| + \frac{|b(s)|}{a^v(s)r(s)} |y_2(s)|^{v-1} \right] ds$$

$$\leq \int_{t_0}^{t} \left(\left| \left(\frac{a^{\frac{1}{p}}}{R(s)} \right)' \right| + \frac{|b(s)|}{a^v(s)r(s)} R^{\beta_1(v-1)}(s) c_2^{\frac{q-p}{p+1}} \right) ds$$

$$\leq c_5 \int_{0}^{t} \left(\frac{|a'(s)|}{pa(s)r(s)} + \frac{|b(s)|}{a^v(s)r(s)} R^{\beta_1(v-1)}(s) \right) ds$$

$$+ \int_{t_0}^{t} \frac{|g(s)|}{R^{\beta}(s)} ds \leq M_2 + \frac{c_3}{2M_1} \int_{t_0}^{t} R^{-\beta}(s)\, ds \tag{4.3.31}$$

for $t \geq t_0$, where $M_2 = c_5 \int_0^\infty \left(\frac{|a'(s)|}{pa(s)r(s)} + \frac{|b(s)|}{a^v(s)r(s)} R^{\beta_1(v-1)}(s) \right) ds < \infty$ by (4.3.19). From (4.3.31) together with (4.3.24), (4.3.26), and (4.3.27), inequality (4.3.30) holds with $m = \frac{c_3}{2}$ and $J_1(t) = M_1 M_2$. Thus, (4.3.30) holds with $m = \frac{c_3}{2}$ if either (4.3.18) or (4.3.19) holds, and J_1 is bounded for $t \geq t_0$. Moreover,

$$-J_1(t) - \frac{c_3}{2} \int_{t_0}^{t} R^{-\beta}(s)\, ds \leq \int_{t_0}^{t} |y(s)|^{\lambda+1} ds - \int_{t_0}^{t} \frac{|y_2(s)|^{\delta}}{R(s)} ds + \frac{y(t)\, y_2(t)}{r(t)}$$

$$\leq J_1(t) + \frac{c_3}{2} \int_{t_0}^{t} R^{-\beta}(s)\, ds \tag{4.3.32}$$

for $t \geq t_0$. Adding the left-hand inequalities in (4.3.29) and (4.3.32) gives

$$2 \int_{t_0}^{t} |y(s)|^{\lambda+1} ds + \frac{y(t)\, y_2(t)}{r(t)} \geq \frac{c_3}{2} \int_{t_0}^{t} R^{-\beta}(s)\, ds - J_1(t) \to \infty \tag{4.3.33}$$

as $t \to \infty$, and subtracting gives

$$2 \int_{t_0}^{t} \frac{|y_2(s)|^{\delta}}{R(s)} ds - \frac{y(t)\, y_2(t)}{r(t)} \geq \frac{c_3}{2} \int_{t_0}^{t} R^{-\beta}(s)\, ds - J_1(t) \to \infty \tag{4.3.34}$$

as $t \to \infty$.

If y' is oscillatory, let $\{t_k\}_{k=1}^{\infty} \to \infty$ be a sequence of zeros of y'. Then letting $t = t_k$ in (4.3.33) and (4.3.34), it is clear that the conclusion of the lemma holds.

Let y' be nonoscillatory. Then either

$$y(t)\, y_2(t) > 0 \quad \text{for large } t \tag{4.3.35}$$

or

$$y(t)\, y_2(t) < 0 \quad \text{for large } t. \tag{4.3.36}$$

We first prove (4.3.21). It clearly holds if (4.3.35) does. So suppose (4.3.36) holds. Then $y(t)\, y_2(t)\, r^{-1}(t) < 0$ for large t and (4.3.33) gives us the contradiction. Hence, (4.3.21) holds.

Finally, we prove (4.3.22). From (4.3.34), (4.3.22) holds if (4.3.35) does. Let (4.3.36) hold and assume that (4.3.22) does not. Then (4.3.34) implies

$$\lim_{t\to\infty} \frac{y(t)\, y_2(t)}{r(t)} = -\infty.$$

In view of (4.3.24) and (4.3.36), $y(t)\, y_2(t) \geq -M_1$, so $\lim_{t\to\infty} r(t) = 0$. This contradicts the assumptions of the lemma and completes the proof. □

Remark 4.3.2. Note that Lemmas 4.3.1, 4.3.2, and 4.3.6 hold without the restriction that $\lambda \leq p$.

4.4 The Sub-Half-Linear Case: Main Results

In this section we give our limit-point and limit-circle results for the sub-half-linear case of equation(4.1.1), namely for equation (4.3.1).

Theorem 4.4.1. *Let $b \geq 0$ for large t and assume that* (4.2.2) *and* (4.3.10) *hold. In addition, if $q > \lambda$, assume that* (4.3.11) *also holds. Then* (4.3.1) *is of the strong nonlinear limit-circle type if and only if*

$$\int_0^\infty R^{-\beta}(t)\, dt < \infty. \tag{4.4.1}$$

Proof. Let y be a nontrivial solution of (4.3.1). By Remark 4.3.1, y is defined on $\mathbb{R}_+$. The hypotheses of Lemmas 4.3.3 and 4.3.4 are satisfied, so there are constants c and c_1 such that

$$0 < c \leq F(t) \leq c_1$$

on $\mathbb{R}_+$. Hence, from this and (4.2.3),

$$c\int_0^\infty R^{-\beta}(t)\,dt \le \gamma\int_0^\infty |y(t)|^{\lambda+1}\,dt + \int_0^\infty \frac{|y_2(t)|^\delta}{R(t)}\,dt$$
$$= \int_0^\infty F(t)R^{-\beta}(t)\,dt \le c_1\int_0^\infty R^{-\beta}(t)\,dt.$$

The conclusion of the theorem then follows from (4.4.1). □

Theorem 4.4.2. *Let* (4.2.2), (4.3.6), *and either* (4.3.18) *or* (4.3.19) *hold. If*

$$\int_0^\infty R^{-\beta}(t)\,dt = \infty,$$

then (4.3.1) *is of the nonlinear limit-point type.*

Proof. The hypotheses of Lemmas 4.3.5 and 4.3.6 are satisfied, so if y is a solution given by Lemma 4.3.5, then (4.3.21) holds, and the conclusion follows. □

Theorem 4.4.3. *Let $b \ge 0$ for large t and let conditions* (4.2.2), (4.3.10), *and either* (4.3.18) *or* (4.3.19) *hold. In addition, if $q > \lambda$, assume that* (4.3.11) *holds. If*

$$\int_0^\infty R^{-\beta}(t)\,dt = \infty,$$

then every nontrivial solution of (4.3.1) *is of the nonlinear limit-point type. If, moreover, r does not tend to zero as $t \to \infty$, then* (4.3.1) *is of the strong nonlinear limit-point type.*

Proof. Note that the hypotheses of Lemmas 4.3.3, 4.3.4, and 4.3.6 are satisfied. Let y be a nontrivial solution of (4.3.1). Then Remark 4.3.1 implies y is defined on $\mathbb{R}_+$, and by Lemmas 4.3.3 and 4.3.4, there are positive constants C_1 and C_2 such that

$$0 < C_1 \le F(t) \le C_2 \quad \text{on } \mathbb{R}_+.$$

Thus, by Lemma 4.3.6, (4.3.21) holds, and if r does not tend to zero as $t \to \infty$, then (4.3.22) holds. This proves the theorem. □

4.5 The Super-Half-Linear Case: Preliminary Lemmas

We are in the super-half-linear case in this section so we are considering the equation

$$\left(a(t)|y'|^{p-1}y'\right)' + b(t)|y'|^{q-1}y' + r(t)|y|^{\lambda-1}y = 0, \quad 0 < q \le p \le \lambda. \tag{4.5.1}$$

Remark 4.5.1. Under the covering assumptions here, the functions a, b, and r are smooth enough so that all solutions of (4.5.1) are defined for large t (see [48, Theorem 2(i)]). Moreover, all nontrivial solutions of (4.5.1) are proper if either $b \le 0$ on $\mathbb{R}_+$ or $q = p$ (see [48, Theorem 4]).

In addition to the constants α, β, γ, α_1, β_1, and ω defined in Section 4.2, we will need the constants

$$\beta_3 = \frac{(\lambda+1)(p+1)}{\lambda(p-q)+(\lambda-q)} \quad \text{for either } p > q \text{ or } \lambda > q,$$

$$\omega_3 = \frac{q+1}{p+1}, \quad \omega_4 = \frac{1}{\lambda+1} + \frac{q}{p+1},$$

$$v = \frac{q}{p} \le 1, \quad v_2 = \frac{q-p}{(\lambda+2)p+1}.$$

Notice that in this case, $\omega_4 \le \omega_3$, and $\omega \le 1$.

For this and the super-half-linear case, we will need the following lemmas.

Lemma 4.5.1. *Let either $p = q$ or $b \le 0$ on $\mathbb{R}_+$. Then for every nontrivial solution y of* (4.5.1) *we have $F(t) > 0$ for large t.*

Proof. Let y be a nontrivial solution of (4.5.1). Then (4.2.3) implies the existence of $T \ge 0$ such that $F(T) > 0$. Suppose, to the contrary, that $F(t_0) = 0$ for some $t_0 > T$. Then (4.2.3) implies $y(t_0) = y'(t_0) = 0$ and so equation (4.5.1) has the solution $\bar{y}$ defined by

$$\bar{y}(t) = y(t) \quad \text{for } t \in [0, t_0] \quad \text{and} \quad \bar{y}(t) = 0 \quad \text{for } t \ge t_0.$$

But this contradicts Remark 4.5.1 and proves the lemma. □

Lemma 4.5.2. *Let y be a solution of* (4.5.1).

(i) *If $t \in \mathbb{R}_+$, we have*

$$|y(t)| \le \gamma^{-\frac{1}{\lambda+1}} R^{-\beta_1}(t) F^{\frac{1}{\lambda+1}}(t) \quad \text{and} \quad |y_2(t)| \le R^{\beta_1}(t) F^{\frac{p}{p+1}}(t). \tag{4.5.2}$$

(ii) *If* $0 \le \tau < t$, *we have*

$$F(t) = F(\tau) - \alpha g(\tau)\, y(\tau)\, y_2(\tau) + \alpha g(t)\, y(t)\, y_2(t)$$
$$-\alpha \int_\tau^t g'(s)\, y(s)\, y_2(s)\, ds - \int_\tau^t \Big[\delta R^{-\alpha}(s)|y_2(s)|^{\frac{1}{p}} \operatorname{sgn} y_2(s)$$
$$-\alpha g(s)\, y(s)\Big] \frac{b(s)}{a^v(s)} |y_2(s)|^{v-1} y_2(s)\, ds \tag{4.5.3}$$

and

$$\left| \int_\tau^t \Big[\delta R^{-\alpha}(s)|y_2(s)|^{\frac{1}{p}} \operatorname{sgn} y_2(s) - \alpha g(s)\, y(s)\Big] \frac{b(s)}{a^v(s)} |y_2(s)|^{v-1} y_2(s)\, ds \right|$$
$$\le \delta \int_\tau^t \frac{|b(s)|}{a^v(s)} R^{v_2}(s)\, F^{\omega_3}(s)\, ds + \gamma_1 \int_\tau^t \frac{|b(s)|}{a^v(s)} R^{v_2}(s)\, F^{\omega_4}(s)\, ds. \tag{4.5.4}$$

Proof. Let y be a solution of (4.5.1). Then it is a solution of the equation

$$\big(a(t)|z'|^{p-1}z'\big)' + r(t)|z|^{\lambda-1}z = e(t) \tag{4.5.5}$$

with $e(t) = -b(t)|y'(t)|^{q-1}y'(t) = -\frac{b(t)}{a^v(t)}|y_2(t)|^{v-1}y_2(t)$. The expressions (4.5.2) and (4.5.3) follow from Lemma 3.9.1 applied to (4.5.5). Inequality (4.5.4) follows from (4.5.2). □

The next two lemmas give us sufficient conditions for the boundedness of F from above and from below by positive constants.

Lemma 4.5.3. *In addition to* (4.2.2), *assume that*

$$\int_0^\infty \frac{|b(t)|}{a^v(t)} R^{v_2}(t)\, dt < \infty \tag{4.5.6}$$

and one of the following conditions holds:
(i) $\lambda = p = q$; *or*
(ii) $b \le 0$ *for large* t *and*

$$\liminf_{t\to\infty} R^{-\beta}(t) \left\{ \int_t^\infty \left[|g'(s)| + \frac{|b(s)|}{a^v(s)} R^{v_2}(s) \right] ds \right\}^{\beta_3}$$
$$\times \exp\left\{ \int_0^t \frac{R'_+(s)}{R(s)}\, ds \right\} = 0; \quad \textit{or} \tag{4.5.7}$$

(iii) $p = q < \lambda$ *and* (4.5.7) *holds.*
Then for any nontrivial solution y *of equation* (4.5.1) *defined on* $\mathbb{R}_+$, *the function* F *is bounded from below for large* t *by a positive constant depending on* y.

Proof. Suppose, to the contrary, that there is a nontrivial solution of (4.5.1) such that

$$\liminf_{t\to\infty} F(t) = 0.$$

By Lemma 4.5.1, $F(t) > 0$ for large t. Let $\bar{t} \in \mathbb{R}_+$ be such that $F(t) > 0$ for $t \geq \bar{t}$ and

$$2\alpha_1 \sup_{s\in[\bar{t},\infty)} |g(s)| + \alpha_1 \int_{\bar{t}}^{\infty} |g'(s)|\, ds + \gamma_2 \int_{\bar{t}}^{\infty} \frac{|b(s)|}{a^v(s)} R^{v_2}(s)\, ds \leq \frac{1}{4}; \tag{4.5.8}$$

the existence of such a $\bar{t}$ follows from (4.2.2) and (4.5.6). Then, for any $t_0 \geq \bar{t}$ such that $F(t_0) \leq 1$, there exist τ and σ such that $t_0 \leq \sigma < \tau$ and

$$2F(\tau) = F(\sigma) = F(t_0) > 0 \quad \text{and} \quad F(\tau) \leq F(t) \leq F(\sigma) \tag{4.5.9}$$

for $\sigma \leq t \leq \tau$. Then (4.5.2) implies

$$|y(t)\, y_2(t)| \leq \gamma^{-\frac{1}{\lambda+1}} F^{\omega}(t) \tag{4.5.10}$$

on $\mathbb{R}_+$. From (4.5.3) (with $\tau = \sigma$ and $t = \tau$), inequalities (4.5.4) and (4.5.10), and the fact that $F(\sigma) \leq 1$, we have

$$\begin{aligned} \frac{F(\sigma)}{2} &= F(\sigma) - F(\tau) \\ &\leq \left[\alpha_1 |g(\tau)| + \alpha_1 |g(\sigma)| + \alpha_1 \int_{\sigma}^{\infty} |g'(s)|\, ds\right] F^{\omega}(\sigma) \\ &\quad + \int_{\sigma}^{\infty} \frac{|b(s)|}{a^v(s)} R^{v_2}(s)\, ds (\gamma_1 F^{\omega_4}(\sigma) + \delta F^{\omega_3}(\sigma)). \end{aligned} \tag{4.5.11}$$

Since $\min(\omega, \omega_3, \omega_4) = \omega_4 \leq 1$, (4.5.8) and (4.5.11) imply

$$F(\sigma) \leq \frac{1}{2} F^{\omega_4}(\sigma). \tag{4.5.12}$$

It is easy to see that $\omega_4 = 1$ if and only if $p = \lambda = q$.
In this case (4.5.12) and $F > 0$ give us a contradiction and the statement of the lemma holds in case (i).

Suppose that $p = \lambda = q$ does not hold; this implies $\omega_4 \neq 1$. Since g is of bounded variation and $\lim_{t\to\infty} g(t) = 0$, we see that

$$|g(\sigma)| = |g(\sigma) - g(\infty)| \leq \int_\sigma^\infty |g'(s)|\, ds. \tag{4.5.13}$$

From this, (4.5.9), and (4.5.11), we obtain

$$F(\sigma) \leq 2\left\{3\alpha_1 \int_\sigma^\infty |g'(s)|\, ds + \gamma_2 \int_\sigma^\infty \frac{|b(s)|}{a^v(s)} R^{v_2}(s)\, ds\right\} F^{\omega_4}(\sigma),$$

or

$$F(t_0) = F(\sigma) \leq K\left\{\int_\sigma^\infty \left[|g'(s)| + \frac{|b(s)|}{a^v(s)} R^{v_2}(s)\right] ds\right\}^{\beta_3}$$

with $K = (6\alpha_1 + 2\gamma_2)^{\beta_3}$. Hence,

$$F(t) \leq K\left\{\int_t^\infty \left[|g'(s)| + \frac{|b(s)|}{a^v(s)} R^{v_2}(s)\right] ds\right\}^{\beta_3} \tag{4.5.14}$$

for all $t \geq \bar{t}$ such that $F(t) \leq 1$. At the same time, (4.5.12) implies $F(t) \leq 2^{-\beta_3} < 1$ for these values of t. Thus, (4.5.14) holds for all $t \geq \bar{t}$ and so $\lim_{t\to\infty} F(t) = 0$.

Now in cases (ii) and (iii) we can actually estimate a bound from below on F. Let

$$Z(t) = F(t) R^{-\beta}(t). \tag{4.5.15}$$

Then (4.2.3) and (4.5.15) imply

$$Z'(t) = \left(R^{-1}(t)\right)' |y_2(t)|^\delta - \delta r^{-1}(t)\, b(t) |y'(t)|^{q+1}. \tag{4.5.16}$$

Define

$$S(t) = \begin{cases} -(R^{-1}(t))'_-, & \text{in case (ii)}, \\ -(R^{-1}(t))'_- - \delta \dfrac{b_+(t)}{r(t)} a^{-\delta}(t), & \text{in case (iii)}. \end{cases}$$

Suppose case (ii) holds. Then (4.5.15) and (4.5.16) imply

$$Z'(t) \geq S(t)\, R(t)\, Z(t). \tag{4.5.17}$$

Suppose case (iii) holds. Then (4.5.15) and (4.5.16) imply

$$Z'(t) \geq -\big(R^{-1}(t)\big)'_-|y_2(t)|^\delta - \delta\frac{b_+(t)}{r(t)}a^{-\delta}(t)|y_2(t)|^\delta$$
$$\geq S(t)\,R(t)\,Z(t).$$

Hence, (4.5.17) holds in both cases (ii) and (iii) for $t \geq \bar{t}$. So

$$Z(t) \geq Z(\bar{t})\exp\left\{\int_{\bar{t}}^{t} S(\sigma)\,R(\sigma)\,d\sigma\right\},$$

and in view of (4.5.14) and the fact that $S(t) \leq 0$,

$$Z(\bar{t})R^\beta(t)\exp\left\{\int_{\bar{t}}^{t} S(\sigma)\,R(\sigma)\,d\sigma\right\}$$
$$\leq F(t) \leq K\left\{\int_t^\infty \left[|g'(s)| + \frac{|b(s)|}{a^v(s)}R^{v_2}(s)\right]ds\right\}^{\beta_3},$$

which contradicts (4.5.7). Notice that

$$\exp\left\{-\int_{\bar{t}}^{t} S(\sigma)\,R(\sigma)\,d\sigma\right\} \leq C\exp\int_0^t \frac{R'_+(s)}{R(s)}\,ds$$

with $C = 1$ in case (ii) and $C = \exp\{\delta\int_0^\infty \frac{|b(s)|}{a(s)}\,ds\}$ in case (iii). □

Lemma 4.5.4. *Let* (4.2.2) *and* (4.5.6) *hold. Then for every solution of* (4.5.1), *the function* F *is bounded.*

Proof. Let y be a solution of (4.5.1) and let $\bar{t} \geq 0$ be such that (4.5.8) holds. Suppose that F is not bounded, i.e.,

$$\limsup_{t\to\infty} F(t) = \infty.$$

Then for any $t_0 \geq \bar{t}$ with $F(t_0) \geq 1$, there exist σ and τ such that $t_0 \leq \sigma < \tau$, $\frac{F(\tau)}{2} = F(\sigma) = F(t_0)$, and

$$1 \leq F(\sigma) \leq F(t) \leq F(\tau) \quad \text{for } \sigma \leq t \leq \tau. \tag{4.5.18}$$

Setting $\tau = \sigma$ and $t = \tau$ in (4.5.2)–(4.5.4), we have (4.5.10) and

$$\frac{F(\tau)}{2} = F(\tau) - F(\sigma) \leq \left[\alpha_1|g(\sigma)| + \alpha_1|g(\tau)| + \alpha_1\int_\sigma^\tau |g'(s)|\,ds\right]F^\omega(\tau)$$
$$+\int_\sigma^\infty \frac{|b(s)|}{a^v(s)}R^{v_2}(s)\,ds(\gamma_1 F^{\omega_4}(\tau) + \delta F^{\omega_3}(\tau)).$$

Since $\max(\omega, \omega_4, \omega_3) \le 1$, (4.5.8) and (4.5.18) imply

$$F(\tau) \le 2\left[2\alpha_1 \max_{s\in[\bar{t},\infty)}|g(s)| + \alpha_1 \int_{\bar{t}}^{\infty} |g'(s)|\,ds \right.$$
$$\left. + \gamma_2 \int_{\bar{t}}^{\infty} \frac{|b(s)|}{a^v(s)} R^{v_2}(s)\,ds\right] F(\tau) \le \frac{F(\tau)}{2}.$$

This contradiction proves that F is bounded. □

Lemma 4.5.5. *Let* (4.2.2) *and* (4.5.6) *hold. Then there exist a solution* y *of* (4.5.1) *and positive constants* c_1 *and* c_2 *and* $t_0 \ge 0$ *such that*

$$0 < c_1 \le F(t) \le c_2 \quad \textit{for } t \in [t_0, \infty). \tag{4.5.19}$$

Proof. Condition (4.2.2) implies that g is bounded, so we can choose $t_0 \in \mathbb{R}_+$ such that

$$M = \frac{1}{4}[3\alpha_1 + \delta + \gamma_1]^{-1},$$
$$|g(t)| \le M \quad \text{for } t \ge t_0,$$
$$\int_{t_0}^{\infty} |g'(s)|\,ds \le M, \quad \text{and} \quad \int_{t_0}^{\infty} \frac{|b(s)|}{a^v(s)} R^{v_2}(s)\,ds \le M.$$

Consider a solution y of (4.5.1) such that $F(t_0) = 1$. First, we will show that

$$F(t) \le \frac{3}{2} \quad \text{for } t \ge t_0. \tag{4.5.20}$$

Suppose (4.5.20) does not hold. Then there exist $t_2 > t_1 \ge t_0$ such that

$$F(t_2) = \frac{3}{2}, \quad F(t_1) = 1 \quad \text{and} \quad 1 < F(t) < \frac{3}{2}$$

for $t \in (t_1, t_2)$. Lemma 4.5.2 (with $\tau = t_1$ and $t = t_2$), together with the facts that $\omega \le 1$ and $\omega_4 \le \omega_3 \le 1$, implies

$$\frac{1}{2} = \frac{F(t_1)}{2} = F(t_2) - F(t_1) \le 3\alpha_1 M F^{\omega}(t_2) + \delta M F^{\omega_3}(t_2) + \gamma_1 M F^{\omega_4}(t_2)$$
$$\le \frac{3}{2} M[3\alpha_1 + \delta + \gamma_1] = \frac{3}{8}.$$

This contradiction shows that (4.5.20) holds.

Now, Lemma 4.5.2 (with $t = t$, $\tau = t_0$) similarly implies

$$|F(t) - 1| \leq 3\alpha_1 M \left(\tfrac{3}{2}\right)^{\omega} + \delta M \left(\tfrac{3}{2}\right)^{\omega_3} + \gamma_1 M \left(\tfrac{3}{2}\right)^{\omega_4}$$
$$\leq \tfrac{3}{2} M[3\alpha_1 + \delta + \gamma_1] < \tfrac{1}{2}$$

for $t \geq t_0$. From this and (4.5.20), we have

$$\tfrac{1}{2} \leq F(t) \leq \tfrac{3}{2} \quad \text{for } t \geq t_0. \tag{4.5.21}$$

Next, we prove that y is defined on $[0, t_0]$. Suppose to the contrary that y is defined on (a, t_0) with $0 \leq a < t_0$ and y cannot be extended to $t = a$. Then $\limsup_{t\to a^+} |y_2(t)| = \infty$. The change of variables $x = t_0 - t$ and $y(t) = Y(x)$ transforms (4.5.1) into

$$(a(t_0 - x)|\dot{Y}|^{p-1}\dot{Y})^{\cdot} - b(t_0 - x)|\dot{Y}|^{q-1}\dot{Y} + r(t_0 - x)|Y|^{\lambda-1}Y = 0, \tag{4.5.22}$$

where "$\cdot$" $= \frac{d}{ds}$, which has the noncontinuable solution $Y(x)$ defined on $[0, t_0 - a)$ and $\limsup_{x\to(t_0-a)^-} |\dot{Y}(x)| = \infty$. This contradicts Remark 4.5.1 applied to equation (4.5.22) and proves that y is defined on $\mathbb{R}_+$. The conclusion of the lemma then follows from (4.5.21). □

Lemma 4.5.6. *Suppose that* (4.2.2) *and* (4.5.6) *hold and*

$$\int_0^\infty R^{-\beta}(t)\, dt = \infty. \tag{4.5.23}$$

In addition, assume that either

$$\lim_{t\to\infty} R^\beta(t) \left[\left| \left(\frac{1}{r(t)} \right)' \right| + \frac{|b(t)|}{a^v(t) r(t)} R^{v_2}(t) \right] = 0 \tag{4.5.24}$$

or

$$\lim_{t\to\infty} R^\beta(t) \left[\frac{|a'(t)|}{a(t)r(t)} + \frac{|b(t)|}{a^v(t) r(t)} R^{v_2}(t) \right] = 0 \tag{4.5.25}$$

holds. If y is a solution of (4.5.1) *with*

$$c_1 \leq F(t) \leq c_2 \tag{4.5.26}$$

for large t and some positive constants c_1 and c_2, then

$$\int_0^\infty |y(t)|^{\lambda+1} dt = \infty. \tag{4.5.27}$$

Moreover, if

$$\limsup_{t\to\infty} r(t) \int_0^t R^{-\beta}(s)\, ds = \infty, \tag{4.5.28}$$

then

$$\int_0^\infty \frac{|y_2(t)|^\delta}{R(t)}\, dt = \infty, \tag{4.5.29}$$

i.e., y is of the strong nonlinear limit-point type.

Proof. Let y be a nontrivial solution of (4.5.1) satisfying (4.5.26) on $[T, \infty) \subset \mathbb{R}_+$. Then, (4.5.2) and (4.5.26) imply the existence of M_1 such that

$$\begin{aligned} |y(t)y_2(t)| &\le \gamma^{-\frac{1}{\lambda+1}} c_2^{\omega} \le M_1, \\ |y(t)|\, |y_2(t)|^v &\le \gamma^{-\frac{1}{\lambda+1}} c_2^{\omega_4} R^{v_2}(t) \le M_1 R^{v_2}(t) \end{aligned} \tag{4.5.30}$$

for $t \ge T$.

Let $t_0 \ge T$ be such that

$$|g(t)| \le c_1 \min(1, \gamma)(4M_1)^{-1} \tag{4.5.31}$$

holds for $t \ge t_0$. It follows from (4.2.1) that

$$\begin{aligned} \int_{t_0}^t |y(s)|^{\lambda+1} ds &= -\int_{t_0}^t \frac{y(s)y_2'(s)}{r(s)}\, ds - \int_{t_0}^t \frac{b(s)}{a^v(s)r(s)}\, y(s)|y_2(s)|^v \operatorname{sgn} y_2(s)\, ds \\ &= -\frac{y(t)y_2(t)}{r(t)} + D + \int_{t_0}^t \frac{|y_2(s)|^\delta}{R(s)}\, ds + J(t), \end{aligned} \tag{4.5.32}$$

where $D = y(t_0)y_2(t_0)r^{-1}(t_0)$ and

$$J(t) = \int_{t_0}^t \left[\left(\frac{1}{r(s)} \right)' y(s)y_2(s) - \frac{b(s)}{a^v(s)r(s)} y(s)\, |y_2(s)|^v \operatorname{sgn} y_2(s) \right] ds. \tag{4.5.33}$$

From (4.5.30) and (4.5.33)

$$|J(t)| \leq M_1 \int_{t_0}^{t} \left[\left| \left(\frac{1}{r(s)} \right)' \right| + \frac{|b(s)|}{a^v(s)r(s)} R^{v_2}(s) \right] ds. \tag{4.5.34}$$

Moreover, (4.2.3) and (4.5.26) imply

$$c_3 \int_{t_0}^{t} R^{-\beta}(s)ds \leq \int_{t_0}^{t} |y(s)|^{\lambda+1} ds + \int_{t_0}^{t} \frac{|y_2(s)|^{\delta}}{R(s)} ds \tag{4.5.35}$$

for $t \geq t_0$ with $c_3 = c_1 \min(1, \gamma)$.

Suppose (4.5.24) holds. By l'Hôpital's rule, there exists $t_1 > t_0$ such that

$$\int_{t_0}^{t} \left[\left| \left(\frac{1}{r(s)} \right)' \right| + \frac{|b(s)|}{a^v(s)r(s)} R^{v_2}(s) \right] ds \leq \frac{c_3}{2M_1} \int_{t_0}^{t} R^{-\beta}(s)\, ds \tag{4.5.36}$$

for $t \geq t_1$. From this, (4.5.32), and (4.5.34), we have

$$\left| \int_{t_0}^{t} |y(s)|^{\lambda+1} ds - \int_{t_0}^{t} \frac{|y_2(s)|^{\delta}}{R(s)} ds + \frac{y(t)y_2(t)}{r(t)} \right| \leq |D| + \frac{c_3}{2} \int_{t_0}^{t} R^{-\beta}(s)\, ds \tag{4.5.37}$$

for $t \geq t_1$.

Now let (4.5.25) hold. Applying l'Hôpital's rule, there exists $t_1 > t_0$ such that

$$\int_{t_0}^{t} \left(\frac{|a'(s)|}{pa(s)r(s)} + \frac{|b(s)|}{a^v(s)r(s)} R^{v_2}(s) \right) ds \leq \frac{c_3}{4M_1} \int_{t_0}^{t} R^{-\beta}(s)\, ds \tag{4.5.38}$$

for $t \geq t_1$. Then using (4.5.31), (4.5.38) and the fact that $\frac{1}{r} = \frac{a^{1/p}}{R}$, we have

$$\begin{aligned} &\int_{t_0}^{t} \left[\left| \left(\frac{1}{r(s)} \right)' \right| + \frac{|b(s)|}{a^v(s)r(s)} R^{v_2}(s) \right] ds \\ &\leq \int_{t_0}^{t} \left(\frac{|a'(s)|}{pa(s)r(s)} + \frac{|b(s)|}{a^v(s)r(s)} R^{v_2}(s) \right) ds + \int_{t_0}^{t} \frac{|g(s)|}{R^{\beta}(s)} ds \\ &\leq \frac{c_3}{2M_1} \int_{t_0}^{t} R^{-\beta}(s)\, ds \end{aligned}$$

for $t \geq t_1$. This together with (4.5.32) and (4.5.34) implies that inequality (4.5.37) again holds. Thus, (4.5.37) holds if either (4.5.24) or (4.5.25) does. Moreover,

$$-|D| - \frac{c_3}{2}\int_{t_0}^{t} R^{-\beta}(s)\,ds \leq \int_{t_0}^{t} |y(s)|^{\lambda+1}ds - \int_{t_0}^{t} \frac{|y_2(s)|^{\delta}}{R(s)}\,ds + \frac{y_1(t)y_2(t)}{r(t)}$$

$$\leq |D| + \frac{c_3}{2}\int_{t_0}^{t} R^{-\beta}(s)\,ds \tag{4.5.39}$$

for $t \geq t_1$. Adding (4.5.35) and the left-hand inequality in (4.5.39) gives

$$2\int_{t_0}^{t} |y(s)|^{\lambda+1}ds + \frac{y(t)y_2(t)}{r(t)} \geq \frac{c_3}{2}\int_{t_0}^{t} R^{-\beta}(s)\,ds - |D| \longrightarrow \infty \tag{4.5.40}$$

as $t \to \infty$, while adding (4.5.35) and the right-hand inequality in (4.5.39) gives

$$2\int_{t_0}^{t} \frac{|y_2(s)|^{\delta}}{R(s)}\,ds - \frac{y(t)y_2(t)}{r(t)} \geq \frac{c_3}{2}\int_{t_0}^{t} R^{-\beta}(s)\,ds - |D| \longrightarrow \infty \tag{4.5.41}$$

as $t \to \infty$.

If y' is oscillatory, let $\{t_k\}_{k=1}^{\infty} \to \infty$ be a sequence of zeros of y'. Then letting $t = t_k$ in (4.5.40) and (4.5.41), we see that y is a strong nonlinear limit-point type solution of (4.5.1).

If y' is nonoscillatory, then either

$$y(t)y_2(t) > 0 \tag{4.5.42}$$

or

$$y(t)y_2(t) < 0 \tag{4.5.43}$$

for large t. First we show that (4.5.27) holds. Clearly (4.5.27) holds if (4.5.42) does, so suppose (4.5.43) holds. Then $y(t)y_2(t)r^{-1}(t) < 0$ for large t and (4.5.27) follows from (4.5.40).

Finally, assume (4.5.28) holds. From (4.5.28) and (4.5.30), it follows that there is a $t_2 \geq t_1$ such that

$$\left|\frac{y(t)y_2(t)}{r(t)}\right| \leq \frac{M_1}{r(t)} \leq \frac{c_3}{4}\int_{0}^{t} R^{-\beta}(s)\,ds$$

for $t \geq t_2$; hence, (4.5.29) follows from this and (4.5.41). □

Remark 4.5.2. Lemma 4.5.6 actually holds for all positive p, λ, and q regardless of their relative size.

4.6 The Super-Half-Linear Case: Main Results

In this section, we present our main results for equation (4.5.1).

Theorem 4.6.1. *Assume that* (4.2.2) *and* (4.5.6) *holds. Then* (4.5.1) *is of the strong nonlinear limit-circle type if and only if*

$$\int_0^\infty R^{-\beta}(t)\,dt < \infty. \tag{4.6.1}$$

Proof. Suppose (4.6.1) holds and let y be any nontrivial solution of (4.5.1) defined on $\mathbb{R}_+$. Then, by Lemma 4.5.4, there is a positive constant c such that $0 \leq F(t) \leq c$. Hence, from this and (4.2.3),

$$\begin{aligned} 0 \leq \gamma \int_0^\infty |y(t)|^{\lambda+1}dt + \int_0^\infty \frac{|y_2(t)|^\delta}{R(t)}\,dt \\ = \int_0^\infty F(t)R^{-\beta}(t)\,dt \ \leq c\int_0^\infty R^{-\beta}(t)\,dt < \infty, \end{aligned}$$

so y is of the strong nonlinear limit-circle type. Thus, (4.5.1) is of the strong nonlinear limit-circle type.

Now suppose that (4.6.1) does not hold, i.e.,

$$\int_0^\infty R^{-\beta}(t)\,dt = \infty.$$

Let y be a solution of (4.5.1) given by Lemma 4.5.5. Then there is $c_1 > 0$ such that $c_1 \leq F(t)$ on $[t_0, \infty) \subset \mathbb{R}_+$. Hence, from this and (4.2.3)

$$\begin{aligned} \gamma \int_{t_0}^\infty |y(t)|^{\lambda+1}dt + \int_{t_0}^\infty \frac{|y_2(t)|^\delta}{R(t)}\,dt &= \int_{t_0}^\infty F(t)R^{-\beta}(t)\,dt \\ &\geq c_1 \int_{t_0}^\infty R^{-\beta}(t)\,dt = \infty. \end{aligned}$$

Thus, either

$$\int_0^\infty |y(t)|^{\lambda+1}dt = \infty \quad \text{or} \quad \int_0^\infty \frac{|y_2(t)|^\delta}{R(t)}\,dt = \infty,$$

and so y and equation (4.5.1) are not of the strong nonlinear limit-circle type. $\square$

Theorem 4.6.2. *Let* (4.2.2), (4.5.6) *and either* (4.5.24) *or* (4.5.25) *holds. If*

$$\int_0^\infty R^{-\beta}(t)\,dt = \infty, \tag{4.6.2}$$

then (4.5.1) *is of the nonlinear limit-point type.*

Proof. The hypothesis of Lemmas 4.5.5 and 4.5.6 are satisfied, so if y is a solution given by Lemma 4.5.5, then (4.5.27) holds and the conclusion follows. □

Theorem 4.6.3. *Let conditions* (4.2.2), (4.5.6) *and either* (4.5.24) *or* (4.5.25) *hold. Assume, in addition, that either:* (i) $\lambda = p = q$; (ii) $b \leq 0$ *for large* t *and* (4.5.7) *holds; or* (iii) $p = q < \lambda$ *and* (4.5.7) *holds. If* (4.6.2) *holds, then every nontrivial solution of* (4.5.1) *is of the nonlinear limit-point type. If, moreover,* (4.5.28) *holds, then* (4.5.1) *is of the strong nonlinear limit-point type.*

Proof. Note that the hypotheses of Lemmas 4.5.3–4.5.6 are satisfied. Let y be a nontrivial solution of (4.5.1) defined on $\mathbb{R}_+$. Then by Lemmas 4.5.3 and 4.5.4, there are positive constants c_1 and c_2 such that

$$0 < c_1 \leq F(t) \leq c_2 \quad \text{for large } t.$$

Thus, by Lemma 4.5.6, (4.5.27) holds; moreover, if (4.5.28) holds, then so does (4.5.29). The existence of a nontrivial solution of (4.5.1) is given by Lemma 4.5.5. This proves the theorem. □

Remark 4.6.1. If R is nondecreasing for large t, then condition (4.5.7) becomes

$$\liminf_{t\to\infty} R^{\alpha}(t) \left\{ \int_t^\infty \left[|g'(s)| + \frac{|b(s)|}{a^v(s)} R^{v_2}(s) \right] ds \right\}^{\beta_3} = 0.$$

If R is nonincreasing for large t, then (4.5.7) becomes

$$\liminf_{t\to\infty} R^{-\beta} \left\{ \int_t^\infty \left[|g'(s)| + \frac{|b(s)|}{a^v(s)} R^{v_2}(s) \right] ds \right\}^{\beta_3} = 0.$$

4.7 The Case $p = q$

In this section we consider the special case of equation (4.1.1) with $p = q$, namely,

$$\left(a(t)|y'|^{p-1}y'\right)' + b(t)|y'|^{p-1}y' + r(t)|y|^{\lambda-1}y = 0. \tag{4.7.1}$$

An important condition used in the study of equations (4.3.1) and (4.5.1) is (4.3.6), which takes the form

$$\int_0^\infty \frac{|b(t)|}{a(t)}dt < \infty$$

for equation (4.7.1). It is possible to remove this condition when studying equation (4.7.1) by using the technique contained in the following lemma. A direct computation proves the lemma.

Lemma 4.7.1. *Equation* (4.7.1) *and the equation*

$$(\bar{a}(t)|y'|^{p-1}y')' + \bar{r}(t)|y|^{\lambda-1}y = 0 \tag{4.7.2}$$

are equivalent where

$$\bar{a}(t) = a(t)\exp\left\{\int_0^t \frac{b(s)}{a(s)}ds\right\} \quad \textit{and} \quad \bar{r}(t) = r(t)\exp\left\{\int_0^t \frac{b(s)}{a(s)}ds\right\}.$$

That is, every solution of (4.7.1) *is a solution of* (4.7.2) *and vice versa.*

Based on this lemma, results for equation (4.7.1) can be obtained by combining Lemma 4.7.1 and known results for equation (4.7.2), such as those that can be found, for example, in Chapter 2 or Chapter 3. Here we only state a sample of the many such possible results.

Define

$$\bar{R}(t) = \bar{a}^{\frac{1}{p}}(t)\bar{r}(t) = R(t)\exp\left\{\delta\int_0^t \frac{b(s)}{a(s)}ds\right\},$$

$$\bar{g}(t) = -\frac{\bar{a}^{\frac{1}{p}}(t)\bar{R}'(t)}{\bar{R}^{\alpha+1}(t)} = -\frac{a^{\frac{1}{p}}(t)}{R^{\alpha+1}(t)}\left[R'(t) + \delta\frac{b(t)}{a(t)}R(t)\right]$$

$$\times \exp\left\{\frac{\lambda - p}{(\lambda+2)p+1}\int_0^t \frac{b(s)}{a(s)}ds\right\},$$

$$\omega_5 = \frac{(\lambda+1)(p+1)}{p-\lambda} \quad \text{for } p > \lambda, \quad \text{and} \quad \omega_5 = \infty \quad \text{for } p = \lambda.$$

Theorem 4.7.1. *Assume that*

$$\lim_{t\to\infty} \bar{g}(t) = 0, \qquad \int_0^\infty |\bar{g}'(s)|\, ds < \infty, \tag{4.7.3}$$

and either (i) $\lambda = p$ *or* (ii) $\lambda < p$ *and*

$$\liminf_{t\to\infty} \bar{R}^\beta(t) \left(\int_t^\infty |\bar{g}'(s)|\, ds \right)^{\omega_5} \exp\left\{ \int_0^t (\bar{R}^{-1}(\sigma))'_+ \bar{R}(\sigma)\, d\sigma \right\} = 0.$$

Then (4.7.1) *is of the strong nonlinear limit-circle type if and only if*

$$\int_0^\infty \bar{R}^{-\beta}(\sigma)\, d\sigma < \infty.$$

Proof. The conclusion follows from Theorem 3.9.4 applied to (4.7.2) and Lemma 4.7.1. □

Our next result follows from Theorem 4.4.2 being applied to (4.7.2) and Lemma 4.7.1.

Theorem 4.7.2. *Let* $\lambda \le p$, *condition* (4.7.3) *and either*

$$\int_0^\infty \left| \left(\frac{1}{\bar{r}(t)} \right)' \right| dt < \infty \quad or \quad \int_0^\infty \frac{|\bar{a}'(r)|}{\bar{a}(t)\bar{r}(t)}\, dt < \infty$$

hold. If

$$\int_0^\infty \bar{R}^{-\beta}(\sigma) d\sigma = \infty,$$

then equation (4.7.1) *is of the nonlinear limit-point type.*

The next theorem is a strong nonlinear limit-point result for equation (4.7.1).

Theorem 4.7.3. *Assume that* $\lambda \le p$, (4.7.3) *holds, and*

$$\lim_{t\to\infty} \frac{a'(t) + b(t)}{a^{1-\frac{\beta}{p}}(t) r(t)} \exp\left\{ \frac{1}{\omega_3} \int_0^t \frac{b(s)}{a(s)}\, ds \right\} = 0.$$

If $\int_0^\infty \bar{R}^{-\beta}(\sigma)\, d\sigma = \infty$, *then equation* (4.3.1) *is of the nonlinear limit-point type. If, in addition,* r *does not tend to zero as* $t \to \infty$, *then equation* (4.7.1) *is of the strong nonlinear limit-point type.*

Proof. This result follows from Theorem 3.9.5 and Lemma 4.7.1 above. Note that Theorem 3.9.5 is proved for $r(t) \ge r_0 > 0$ for $t \in \mathbb{R}_+$, but it is easy to see from (3.9.53) and the end of its proof that Theorem 3.9.5 holds as long as r does not tend to zero as $t \to \infty$. □

Theorem 4.7.4. (i) *Let $\lambda > p$ and*

$$\int_0^\infty \bar{R}^{-\beta}(t) \left(\int_0^t |\bar{g}'(s)|\, ds + 1 \right)^{\frac{(\lambda+1)(p+1)}{\lambda-p}} dt < \infty.$$

Then equation (4.7.1) *is of the nonlinear limit-circle type.*

(ii) *Assume that $\lambda \geq p$,*

$$\lim_{t\to\infty} \bar{g}(t) = 0 \quad \text{and} \quad \int_0^\infty |\bar{g}'(t)|\, dt < \infty. \tag{4.7.4}$$

Then (4.7.1) *is of the strong nonlinear limit-circle type if and only if*

$$\int_0^\infty \bar{R}^{-\beta}(t)\, dt < \infty.$$

Proof. (i) If $\bar{g}$ is not identically a constant on $\mathbb{R}_+$ the result follows from Lemma 4.7.1 and from Theorem 2.4.1 applied to (4.7.2). If $\bar{g} \equiv$ const. on $\mathbb{R}_+$, then the statement follows from Theorem 4.6.1 applied to (4.7.2) and Lemma 4.7.1. Part (ii) follows from Lemma 4.7.1 and Theorem 4.6.1 applied to (4.7.2). □

The next result follows from Lemma 4.7.1 and Theorem 4.6.2 applied to equation (4.7.2).

Theorem 4.7.5. *Let $\lambda \geq p$,* (4.7.4) *hold, and assume that either*

$$\lim_{t\to\infty} R^\beta(t) \left(\frac{|r'(t)|}{r^2(t)} + \frac{|b(t)|}{a(t)r(t)} \right) \exp\left\{ \frac{\lambda - p}{(\lambda+2)p+1} \int_0^t \frac{b(s)}{a(s)}\, ds \right\} = 0 \tag{4.7.5}$$

or

$$\lim_{t\to\infty} R^\beta(t) \frac{|a'(t)| + |b(t)|}{a(t)r(t)} \exp\left\{ \frac{\lambda - p}{(\lambda+2)p+1} \int_0^t \frac{b(s)}{a(s)}\, ds \right\} = 0. \tag{4.7.6}$$

If

$$\int_0^\infty (\bar{R}(t))^{-\beta}\, dt = \infty, \tag{4.7.7}$$

then (4.7.1) *is of the nonlinear limit-point type.*

Our last theorem in this chapter is a strong nonlinear limit-point result for equation (4.7.1).

Theorem 4.7.6. *Assume that* (4.7.4), (4.7.7) *and either* (4.7.5) *or* (4.7.6) *hold. In addition, assume that either:* (i) $\lambda = p$; (ii) $b \leq 0$ *on* $\mathbb{R}_+$ *and*

$$\liminf_{t\to\infty} (\bar{R}(t))^{-\beta} \left\{ \int_t^\infty |\bar{g}'(s)|\, ds \right\}^{\frac{(\lambda+1)(p+1)}{\lambda-p}} \exp\left\{ \frac{\bar{R}'_+(s)}{\bar{R}(s)}\, ds \right\} = 0; \tag{4.7.8}$$

or (iii) $\lambda > p$ *and* (4.7.8) *holds. Then every nontrivial solution of* (4.7.1) *is of the nonlinear limit-point type. If, moreover,*

$$\limsup_{t\to\infty} \bar{r}(t) \int_0^t \left(\bar{R}(s)\right)^{-\beta} ds = \infty,$$

then (4.7.1) *is of the strong nonlinear limit-point type.*

Proof. This follows from Lemma 4.7.1 and Theorem 4.6.3 applied to (4.5.34). □

4.8 Examples and Discussion

Example 4.8.1. Consider the equation

$$\left(|y'|^{p-1}y'\right)' + b(t)|y'|^{q-1}y' + t^\sigma |y|^{\lambda-1}y = 0, \quad t \geq 1. \tag{4.8.1}$$

Assume that $\lambda \leq p \leq q$, $\sigma\alpha > -1$, and $\beta\sigma \leq 1$. If b satisfies either

$$\sigma \geq 0 \quad \text{and} \quad \int_1^\infty t^{\sigma v_1} |b(t)|\, dt < \infty,$$

or

$$\sigma < 0 \quad \text{and} \quad \int_1^\infty t^{\sigma(v_1-1)} |b(t)|\, dt < \infty,$$

then the conditions of Theorem 4.4.2 are satisfied, so equation (4.8.1) is of the nonlinear limit-point type.

Example 4.8.2. Consider the special case of equation (4.8.1) with $p = 1$, namely,

$$y'' + b(t)|y'|^{q-1}y' + t^\sigma |y|^{\lambda-1}y = 0, \quad t \geq 1. \tag{4.8.2}$$

Then we have $0 < \lambda \leq 1 \leq q$, so if

$$-\frac{\lambda+3}{2} < \sigma \leq \frac{\lambda+3}{\lambda+1}$$

and either

$$\sigma \geq 0 \quad \text{and} \quad \int_1^\infty t^{\frac{\sigma(q-1)}{\lambda+3}} |b(t)|\, dt < \infty$$

or

$$\sigma < 0 \quad \text{and} \quad \int_1^\infty t^{\frac{q-\lambda-4}{\lambda+3}\sigma} |b(t)|\, dt < \infty,$$

holds, equation (4.8.2) is of the nonlinear limit-point type.

Example 4.8.3. Consider the equation

$$y'' + b(t)y' + t^\sigma y = 0, \quad t \ge 1, \tag{4.8.3}$$

with $b(t) \ge 0$. Note that here $p = q = \lambda = 1$. Assume that

$$\int_1^\infty b(t)\, dt < \infty.$$

Then by Theorem 4.4.1, equation (4.8.3) is of the strong nonlinear limit-circle type if and only if $\sigma > 2$. By Theorem 4.4.3, equation (4.8.3) is of the strong nonlinear limit-point type if $0 < \sigma \le 2$. It is worth noting that this agrees with the well-known limit-circle criteria

$$\int_0^\infty r^{-\frac{1}{2}}(t)\, dt < \infty$$

of Dunford and Schwartz [72, p. 1414] (also see the discussion in [25]).

For our next example, we consider the case where $p = q = \lambda$. It may be convenient to refer to this case as the *fully half-linear* equation.

Example 4.8.4. Consider the equation

$$\left(t^a |y'|^{\epsilon-1} y'\right)' + t^b |y'|^{\epsilon-1} y' + t^\sigma |y|^{\epsilon-1} y = 0, \quad t \ge 1, \tag{4.8.4}$$

where $\epsilon > 0$. If $\sigma + \epsilon + 1 > a > \max\{b+1, \epsilon(1-\sigma)+1\}$, then equation (4.8.4) is of the strong nonlinear limit-circle type by Theorem 4.4.1. On the other hand, if $\sigma > 0$ and $b + 1 < a < \min\{\sigma + \epsilon + 1, \epsilon(1-\sigma)+1\}$, then equation (4.8.4) is of the strong nonlinear limit-point type by Theorem 4.4.3.

Example 4.8.5. Consider the equation

$$y'' + t^s y' + t^\sigma |y|^{\lambda-1} y = 0, \quad t \ge 1, \tag{4.8.5}$$

with $s \in \mathbb{R}$ and $\lambda \ge 1$. We have the following results.

(i) If $s < -1$ and $\sigma > \frac{\lambda+3}{\lambda+1}$, then equation (4.8.5) is of the strong nonlinear limit-circle type by Theorem 4.6.1.

(ii) If $s < -1$ and $-\frac{\lambda+3}{2} < \sigma \le \frac{\lambda+3}{\lambda+1}$, then equation (4.8.5) is of the strong nonlinear limit-point type by Theorem 4.6.2.

(iii) If $\lambda = 1$, $s < -1$, and $-2 < \sigma \le 2$, then (4.8.5) is of the strong nonlinear limit-point type by Theorem 4.6.3(i).

(iv) If $\lambda > 1$, $s < -1 - \frac{\sigma(\lambda-1)}{(\lambda+3)(\lambda+1)}$, and $0 < \sigma \le \frac{\lambda+3}{\lambda+1}$, then (4.8.5) is of the strong nonlinear limit-point type by Theorem 4.6.3(iii).

In our next example we have that $b(t)$ in equation (4.3.1) (or (4.7.1)) is negative.

Example 4.8.6. Consider the equation

$$y'' - t^s y' + t^\sigma |y|^{\lambda-1} y = 0, \quad t \ge 1, \tag{4.8.6}$$

with $s \in \mathbb{R}$, $\lambda \ge 1$. Calculations show the following.

(i) Equation (4.8.6) is of the nonlinear limit-circle type if:

(a) $s < -1$ and $\sigma > \frac{\lambda+3}{\lambda+1}$ (by Theorem 4.6.1);

(b) $s = -1$ and $\sigma > -\frac{\lambda-1}{\lambda+1}$ (by Theorem 4.7.4(ii)).

(ii) Equation (4.8.6) is of the nonlinear limit-point type if:

(c) $s < -1$ and $-\frac{\lambda+3}{2} < \sigma \le \frac{\lambda+3}{\lambda+1}$ (by Theorem 4.6.2);

(d) $s = -1$ and $-\frac{2(\lambda+1)}{\lambda+3} < \sigma \le -\frac{\lambda-1}{\lambda+1}$ (by Theorem 4.7.5);

(e) $s > -1$ and $\lambda > 1$ (by Theorem 4.7.5);

(f) $s > -1$, $\lambda = 1$, and $\sigma > \max\{2s, s\}$ (by Theorem 4.7.5).

The limit-point/limit-circle problem for the linearly damped equation

$$(a(t)y')' + b(t)y' + r(t)y^\lambda = 0$$

with $b(t) \ge 0$ was studied in [153] with $\lambda \le 1$ being the ratio of odd positive integers and in [176] with $\lambda \ge 1$ an odd integer. The results in both of these papers tend to be modifications of results in [98, 99, 104] to accommodate the presence of a damping term. To compare the results in [153, 176] to those here, consider the equation

$$y'' + t^s y' + t^\sigma |y|^{\lambda-1} y = 0, \quad t \ge 1, \tag{4.8.7}$$

with $s \in \mathbb{R}$, $\sigma \ge 0$, and $0 < \lambda \le 1$. We have the following results.

(i) Equation (4.8.7) is of the nonlinear limit-circle type if:

(a) $\lambda = 1$ and $s > -1$ (by Theorem 4.7.1);

(b) $\lambda < 1$ and $-1 < s < \frac{\sigma}{2}$ (by Theorem 4.7.1);

(c) $s = -1$ and $\sigma > \frac{1-\lambda}{\lambda+1}$ (by Theorem 4.7.1);

(d) $\lambda = 1$, $s < -1$, and $\sigma > \frac{\lambda+3}{\lambda+1}$ (by Theorem 4.4.1);

(e) $\lambda < 1$, $\sigma > \frac{\lambda+3}{\lambda+1}$, and $s < -1 - \sigma\frac{1-\lambda}{\lambda+3}$ (by Theorem 4.4.1);

(ii) Equation (4.8.7) is of the nonlinear limit-point type if:

(f) $s = -1$ and $\sigma \leq \frac{1-\lambda}{\lambda+1}$ (by Theorem 4.7.2);

(g) $s < -1$ and $\sigma \leq \frac{\lambda+3}{\lambda+1}$ (by Theorem 4.4.2);

Now by [153, Corollary 2.3], equation (4.8.7) is of the nonlinear limit-circle type if $s \leq -\frac{\sigma(1-\lambda)}{2(\lambda+3)}$ and $\sigma > \frac{\lambda+3}{\lambda+1}$. The nonlinear limit-point result [153, Theorem 2.6] does not apply to equation (4.8.7). This shows that our results substantially extend the ones in [153] in the case of nonlinear limit-circle type results and are new in the case of nonlinear limit-point results. The results in [153] follow from ours if $s \geq -1$ and for $s < -1$ and $\lambda = 1$. There are errors in the proofs of the results in [176].

In particular, for equation (4.8.7) with $\lambda = 1$, i.e.,

$$y'' + t^s y' + t^\sigma y = 0, \quad t \geq 1, \tag{4.8.8}$$

the results in [153,176] show that (4.8.8) is of the nonlinear limit-circle type if $s \leq 0$ and $\sigma > 2$ and by results in the present paper (4.8.8) is of the nonlinear limit-point type if and only if any one of the following conditions holds:

(h) $s < -1$ and $\sigma > 2$;
(i) $s = -1$ and $\sigma > 0$;
(j) $s > -1$ and σ is arbitrary.

Hence, the results in [153, 176] follow from ours, and our results are substantially better; note that we obtain necessary and sufficient condition for (4.8.8) to be of the nonlinear limit-circle type.

Consider once again equation (4.8.7) but now with $s, \sigma \in \mathbb{R}$ and $\lambda \geq 1$. The following results hold.

(iii) Equation (4.8.7) is of the nonlinear limit-circle type if:

(k) $s < -1$ and $\sigma > \frac{\lambda+3}{\lambda+1}$ (by Theorem 4.7.4(ii));

(l) $s = -1$ and $\sigma > -\frac{\lambda-1}{\lambda+1}$ (by Theorem 4.7.4(i) if $\lambda > 1$ and by Theorem 4.7.4(ii) if $\lambda = 1$);

(m) $s > -1$, $\lambda > 1$, and $\sigma > \frac{\lambda+3}{\lambda+1}$ (by Theorem 4.7.4(i));

(n) $s > -1$, $\lambda = 1$ and $s < \frac{\sigma}{2}$ (by Theorem 4.7.4(ii)).

(iv) Equation (4.8.7) is of the nonlinear limit-point type if:

(o) $s < -1$ and $-\frac{\lambda+3}{\lambda+1} < \sigma \leq \frac{\lambda+3}{\lambda+1}$ (by Theorem 4.7.5);

(p) $s = -1$ and $-2 < \sigma \leq -\frac{\lambda-1}{\lambda+1}$ (by Theorem 4.7.5).

By [176, Corollary 2.1], equation (4.8.7) is of the nonlinear limit-circle type if either (i) $\lambda = 1$, $\sigma > 2$, and $s < 0$, or (ii) $\lambda > 1$, $\frac{\lambda+3}{\lambda+1} < \sigma < \frac{2(\lambda+3)}{\lambda-1}$, and $s < -\sigma\frac{\lambda-1}{2(\lambda+3)}$. This shows that, in the case of nonlinear limit-circle type results, the results in [176] are a special case of those here.

Remark 4.8.1. Theorem 2.3 in [176] appears to show that equation (4.8.7) is of the nonlinear limit-point type if $-1 < \sigma \leq \frac{\lambda+3}{\lambda+1}$, $s \leq -\sigma\frac{\lambda-1}{2(\lambda+3)}$, and $s < \frac{\sigma-1}{2}$. We have a contradiction to our case (l) above if $s = -1$, $\lambda = 1$, and $\sigma = 1$. The proof of [176, Theorem 2.3] is incorrect; in their expression for $\dot{V}$ the term "$-p(t)[a(t)r(t)]^{\beta-2\alpha}x^{2k}(s)$" is missing.

Chapter 5

Higher Order Equations

5.1 Introduction

In this chapter, we examine limit-point, limit-circle, and other asymptotic properties of solutions of some higher order equations. We begin with equations of order $2n$ and then take a closer look at equations of order 4.

5.2 Equations of Order $2n$

Here we consider the even-order differential equation

$$\left(a(t)|y^{(n)}|^{p-1}y^{(n)}\right)^{(n)} = r(t)|y|^{\lambda-1}y \tag{5.2.1}$$

where $n \geq 2$, a and r are continuous functions, $p > 0$, $\lambda > 0$, $a(t) > 0$, and there exists $\alpha \in \{0, 1\}$ such that $\alpha + n$ is odd and $(-1)^{\alpha}r(t) > 0$ on $\mathbb{R}_+$.

Definition 5.2.1. A function $y \in C^n\left([0,\tau)\right), \tau \leq \infty$, is called a *solution* of (5.2.1) if

$$a(t)|y^{(n)}(t)|^{p-1}y^{(n)}(t) \in C^n\left([0,\tau)\right)$$

and (5.2.1) holds.

We examine solutions on their maximal interval of existence $[0, \tau)$, $\tau \leq \infty$. If $\tau < \infty$, then y is said to be *noncontinuable.* A continuable solution y of equation (5.2.1) is *proper* if it is nontrivial in any neighborhood of ∞, and we let $\mathcal{S}$ denote the set of all continuable solutions of (5.2.1).

It can easily be seen that (5.2.1) is equivalent to the nonlinear system

$$\begin{aligned} y_i' &= y_{i+1}, \qquad \text{for } i = 1, 2, \ldots, n-1, \\ y_n' &= a^{-\frac{1}{p}}(t)|y_{n+1}|^{\frac{1}{p}} \operatorname{sgn} y_{n+1}, \\ y_j' &= y_{j+1}, \qquad \text{for } j = n+1, n+2, \ldots, 2n-1, \\ y_{2n}' &= r(t)|y_1|^{\lambda} \operatorname{sgn} y_1. \end{aligned} \tag{5.2.2}$$

The relation between a solution y of (5.2.1) and $(y_1, \ldots, y_{2n})$ of (5.2.2) is

$$\begin{aligned} y_i &= y^{(i-1)} \qquad \text{for } i = 1, 2, \ldots, n; \\ y_{n+1} &= a(t)|y^{(n)}|^{p-1} y^{(n)}, \\ y_j &= y_{n+1}^{(j-n-1)} \quad \text{for } j = n+2, \ldots, 2n. \end{aligned} \tag{5.2.3}$$

Whenever we consider a solution y of (5.2.1), we will make use of the relationships in (5.2.3) without further mention.

Definition 5.2.2. Let y be a proper solution of (5.2.1). Then y is called oscillatory if there exists a sequence $\{t_k\}_{k=1}^{\infty}$ of zeros of y tending to ∞. Otherwise, y is said to be nonoscillatory. In particular, a solution $y \equiv 0$ in a neighborhood of ∞ is nonoscillatory.

The definitions of nonlinear limit-point and limit-circle solutions of the higher order equation (5.2.1) are the same as they are in the second-order case. But there is a difference in the definitions for strong nonlinear limit-point and limit-circle solutions.

Definition 5.2.3. A continuable solution y of (5.2.1) is said to be of the *strong nonlinear limit-circle* type if

$$\int_0^{\infty} |y(\sigma)|^{\lambda+1}\, d\sigma < \infty \quad \text{and} \quad \int_0^{\infty} \frac{a(\sigma)}{|r(\sigma)|} |y^{(n)}(\sigma)|^{p+1}\, d\sigma < \infty.$$

Equation (5.2.1) is said to be of the *strong nonlinear limit-circle* type if every continuable solution is of the strong nonlinear limit-circle type.

Definition 5.2.4. A continuable solution y of (5.2.1) is said to be of the *strong nonlinear limit-point* type if

$$\int_0^{\infty} |y(\sigma)|^{\lambda+1}\, d\sigma = \infty \quad \text{and} \quad \int_0^{\infty} \frac{a(\sigma)}{|r(\sigma)|} |y^{(n)}(\sigma)|^{p+1}\, d\sigma = \infty.$$

Equation (5.2.1) is said to be of the *strong nonlinear limit-point* type if equation (5.2.1) has proper solutions and every one of these is of the strong nonlinear limit-point type.

Due to the nature of our discussion in this and the next section of this chapter, it will be convenient to classify the solution set $\mathcal{S}$ of continuable solutions of equation (5.2.1) in the following way.

Definition 5.2.5. Denote by $\mathcal{O} \subset \mathcal{S}$ (respectively, $\mathcal{N} \subset \mathcal{S}$) the set of all oscillatory (respectively, nonoscillatory) solutions of (5.2.1). Let $\mathcal{N}_0$ be the set of solutions of (5.2.1) that are trivial in a neighborhood of ∞. Let $\mathcal{N}_1 \subset \mathcal{N}$ be the set of solutions of (5.2.1) such that

$$\lim_{t\to\infty} |y_i(t)| = \infty \text{ and } y_i(t)y_1(t) > 0 \text{ for large } t \text{ and } i = 1, 2, \ldots, 2n,$$

and let $\mathcal{N}_2 \subset \mathcal{N}$ be the set of proper solutions of (5.2.1) such that

$$\lim_{t\to\infty} |y_i(t)| = 0 \text{ and } (-1)^{i+1} y_i(t)y_1(t) > 0 \text{ for large } t \text{ and } i = 1, 2, \ldots, 2n.$$

It is possible to give a more refined classification of the nonoscillatory solutions of equation (5.2.1), but that will not be needed here.

Lemma 5.2.1. *Let*

$$\int_0^\infty a^{-\frac{1}{p}}(s)\, ds = \infty \quad \text{and} \quad \int_0^\infty |r(s)|\, ds = \infty\,. \tag{5.2.4}$$

Then $\mathcal{S} = \mathcal{O} \cup \mathcal{N}_0 \cup \mathcal{N}_1 \cup \mathcal{N}_2$ *if* $\alpha = 0$ *or* $\mathcal{S} = \mathcal{O} \cup \mathcal{N}_0$ *if* $\alpha = 1$.

The above lemma follows from [5, Theorem 1]; the following lemma is a consequence of [6, Theorems 2 and 3].

Lemma 5.2.2. *Let* $y \in \mathcal{O}$. *Then there is* $T \in \mathbb{R}_+$ *such that all zeros of* y_i, $i = 1, 2, \ldots, 2n$, *on* $[T, \infty)$ *are simple; if* $i \in \{1, \ldots, 2n\}$, *then between two consecutive zeros of* y_i *lying in* $[T, \infty)$ *there exists one and only one zero of* y_{i+1}. *We set* $y_{2n+1} = y_1$.

For any solution y of (5.2.1), we define the function F by

$$F(t) = \sum_{i=0}^{n-1} (-1)^{i+\alpha} y_{2n-i}(t) y_{i+1}(t)\,. \tag{5.2.5}$$

Lemma 5.2.3. *Let y be a solution of (5.2.1) defined on $[0,\tau)$. Then*

$$F'(t) = (-1)^{\alpha} r(t)|y(t)|^{\lambda+1} + (-1)^{\alpha+n-1} a(t)|y^{(n)}(t)|^{p+1} \geq 0, \quad t \in [0,\tau),$$

and so F is a nondecreasing function for any solution y of (5.2.1).

Proof. This result can be obtained by a direct computation or it can be seen as a consequence of [6, Lemma 5]. The nonnegativity follows from the fact that $\alpha + n$ is odd. □

Lemma 3 yields the following classification of oscillatory solutions.

Definition 5.2.6. We let $\mathcal{O}_1 \subset \mathcal{O}$ $(\mathcal{O}_2 \subset \mathcal{O})$ be the set of oscillatory solutions for which $\lim_{t\to\infty} F(t) = \infty (\lim_{t\to\infty} F(t) < \infty)$.

Lemma 5.2.4. *Let $[a,b] \subset \mathbb{R}_+$ and a_0, a_1, and r_1 be positive constants such that*

$$a_0 \leq a(t) \leq a_1 \quad \textit{and} \quad |r(t)| \leq r_1 \quad \textit{on } [a,b]. \tag{5.2.6}$$

Let y be a solution of (5.2.2) such that $y_i, i = 1, 2, \ldots, 2n$, have zeros on $[a,b]$ and let

$$\nu_i = \max_{a \leq t \leq b} |y_i(t)|, \quad i = 1, 2, \ldots, 2n.$$

Then,

$$\nu_i \leq C\nu_1^{m_i}, \quad i = 2, 3, \ldots, 2n, \tag{5.2.7}$$

where C is a positive constant depending on a_0, a_1, r_1, and n,

$$m_i = 1 + \frac{(\lambda - p)(i-1)}{pn+n} > 0 \quad \textit{for } i = 2, 3, \ldots, n, \tag{5.2.8}$$

and

$$m_i = p + (\lambda - p)\frac{pn + i - n - 1}{pn+n} > 0 \quad \textit{for } i = n+1, \ldots, 2n. \tag{5.2.9}$$

Proof. Let $\nu_{2n+1} = \max_{a\leq t\leq b} |y'_{2n}(t)|$. We can choose intervals $J_i \subset [a,b]$, $i = 2, \ldots, 2n$, such that

$$\min_{t\in J_i} |y_i(t)| = 0 \quad \text{and} \quad \nu_i = \max_{t\in J_i} |y_i(t)|,$$

where ν_i and the zero of y_i occur at the endpoints of the interval J_i and y_i does not change its sign on J_i, $i \in \{2, \ldots, 2n\}$.

First, we prove that

$$\nu_i \leq C_i \nu_1^{s_i} \nu_{i+1}^{\gamma_i}, \quad i = 2, \ldots, 2n, \tag{5.2.10}$$

where

$$s_i = \frac{1}{i}, \quad \gamma_i = \frac{i-1}{i}, \quad C_i = 2^{i-1} \quad \text{for } 2 \le i \le n-1,$$

$$s_n = \frac{1}{n}, \quad \gamma_n = \frac{n-1}{pn}, \quad C_n = 2^{(n-1)} a_0^{-\frac{n-1}{pn}},$$

$$s_i = \frac{p}{pn+i-n}, \quad \gamma_i = \frac{pn+i-n-1}{pn+i-n}, \quad \text{for } i = n+1, \ldots, 2n,$$

$$C_{n+1} = \left(\frac{2p+2}{p} a_1^{\frac{1}{p}} C_n\right)^{\gamma_{n+1}}, \quad C_i = (4C_{i-1})^{\gamma_i}, \quad \text{for } i = n+2, \ldots, 2n. \tag{5.2.11}$$

Let either $1 < j \le n-1$ or $n+1 < j \le 2n$. Then (5.2.2) and (5.2.3) yield

$$\begin{aligned} \nu_j^2 &= 2\int_{J_j} \left|y_j(s)\, y_j'(s)\right| ds = 2\int_{J_j} \left|y_{j-1}'(s)\, y_{j+1}(s)\right| ds \\ &\le 4\,\nu_{j-1}\nu_{j+1}; \end{aligned} \tag{5.2.12}$$

hence, (5.2.10) holds for $i = 2$. For $i = 3, 4, \ldots, n-1$, (5.2.10) can be proved by mathematical induction using (5.2.12), or it can be seen from the proof of [122, Lemma 5.2]. Furthermore, (5.2.2), (5.2.3), (5.2.6), and (5.2.10) with $i = n-1$ yield

$$\begin{aligned} \nu_n^2 &= 2\int_{J_n} \left|y_n(s)\, y_n'(s)\right| ds = 2\int_{J_n} \left|y_{n-1}'(s)\right| a^{-\frac{1}{p}}(s) \left|y_{n+1}(s)\right|^{\frac{1}{p}} ds \\ &\le 4a_0^{-\frac{1}{p}} \nu_{n-1} \nu_{n+1}^{\frac{1}{p}} \le 4a_0^{-\frac{1}{p}} 2^{n-2} \nu_1^{\frac{1}{n-1}} \nu_n^{\frac{n-2}{n-1}} \nu_{n+1}^{\frac{1}{p}}. \end{aligned}$$

Thus,

$$\nu_n^{\frac{n}{n-1}} \le 2^n a_0^{-\frac{1}{p}} \nu_1^{\frac{1}{n-1}} \nu_{n+1}^{\frac{1}{p}} \tag{5.2.13}$$

and (5.2.10) holds for $i = n$.

The conclusion for $i > n$ can be proved using induction. From (5.2.2) and (5.2.10) for $i = n$, we have

$$\begin{aligned} \nu_{n+1}^{\frac{p+1}{p}} &= \frac{p+1}{p} \int_{J_{n+1}} \left|y_{n+1}^{\frac{1}{p}}(s)\, y_{n+1}'(s)\right| ds \\ &= \frac{p+1}{p} \int_{J_{n+1}} a^{\frac{1}{p}}(s) \left|y_n'(s)\, y_{n+2}(s)\right| ds \\ &\le 2\frac{p+1}{p} a_1^{\frac{1}{p}} \nu_n\, \nu_{n+2} \le 2\frac{p+1}{p} a_1^{\frac{1}{p}} C_n \nu_1^{s_n} \nu_{n+1}^{\frac{n-1}{pn}} \nu_{n+2}, \end{aligned}$$

or

$$\nu_{n+1}^{(p+1-\frac{n-1}{n})/p} \le 2\frac{p+1}{p} a_1^{\frac{1}{p}} C_n \nu_1^{s_n} \nu_{n+2},$$

and hence (5.2.10) holds for $i = n+1$.

Suppose that (5.2.10) holds for $i = n+1, n+2, \ldots, j-1 \le 2n-1$. Then (5.2.10) and (5.2.12) yield

$$\nu_j^2 \le 4\nu_{j-1}\nu_{j+1} \le 4C_{j-1}\nu_1^{s_{j-1}} \nu_j^{\gamma_{j-1}} \nu_{j+1},$$

so

$$\nu_j^{\frac{pn+j-n}{pn+j-n-1}} \le 4C_{j-1}\nu_1^{\frac{p}{pn+j-n-1}} \nu_{j+1};$$

hence, (5.2.10) holds for $i = j$. Thus, (5.2.10) holds for all $i = 2, 3, \ldots, 2n$.

Now using (5.2.10) and induction, we see that

$$\nu_i \le \bar{C}_i \nu_1^{d_i} \nu_{2n+1}^{l_i} \tag{5.2.14}$$

with

$$\begin{aligned} d_i &= s_i + d_{i+1}\gamma_i, \qquad i = 2n-1, 2n-2, \ldots, 2, \\ d_{2n} &= s_{2n}, \end{aligned} \tag{5.2.15}$$

$$l_i = \prod_{j=i}^{2n} \gamma_j, \tag{5.2.16}$$

$$\bar{C}_i = C_i \bar{C}_{i+1}^{\gamma_i}, \quad \bar{C}_{2n} = C_{2n}.$$

It follows from (5.2.11) and (5.2.16) that

$$l_i = \frac{i-1}{pn+n} \quad \text{for } i = 2, 3, \ldots, n,$$

and

$$l_i = \frac{pn+i-n-1}{pn+n} \quad \text{for } i = n+1, \ldots, 2n.$$

An induction argument and (5.2.15) show that

$$d_i = p\left(1 - \frac{pn+i-n-1}{pn+n}\right) \quad \text{for } i = 2n, \ldots, n+1,$$

$$d_n = 1 - \frac{p(n-1)}{pn+n} > 0,$$

and

$$d_i = 1 - \frac{p(i-1)}{pn+n} > 0, \quad i = n-1, \ldots, 2.$$

Finally, if we let $\mathcal{C}_0 = \max_{1 \le i \le 2n} \bar{\mathcal{C}}_i$, then (5.2.2), (5.2.8), (5.2.9), and (5.2.14) yield

$$\nu_i \le \mathcal{C}_0 \, \nu_1^{d_i} \, r_1^{l_i} \, \nu_1^{\lambda l_i} = \mathcal{C}_0 \, r_1^{l_i} \, \nu_1^{m_i},$$

and so (5.2.7) holds with $\mathcal{C} = \mathcal{C}_0 \max_{1 \le i \le 2n} r_1^{l_i}$. □

Lemma 5.2.5. *Let $y \in \mathcal{S}$ be such that $\lim_{t\to\infty} F(t) < \infty$. Assume that there are positive constants a_0, a_1, r_0, and r_1 such that*

$$a_0 \le a(t) \le a_1 \quad \text{and} \quad r_0 \le |r(t)| \le r_1 \quad \text{for } t \in \mathbb{R}_+. \tag{5.2.17}$$

Then,

$$\lim_{t\to\infty} y_j(t) = 0 \quad \text{for } j = 1, \ldots, 2n \tag{5.2.18}$$

and

$$\lim_{t\to\infty} F(t) = 0. \tag{5.2.19}$$

Proof. Since (5.2.17) implies that (5.2.4) holds, $\mathcal{S} = \mathcal{O} \cup \mathcal{N}_0 \cup \mathcal{N}_1 \cup \mathcal{N}_2$. Let $y \in \mathcal{S}$ and $\lim_{t\to\infty} F(t) = \mathcal{C} \in (-\infty, \infty)$. Then Lemma 5.2.3 yields

$$\int_0^\infty |r(t)| \, |y(t)|^{\lambda+1} dt < \infty \quad \text{and} \quad \int_0^\infty a(t) |y^{(n)}(t)|^{p+1} dt < \infty. \tag{5.2.20}$$

From this and (5.2.17), we have $y \in L_{\lambda+1}(\mathbb{R}_+)$ and $y^{(n)} \in L_{p+1}(\mathbb{R}_+)$. Hence, Lemma 1.5 in [132] yields (5.2.18) for $j = 1, 2, \ldots, n$. As y tends to zero, then $\lim_{n\to\infty} \nu_1(n) = 0$ where $\nu_1(n) = \max_{n \le t < \infty} |y_1(t)|$. If $y \in \mathcal{O}$, then Lemma 5.2.4 applied on $[n, b]$ with $b \to \infty$ shows that (5.2.18) holds for all j, and so (5.2.19) is a consequence of (5.2.18). If $y \in \mathcal{N}_0 \cup \mathcal{N}_2$, then (5.2.18) and (5.2.19) are immediate. If $y \in \mathcal{N}_1$, then $\lim_{t\to\infty} |y(t)| = \infty$, which contradicts (5.2.17) and (5.2.20). □

Remark 5.2.1. If (5.2.17) holds, then for any solution $y \in \mathcal{S}$, $\lim_{t\to\infty} F(t)$ is either zero or ∞.

Lemma 5.2.6. *Let $\lambda \le p$. Then all solutions of (5.2.1) are continuable.*

Proof. Let y be a noncontinuable solution of (5.2.1) defined on $J = [0, \tau)$ with $\tau < \infty$. Then,

$$\limsup_{t \to \tau^-} |y_{2n}(t)| = \infty. \tag{5.2.21}$$

Furthermore, (5.2.2) yields

$$|y_i(t)| \leq |y_i(0)| + \int_0^t |y_{i+1}(s)|\, ds, \quad i = 1, 2, \ldots, 2n-1,\ i \neq n, \tag{5.2.22}$$

$$|y_n(t)| \leq |y_n(0)| + \int_0^t a^{-\frac{1}{p}}(s) |y_{n+1}(s)|^{\frac{1}{p}}\, ds \tag{5.2.23}$$

and

$$|y_{2n}(t)| \leq |y_{2n}(0)| + \int_0^t |r(s)|\, |y_1(s)|^{\lambda}\, ds \tag{5.2.24}$$

for $t \in J$. Set

$$v_i(t) = \max_{0 \leq s \leq t} |y_i(s)|, \qquad i = 1, 2, \ldots, 2n,$$

and

$$\mathcal{C} = \max\left(\tau \max_{0 \leq s \leq \tau} a^{-\frac{1}{p}}(s),\ \max_{0 \leq s \leq \tau} \big|r(s)\big|,\ \max_{1 \leq i \leq 2n} \big|y_i(0)\big|, \tau\right).$$

Then, (5.2.22) and (5.2.23) yield

$$v_i(t) \leq \mathcal{C} + \mathcal{C} v_{i+1}\,, \quad i = 1, 2, \ldots, 2n-1,\ i \neq n, \tag{5.2.25}$$

and

$$v_n(t) \leq \mathcal{C} + \mathcal{C} v_{n+1}^{\frac{1}{p}}\,. \tag{5.2.26}$$

Hence, using a recursive argument, (5.2.25) yields

$$v_1(t) \leq \mathcal{C}_1 + \mathcal{C}^{n-1} v_n(t)$$

and

$$v_{n+1}(t) \leq \mathcal{C}_1 + \mathcal{C}^{n-1} v_{2n}(t)$$

for $t \in J$, where $\mathcal{C}_1 = \sum_{i=1}^{n-1} \mathcal{C}^i$. From this and (5.2.26), we have

$$\begin{aligned}\big|y_1(t)\big| \leq v_1(t) &\leq \mathcal{C}_1 + \mathcal{C}^{n-1}\big(\mathcal{C} + \mathcal{C} v_{n+1}^{\frac{1}{p}}(t)\big)\\ &\leq \mathcal{C}_1 + \mathcal{C}^n + \mathcal{C}^n \big(\mathcal{C}_1 + \mathcal{C}^{n-1} v_{2n}(t)\big)^{\frac{1}{p}} \leq \mathcal{C}_2 + \mathcal{C}_3 v_{2n}^{\frac{1}{p}}(t)\end{aligned}$$

for $t \in J$, where $\mathcal{C}_2 = \mathcal{C}_1 + \mathcal{C}^n + 2^{\frac{1}{p}}\mathcal{C}^n\mathcal{C}_1^{\frac{1}{p}}$ and $\mathcal{C}_3 = 2^{\frac{1}{p}}\mathcal{C}^n\mathcal{C}^{\frac{n-1}{p}}$. Hence, (5.2.24) yields

$$\begin{aligned} v_{2n}(t) &\leq \mathcal{C} + \mathcal{C}\int_0^t |y_1(s)|^\lambda\, ds \leq \mathcal{C} + \mathcal{C}\int_0^t \left[\mathcal{C}_2 + \mathcal{C}_3 v_{2n}^{\frac{1}{p}}(s)\right]^\lambda ds \\ &\leq \mathcal{C}_4 + \mathcal{C}_5\int_0^t v_{2n}^{\frac{\lambda}{p}}(s)\, ds \end{aligned} \tag{5.2.27}$$

with $\mathcal{C}_4 = \mathcal{C} + 2^\lambda\mathcal{C}\tau\mathcal{C}_2$ and $\mathcal{C}_5 = 2^\lambda\mathcal{C}\mathcal{C}_3^\lambda$. In view of (5.2.21), $\lim_{t\to\tau^-} v_{2n}(t) = \infty$, and so (5.2.27) yields the existence of $t_0 \in [0,\tau)$ such that $v_{2n}(t) \geq 1$ on $[t_0,\tau)$. Thus, from (5.2.27) and the fact that $\lambda/p \leq 1$, we have

$$v_{2n}(t) \leq \mathcal{C}_4 + \mathcal{C}_5\int_{t_0}^t v_{2n}(s)\, ds\,, \quad t \in [t_0,\tau)\,. \tag{5.2.28}$$

An application of Gronwall's inequality shows that v_{2n} is bounded on $[0,\tau)$, which contradicts (5.2.21). □

Remark 5.2.2. The result contained in Lemma 5.2.6 is proved for the case $\lambda \geq 1$ in [137, Theorem].

Our final lemma in this section is due to Medveď and Pekárková [137, Lemma 2.1].

Lemma 5.2.7. *Let $K > 0$ and $m > 1$ be constants, $Q(t)$ be a continuous, nonnegative function on $\mathbb{R}_+$, and let u be a continuous, nonnegative function on $\mathbb{R}_+$ satisfying the inequality*

$$u(t) \leq K + \int_0^t Q(s)u^m(s)\, ds$$

on $[0,\tau), \tau \leq \infty$. If

$$(m-1)K^{m-1}\int_0^\infty Q(s)\, ds < 1,$$

then

$$u(t) \leq K\left[1 - (m-1)K^{m-1}\int_0^t Q(s)\, ds\right]^{\frac{1}{1-m}}, \qquad t \in [0,\tau).$$

5.3 Limit-Point/Limit-Circle Solutions of Even-Order Equations

Now that we have presented our preliminary results, we can turn our attention to the limit-point and limit-circle properties of solutions of (5.2.1). Our first theorem is a strong nonlinear limit-circle type result.

Theorem 5.3.1. *If there exists r_0 such that*

$$0 < r_0 \le r(t), \quad t \in \mathbb{R}_+, \tag{5.3.1}$$

then for every $y \in \mathcal{O}_2 \cup \mathcal{N}_0 \cup \mathcal{N}_2$,

$$\int_0^\infty \left|y(t)\right|^{\lambda+1} dt < \infty \quad \text{and} \quad \int_0^\infty \frac{a(t)}{r(t)} \left|y^{(n)}(t)\right|^{p+1} dt < \infty;$$

i.e., $y \in \mathcal{O}_2 \cup \mathcal{N}_0 \cup \mathcal{N}_2$ is of the strong nonlinear limit circle type.

Proof. If $y \in \mathcal{O}_2 \cup \mathcal{N}_2$, then $F(t)$ is bounded and an integration of $F'(t)$ (see Lemma 5.2.3) yields (5.2.20) and the statement follows from (5.3.1). For $y \in \mathcal{N}_0$, the result is clear. □

The conclusions in our next theorem are of a limit-point type character.

Theorem 5.3.2. *Let $\lambda \le p$, let a_0, a_1, r_0, and r_1 be positive constants, and let $y \in \mathcal{O}_1$.*

(i) *If*

$$a_0 \le a(t) \le a_1 \quad \text{and} \quad \left|r(t)\right| \le r_1 \quad \text{on } \mathbb{R}_+, \tag{5.3.2}$$

then

$$\int_0^\infty \left|y(t)\right|^{\lambda+1} = \infty.$$

(ii) *If $a_0 \le a(t) \le a_1$ and $r_0 \le \left|r(t)\right| \le r_1$ on $\mathbb{R}_+$, then*

$$\int_0^\infty \frac{a(t)}{\left|r(t)\right|} \left|y^{(n)}(t)\right|^{p+1} dt = \infty.$$

Proof. Let $y \in \mathcal{O}_1$.

(i) We will prove that y is unbounded on $\mathbb{R}_+$. Suppose that y is bounded on $\mathbb{R}_+$. Then Lemma 5.2.4 and (5.3.2) imply that $y_i(t)$, $i = 1, 2, \ldots, 2n-1$, are bounded, so $\lim_{t\to\infty} F(t) < \infty$. This contradicts the fact that $y \in \mathcal{O}_1$ and shows that

$$\limsup_{t\to\infty} |y_1(t)| = \infty.$$

Let $\{t_k\}_{k=1}^{\infty}$ and $\{a_k\}_{k=1}^{\infty}$ be increasing sequences with $a_k < t_k < a_{k+1}$, $k = 1, 2, \ldots$, $a_1 \geq T$ where T is from Lemma 5.2.2, $y_1'(t_k) = y_2(t_k) = 0$, y_1 has $2n$ zeros on $[a_k, t_k]$, and

$$\nu_1 = \max_{t\in[a_k,t_k]} |y_1(t)| = |y(t_k)| \quad \text{for } k = 1, 2, \ldots.$$

Then it follows from Lemma 5.2.2 that y_i, $i = 1, \ldots, 2n-1$, have zeros on $[a_k, t_k]$.

Let $\tau_k \in (a_k, t_k)$ be such that $|y_1(\tau_k)| = \frac{1}{2}|y_1(t_k)|$ and $|y_1(t)|$ is increasing on $[\tau_k, t_k)$. The hypotheses of Lemma 5.2.4 are satisfied on $\Delta_k = [\tau_k, t_k]$, so

$$\begin{aligned}\frac{1}{2}|y_1(t_k)| = |y_1(t_k)| - |y_1(\tau_k)| &= \int_{\Delta_k} |y_2(s)|\, ds \\ &\leq \mathcal{C}(t_k - \tau_k)|y_1(t_k)|^{m_2}\end{aligned} \tag{5.3.3}$$

where $\mathcal{C}$ is given by Lemma 5.2.4 and $m_2 = 1 + \frac{\lambda - p}{pn+n} \in (0, 1)$. From this and (5.3.3), we see that there is a positive constant M such that

$$t_k - \tau_k \geq M\,, \quad k = 1, 2, \ldots$$

Hence,

$$\int_0^{\infty} |y(t)|^{\lambda+1} dt \geq \sum_{k=1}^{\infty} \int_{\Delta_k} |y_1(t)|^{\lambda+1} dt \geq \frac{M}{2^{\lambda+1}} \sum_{k=1}^{\infty} |y_1(t_k)|^{\lambda+1} = \infty. \tag{5.3.4}$$

(ii) System (5.2.2) can be rewritten as

$$\begin{aligned} Z &= (Z_1, \ldots, Z_{2n}), \\ Z_i' &= Z_{i+1}\,, \quad i = 1, 2, \ldots, n-1, \\ Z_n' &= r(t)|Z_{n+1}|^{\lambda} \operatorname{sgn} Z_{n+1}, \\ Z_j' &= Z_{j+1}\,, \quad j = n+1, \ldots, 2n-1, \\ Z_{2n}' &= a^{-\frac{1}{p}}(t)|Z_1|^{\frac{1}{p}} \operatorname{sgn} Z_1, \end{aligned} \tag{5.3.5}$$

with $Z_i = y_{n+i}$, $Z_{n+i} = y_i$, $i = 1, 2, \ldots, n$.

Since $1/p < 1/\lambda$ and the proof of Lemma 5.2.4 does not depend on the signs of a and r, Lemma 5.2.4 can be applied to (5.3.5). The remainder of the proof is similar to that of the proof of part (i); in this case, we investigate $Z_1 = y_{n+1}$ instead of y_1, have $\Delta_k = [\tau_k, t_k]$ as an interval where $y_{n+1}'(t_k) = 0$, $|y_{n+1}(\tau_k)| = \frac{1}{2}|y_{n+1}(t_k)|$, $|y_{n+1}|$ is increasing, $\max_{t\in[a_k,t_k]} |y_{n+1}(t)| = |y_{n+1}(t_k)|$, and y_{n+1} has just $2n$ zeros on $[a_k, t_k]$. We can then prove similarly that $\limsup_{t\to\infty} |y_{n+1}(t)| = \infty$.

Lemma 5.2.4 then yields

$$\begin{aligned}\int_0^\infty \frac{a(t)}{|r(t)|}\left|y^{(n)}(t)\right|^{p+1} dt &= \int_0^\infty a^{-\frac{1}{p}}(t)\left|r^{-1}(t)\right|\left|y_{n+1}(t)\right|^{\frac{p+1}{p}} dt \\ &\geq a_1^{-\frac{1}{p}} r_1^{-1} \sum_{k=1}^{\infty} \int_{\Delta_k} \left|y_{n+1}(t)\right|^{\frac{p+1}{p}} dt \\ &\geq 2^{-\frac{p+1}{p}} a_1^{-\frac{1}{p}} r_1^{-1} M \sum_{k=1}^{\infty} \left|y_{n+1}(t_k)\right|^{\frac{p+1}{p}} dt = \infty\end{aligned}$$

for a similarly chosen M. This completes the proof of the theorem. □

Our next result is for solutions in the class $\mathcal{N}_1$.

Theorem 5.3.3. *Let* $\alpha = 0$.

(i) *If* $\lambda \leq p$ *and* (5.2.4) *holds, then* $\mathcal{N}_1 \neq \emptyset$.

(ii) *If* $y \in \mathcal{N}_1$, *then* $\int_0^\infty |y(s)|^{\lambda+1} ds = \infty$.

(iii) *If* $y \in \mathcal{N}_1$ *and*

$$\int_0^\infty a(t)^{-\frac{1}{p}} r^{-1}(t) \left[\int_T^t \int_T^{s_1} \cdots \int_T^{s_{n-1}} r(s_n) \left[\int_T^{s_n} \cdots \int_T^{s_{2n-1}} a^{-\frac{1}{p}}(\sigma) \right.\right.$$
$$\left.\left. \times\, d\sigma\, ds_{2n-1} \ldots ds_{n+1}\right]^{\lambda} ds_n \ldots ds_1\right]^{\frac{p+1}{p}} dt = \infty, \tag{5.3.6}$$

then

$$\int_0^\infty \frac{a(s)}{r(t)}\left|y^{(n)}(s)\right|^{p+1} ds = \infty.$$

Proof. (i) According to Lemma 5.2.6, all solutions of (5.2.1) are defined on $\mathbb{R}_+$. Consider a solution y of (5.2.1) with Cauchy initial conditions

$$y_i(0) = 1, \quad i = 1, 2, \ldots, 2n.$$

Since $r > 0$ on $\mathbb{R}_+$, it follows from (5.2.3) that y is nonoscillatory and all components of y are increasing on $\mathbb{R}_+$. The statement follows from Lemma 5.2.1.

(ii) It follows from Definition 5.2.5 that $|y|$ is bounded from below by a positive constant in a neighborhood of ∞, and so the conclusion is immediate.

(iii) Let $y \in \mathcal{N}_1$ and suppose, for simplicity, that $T \in \mathbb{R}_+$ exists such that $y_i(t) > 0$ for $t \in [T, \infty)$ and $i = 1, 2, \ldots, 2n$. Then,

$$y_i(t) \geq \int_T^t y_i'(s)\,ds = \int_T^t y_{i+1}(s)\,ds, \qquad i \neq n,\ i \neq 2n,$$

$$y_n(t) \geq \int_T^t a^{-\frac{1}{p}}(s) y_{n+1}^{\frac{1}{p}}(s)\,ds,$$

and

$$y_{2n}(t) \geq \int_T^t r(s)\, y_1^{\lambda}(s)\,ds.$$

From this, we obtain

$$y_1(t) \geq \underbrace{\int_T^t \cdots \int_T^s}_{(n-1)\text{-times}} y_n(s)\,ds \ldots ds \geq y_{n+1}^{\frac{1}{p}}(0) \underbrace{\int_T^t \cdots \int_T^s}_{n\text{-times}} a^{-\frac{1}{p}}(s)\,ds \ldots ds,$$

or

$$y_{n+1}(t) \geq \underbrace{\int_T^t \cdots \int_T^s}_{n\text{-times}} r(s) y_1^{\lambda}(s)\,ds \ldots ds.$$

Hence,

$$y_{n+1}(t) \geq y_{n+1}^{\frac{\lambda}{p}}(0) \underbrace{\int_T^t \cdots \int_T^s}_{n\text{- times}} r(s) \Big[\underbrace{\int_T^s \cdots \int_T^s}_{n\text{- times}} a^{-\frac{1}{p}}(s)\,ds \ldots ds \Big]^{\lambda} ds \ldots ds,$$

and the conclusion follows from this and (5.3.6). □

We offer our next theorem as a sort of summary of the above results.

Theorem 5.3.4. *Let $\lambda \leq p$ and let $a_0, a_1, r_0,$ and r_1 be positive constants such that*

$$a_0 \leq a(t) \leq a_1, \quad r_0 \leq \big|r(t)\big| \leq r_1 \quad \textit{on } \mathbb{R}_+.$$

(i) *All solutions of (5.2.1) are continuable and $\mathcal{S} = \mathcal{O}_1 \cup \mathcal{O}_2 \cup \mathcal{N}_0 \cup \mathcal{N}_1 \cup \mathcal{N}_2$. Moreover, if $\alpha = 0$, then $\mathcal{O}_1 \cup \mathcal{N}_1 \neq \emptyset$, and if $\alpha = 1$, then $\mathcal{O}_1 \neq \emptyset$ and $\mathcal{N}_1 = \emptyset$.*

(ii) *For any solution y of (5.2.1), the following statements are equivalent:*

(a) *y is of the nonlinear limit-circle type;*

(b) *y is of the strong nonlinear limit-circle type;*
(c) $y \in \mathcal{O}_2 \cup \mathcal{N}_0 \cup \mathcal{N}_2$.

(iii) *For any solution y of (5.2.1), the following statements are equivalent:*
(a) *y is of the nonlinear limit-point type;*
(b) *y is of the strong nonlinear limit-point type;*
(c) *$y \in \mathcal{O}_1$ in case $\alpha = 1$ and $y \in \mathcal{O}_1 \cup \mathcal{N}_1$ in case $\alpha = 0$.*

Proof. Part (i) follows from Lemma 5.2.6, Lemma 5.2.1, Lemma 5.2.5, and the fact that $F(0) > 0$ implies $y \in \mathcal{O}_1 \cup \mathcal{N}_1$. Parts (ii) and (iii) follow from Theorems 5.3.1, 5.3.2, and 5.3.3. □

The next theorem is a limit-point result. Even though it requires α to be 0, it does not need n to be odd.

Theorem 5.3.5. *Let $\lambda > p$, $\alpha = 0$, and*

$$\int_0^\infty r(t)\, t^{(n-1)\lambda} \left(\int_0^t a^{-\frac{1}{p}}(\sigma)\sigma^{(n-1)/p} d\sigma \right)^\lambda dt < \infty. \tag{5.3.7}$$

Then equation (5.2.1) is of the nonlinear limit-point type.

Proof. Consider a solution y of (5.2.1) with Cauchy initial conditions

$$y_i(0) = y_i > 0, \quad i = 1, 2, \ldots, 2n.$$

From this and (5.2.2) we see that y_i, $i = 1, \ldots, 2n$, are positive and increasing on the maximal interval of existence $J = [0, \tau)$, $\tau \le \infty$. Denote by $u(t) = \max_{0 \le s \le t} y_{2n}(s)$, $t \in J$. It follows from (5.2.2) that

$$\begin{aligned}
y_{2n}(t) &\le y_{2n}(0) + \int_0^t r(s) y_1^\lambda(s)\, ds \\
&\le y_{2n}(0) + \int_0^t r(s) \left(\sum_{i=0}^{n-1} \frac{y_{i+1}}{i!} s^i \right. \\
&\quad \left. + \int_0^s \frac{(s-\sigma)^{n-1}}{(n-1)!} a^{-\frac{1}{p}}(\sigma) y_{n+1}^{\frac{1}{p}}(\sigma)\, d\sigma \right)^\lambda ds \\
&\le y_{2n}(0) + n^\lambda \int_0^t r(s) \left(\sum_{i=0}^{n-1} \frac{y_{i+1}^\lambda}{(i!)^\lambda} s^{\lambda i} \right. \\
&\quad \left. + s^{(n-1)\lambda} \left(\int_0^s a^{-\frac{1}{p}}(\sigma) y_{n+1}^{\frac{1}{p}}(\sigma)\, d\sigma \right)^\lambda \right) ds\,;
\end{aligned} \tag{5.3.8}$$

here we used the inequality

$$(a_1 + \cdots + a_n)^\nu \le n^\nu (a_1^\nu + \cdots + a_n^\nu)$$

for $\nu > 0$ and $a_i \ge 0$, $i = 1, 2, \ldots, n$. By Taylor's theorem, (5.2.2), and (5.2.3), we have

$$\left|y_{n+1}(t)\right| \le \sum_{j=n+1}^{2n-1} \frac{y_j}{(j-n-1)!} t^{j-n-1} + \int_0^t \frac{(t-\sigma)^{n-2}}{(n-2)!} y_{2n}(\sigma)\, d\sigma,$$

and together with (5.3.8), we obtain

$$\begin{aligned} y_{2n}(t) &\le y_{2n}(0) + (n+1)^\lambda \sum_{i=0}^{n-1} \left(\frac{y_{i+1}}{i!}\right)^\lambda \int_0^\infty r(s) s^{\lambda i} ds \\ &+ (n+1)^\lambda \int_0^t r(s) s^{(n-1)\lambda} \\ &\times \left\{ \int_0^s a^{-\frac{1}{p}}(\sigma) n^{\frac{1}{p}} \left[\sum_{j=n+1}^{2n-1} \left(\frac{y_j}{(j-n-1)!} \right)^{\frac{1}{p}} \sigma^{(j-n-1)/p} \right.\right. \\ &+ \left.\left. \left(u(s) \frac{\sigma^{n-1}}{(n-1)!} \right)^{\frac{1}{p}} \right] d\sigma \right\}^\lambda ds, \end{aligned}$$

or

$$\begin{aligned} y_{2n}(t) \le\ & y_{2n}(0) + (n+1)^\lambda \sum_{i=0}^{n-1} \left(\frac{y_{i+1}}{i!}\right)^\lambda \int_0^\infty r(s) s^{\lambda i} ds \\ &+ (n+1)^{2\lambda + \frac{1}{p}} \sum_{j=n+1}^{2n-1} \left(\frac{y_j}{(j-n-1)!} \right)^{\frac{\lambda}{p}} \\ &\times \int_0^\infty r(s) s^{(n-1)\lambda} \left(\int_0^s a^{-\frac{1}{p}}(\sigma) \sigma^{(j-n-1)/p} d\sigma \right)^\lambda ds \\ &+ \frac{(n+1)^{2\lambda+\frac{1}{p}}}{[(n-1)!]^{\lambda/p}} \int_0^t r(s) s^{(n-1)\lambda} u^{\frac{\lambda}{p}}(s) \left(\int_0^s a^{-\frac{1}{p}}(\sigma) \sigma^{(n-1)/p} \right)^\lambda d\sigma\, ds. \end{aligned}$$

Hence,

$$y_{2n}(t) \le K + \int_0^t Q(s)\, u^{\frac{\lambda}{p}}(s)\, ds \quad \text{and} \quad u(t) \le K + \int_0^t Q(s)\, u^{\frac{\lambda}{p}}(s)\, ds$$

with

$$K = y_{2n}(0) + (n+1)^{\lambda} \sum_{i=0}^{n-1} \left(\frac{y_{i+1}}{i!}\right)^{\lambda} \int_0^{\infty} r(s) s^{\lambda i} ds + (n+1)^{2\lambda + \frac{1}{p}}$$

$$\times \sum_{j=n+1}^{2n-1} \left(\frac{y_j}{(j-n-1)!}\right)^{\frac{\lambda}{p}} \int_0^{\infty} r(s) s^{(n-1)\lambda} \left(\int_0^s a^{-\frac{1}{p}} \sigma^{(j-n-1)/p} d\sigma\right)^{\lambda} ds$$

and

$$Q(t) = [(n-1)!]^{-\lambda/p} (n+1)^{2\lambda+\frac{1}{p}} r(t) t^{(n-1)\lambda} \left(\int_0^t a^{-\frac{1}{p}}(\sigma)\, \sigma^{(n-1)/p} d\sigma\right)^{\lambda}.$$

Now, choose y_i, $i = 1, 2, \ldots, 2n$, such that

$$\left(\frac{\lambda}{p} - 1\right) K^{\frac{\lambda}{p}-1} \int_0^{\infty} Q(s)\, ds < 1;$$

it is possible to do this in view of (5.3.7). The hypotheses of Lemma 5.2.7 are satisfied with $m = \lambda/p$, so we have

$$u(t) \leq K \left[1 - \left(\frac{\lambda}{p} - 1\right) K^{\frac{\lambda}{p}-1} \int_0^{\infty} Q(s)\, ds\right]^{\frac{p}{p-\lambda}}, \quad t \in J.$$

From this and from (5.2.2), we have that y_i, $i = 1, 2, \ldots, 2n$, are bounded on every finite subinterval of J. Since J is the maximal interval of existence for y, we have $\tau = \infty$. Hence, y is proper and increasing and the conclusion of the theorem follows. □

Remark 5.3.1. The method of proof used for Theorem 5.3.5 was also used in [137] for another type of differential equation.

It is reasonable to ask if it is possible to obtain results similar to those in Theorem 5.3.3 for $\lambda > p$ or for n even. The answer is affirmative for the special case of (5.2.1) with $p = 1$, $\lambda = 1$, and $a(t) \equiv 1$, i.e., for the equation

$$y^{(2n)} = r(t)|y|^{\lambda-1} y. \tag{5.3.9}$$

We will need the following lemma.

Lemma 5.3.1. *Let n be even, $[T, T_1] \subset \mathbb{R}_+, T \leq T_1 - 2, C_i \in \mathbb{R}$ for $i = 1, 2, \ldots, 2n, \lambda > 1, \mu = \big(2(\lambda+1)n - \lambda + 1\big)/\big(2(\lambda+1)n - \lambda + 1 - 2\frac{\lambda-1}{\lambda+1}\big)$, and*

$$0 < r_0 \leq -r(t) \leq r_1 \quad on\ [T, T_1].$$

Then there exist constants $\delta > 0$ and $K > 0$, which do not depend on T, T_1, r_0, or r_1, such that a solution of the Cauchy initial value problem consisting of equation (5.3.5) and the conditions

$$y^{(i)}(T) = C_{i+1}, \quad i = 0, 1, \ldots, 2n-1,$$

is noncontinuable if

$$\sum_{i=0}^{n-1}(-1)^{i+1}C_{i+1}C_{2n-i} > Kr_0^{-\frac{1}{\mu-1}} r_1^{\frac{(2n-1)(\lambda+1)\mu}{(\mu-1)[2n(\lambda+1)-\lambda+1]}} (T_1 - T)^{-\frac{2\mu}{(\lambda+1)(\mu-1)}}$$

and

$$\sum_{i=0}^{2n-1} |C_{i+1}| > 2^{\frac{2n}{\lambda-1}+1} r_0^{-\frac{\lambda}{\lambda-1}} n(\delta r_0 + r_1)(T_1 - T)^{-\frac{2n}{\lambda-1}}.$$

Proof. This follows from the proof of [61, Theorem 1] (with $\eta_1 = r_0$, $\eta_2 = r_1$, $a = T$, $b = T_1$, and $t_0 = \frac{T_1 - T}{2}$). Observe that $\mu > 1$. □

The following result can then be proved.

Theorem 5.3.6. *Assume that n is even, $\lambda > 1, r_0 < r_1$, and*

$$0 < r_0 \le -r(t) \le r_1 t^{\sigma} \quad \text{for } t \in \mathbb{R}_+,$$

with $\sigma < \frac{2}{(\lambda+1)^2(2n-1)}[2n(\lambda+1) - \lambda + 1]$. Then equation (5.3.9) is of the strong nonlinear limit-circle type and $\mathcal{O}_1 = \emptyset$.

Proof. By Lemma 5.2.1, $\mathcal{S} = \mathcal{O}_1 \cup \mathcal{O}_2 \cup \mathcal{N}_0$. We will prove that $\mathcal{O}_1 = \emptyset$. Suppose $y \in \mathcal{O}_1 \ne \emptyset$. Then Definition 5.2.6 implies $\lim_{t\to\infty} F(t) = \infty$, and so there exists $T \in \mathbb{R}$ such that

$$F(T) = \sum_{i=0}^{n-1}(-1)^{i+1} y^{(i)}(T) y^{(2n-i-1)}(T) > 0.$$

Choose $T_1 \in [T+2, \infty)$ such that

$$F(T) > Kr_0^{-\frac{1}{\mu-1}} \left(r_1 T_1^{\sigma}\right)^{\frac{(2n-1)(\lambda+1)\mu}{(\mu-1)[2n(\lambda+1)-\lambda+1]}} (T_1 - T)^{-\frac{2\mu}{(\lambda+1)(\mu-1)}}$$

and

$$\sum_{i=0}^{2n-1} |y^{(i)}(T)| > 2^{\frac{2n}{\lambda-1}+1} n r_0^{-\frac{\lambda}{\lambda-1}} (\delta r_0 + r_1 T_1^{\sigma})(T_1 - T)^{-\frac{2n}{\lambda-1}},$$

where δ, K, and μ are given in Lemma 5.3.1; this choice is possible since $\mu > 1$,

$$\frac{2}{(\lambda+1)(\mu-1)} > \frac{(2n-1)(\lambda+1)\sigma}{(\mu-1)[2n(\lambda+1)-\lambda+1]},$$

and $\frac{2n}{\lambda-1} > \sigma$. This latter inequality follows from the fact that $\lambda > 1$, $n \geq 2$, and $1/(\lambda+1) < 1/(\lambda-1)$. Now all the assumptions of Lemma 5.3.1 hold with $r_0 = r_0$, $T = T$, $T_1 = T_1$, and $r_1 = r_1 T_1^{\sigma}$, so the solution y is noncontinuable. This contradiction proves that $\mathcal{O}_1 = \emptyset$. Thus, $\mathcal{O} = \mathcal{O}_2$ and the conclusion of the theorem follows from Theorem 5.3.1. □

To illustrate Theorem 5.3.3, consider the equation

$$\left(|y^{(n)}|^{p-1}y^{(n)}\right)^{(n)} = t^{\beta}|y|^{\lambda-1}y, \quad t \geq 1. \tag{5.3.10}$$

If n is odd, $y \in \mathcal{N}_1$, and $\beta > -1 - n\lambda$, then y is a strong limit-point type solution of (5.3.10). As an example of Theorem 5.3.5, again consider equation (5.3.10) but with $\lambda > p$. If $\beta < -1 - \lambda[pn + n - 1]/p$, then equation (5.3.10) is of the nonlinear limit-point type. Finally, note that in Theorem 5.3.6, σ need not be small. For example, if $n = 2$ and $\lambda = 2$, then $0 \leq \sigma < 22/27$, while if $n = 2$ and $\lambda = 1.1$, then $0 \leq \sigma \leq 5/4$.

Remark 5.3.2. We wish like to point out that although we made the covering assumption that $\alpha + n$ is odd, this is actually not needed in Lemmas 5.2.4 and 5.2.6 and Theorems 5.3.2, 5.3.3, and 5.3.5.

5.4 Fourth-Order Equations

In this section, we examine the special case of equation (5.2.1) with $n = 2$, namely,

$$\left(a(t)|y''|^{p-1}y''\right)'' + r(t)|y|^{\lambda-1}y = 0. \tag{5.4.1}$$

Here, $r(t) > 0$. The system (5.2.2) then takes the form

$$\begin{cases} y_1' = y_2, \\ y_2' = a^{-\frac{1}{p}}(t)|y_3|^{\frac{1}{p}}\operatorname{sgn} y_3, \\ y_3' = y_4, \\ y_4' = -r(t)|y_1|^{\lambda}\operatorname{sgn} y_1. \end{cases} \tag{5.4.2}$$

The relation between a solution y of (5.4.1) and $(y_1, \ldots, y_4)$ of (5.4.2) is

$$y_1 = y, \quad y_2 = y', \quad y_3 = a(t)|y''|^{p-1}y'', \quad y_4 = \left(a(t)|y''|^{p-1}y''\right)'. \tag{5.4.3}$$

As before, a proper solution of y (5.4.1) is *oscillatory* if there exists a sequence $\{t_k\}_{k=1}^{\infty}$ of zeros of y tending to ∞. Otherwise, y is called *nonoscillatory*. In particular, a solution $y \equiv 0$ in a neighborhood of ∞ is nonoscillatory. We will also say that a solution y is *strongly oscillatory* if the set of zeros of y has no finite accumulation point in its interval of definition.

As pointed out above, the definitions of nonlinear limit-circle and nonlinear limit-point solutions are the same as they are in the second-order case. Definitions 5.2.3 and 5.2.4 for strong nonlinear limit-circle and strong nonlinear limit-point solution take the following form in the fourth-order case.

Definition 5.4.1. A continuable solution y of (5.4.1) is said to be of the *strong nonlinear limit-circle* type if

$$\int_0^\infty |y(\sigma)|^{\lambda+1}\,d\sigma < \infty \quad \text{and} \quad \int_0^\infty \frac{a(\sigma)}{r(\sigma)}|y''(\sigma)|^{p+1}\,d\sigma < \infty.$$

Equation (5.4.1) is said to be of the *strong nonlinear limit-circle* type if every continuable solution is of the strong nonlinear limit-circle type.

Definition 5.4.2. A continuable solution y of (5.4.1) is said to be of the *strong nonlinear limit-point* type if

$$\int_0^\infty |y(\sigma)|^{\lambda+1}\,d\sigma = \infty \quad \text{and} \quad \int_0^\infty \frac{a(\sigma)}{r(\sigma)}|y''(\sigma)|^{p+1}\,d\sigma = \infty.$$

Equation (5.4.1) is said to be of the *strong nonlinear limit-point* type if equation (5.4.1) has proper solutions and every one of these is of the strong nonlinear limit-point type.

For the purposes of our discussion here, we will give a more refined classification of the solutions in the set $\mathcal{S}$ of continuable solutions of equation (5.4.1) that the one given in Section 5.2.

Denote by $\mathcal{O} \subset \mathcal{S}$ (respectively, $\mathcal{N} \subset \mathcal{S}$) the set of all oscillatory (respectively, nonoscillatory) solutions of (5.4.1). Let $\mathcal{N}_0$ be the set of solutions that are trivial in a neighborhood of ∞.

Definition 5.4.3. (i) Let $\mathcal{O}_1 \subset \mathcal{O}$ be the set of those solutions $y \in \mathcal{O}$ for which there are sequences $\{t_k^i\}_{k=1}^\infty$, $i = 1, 2, 3, 4$, such that

$$\begin{gathered} t_k^1 < t_k^4 < t_k^3 < t_k^2 < t_{k+1}^1, \quad \lim_{k\to\infty} t_k^1 = \infty, \\ y_j(t_k^j) = 0, \quad j = 1, 2, 3, 4, \\ y_i(t_k^j) \neq 0, \quad j = 1, 2, 3, 4, \quad i \neq j, \\ y_i(t)y_1(t) > 0 \quad \text{for } t \in (t_k^1, t_k^i), \quad i = 2, 3, 4, \quad \text{and} \\ y_i(t)y_1(t) < 0 \quad \text{for } t \in (t_k^i, t_{k+1}^1), \quad i = 2, 3, 4. \end{gathered} \tag{5.4.4}$$

(ii) Let $\mathcal{O}_2 \subset \mathcal{O}$ be the set of those solutions $y \in \mathcal{O}$ for which there are sequences $\{t_k^i\}_{k=1}^{\infty}$, $i = 1, 2, 3, 4$, such that

$$\begin{gathered} t_k^1 < t_k^2 < t_k^3 < t_k^4 < t_{k+1}^1, \quad \lim_{k \to \infty} t_k^1 = \infty, \\ y_j(t_k^j) = 0, \quad j = 1, 2, 3, 4, \\ y_i(t_k^j) \neq 0, \quad j = 1, 2, 3, 4, \quad i \neq j, \\ (-1)^i y_i(t) y_1(t) > 0 \quad \text{for } t \in (t_k^1, t_k^i), \quad i = 2, 3, 4, \quad \text{and} \\ (-1)^i y_i(t) y_1(t) < 0 \quad \text{for } t \in (t_k^i, t_{k+1}^1), \quad i = 2, 3, 4. \end{gathered} \tag{5.4.5}$$

Definition 5.4.4. The complete classification of nonoscillatory solutions of (5.4.1) is as follows.

(i) Let $\mathcal{N}_1 \subset \mathcal{N}$ be the set of those solutions $y \in \mathcal{N}$ for which there exists $t_y \in \mathbb{R}_+$ with the property

$$y_i(t) y_1(t) > 0 \quad \text{on } [t_y, \infty) \quad \text{for } i = 2, 3, 4. \tag{5.4.6}$$

(ii) Let $\mathcal{N}_2 \subset \mathcal{N}$ be the set of those solutions $y \in \mathcal{N}$ for which there exists $t_y \in \mathbb{R}_+$ with the property

$$\begin{aligned} y_j(t) y_1(t) &> 0 \quad \text{for } j = 2, 3, \quad \text{and} \\ y_1(t) y_4(t) &< 0 \quad \text{for } t \in [t_y, \infty). \end{aligned}$$

(iii) Let $\mathcal{N}_3 \subset \mathcal{N}$ be the set of those solutions $y \in \mathcal{N}$ for which there exists $t_y \in \mathbb{R}_+$ with the property

$$\begin{aligned} y_i(t) y_1(t) &> 0 \quad \text{for } i = 2, 4, \quad \text{and} \\ y_3(t) y_1(t) &< 0 \quad \text{for } t \in [t_y, \infty). \end{aligned} \tag{5.4.7}$$

(iv) Let $\mathcal{N}_4 \subset \mathcal{N}$ be the set of those solutions $y \in \mathcal{N}$ for which there exists $t_y \in \mathbb{R}_+$ with the property

$$\begin{aligned} y_j(t) y_1(t) &< 0 \quad \text{for } j = 3, 4, \quad \text{and} \\ y_1(t) y_2(t) &> 0 \quad \text{for } t \in [t_y, \infty). \end{aligned}$$

(v) Let $\mathcal{N}_5 \subset \mathcal{N}$ be the set of those solutions $y \in \mathcal{N}$ for which there exists $t_y \in \mathbb{R}_+$ with the property

$$\begin{aligned} y_i(t) y_1(t) &> 0 \quad \text{for } i = 3, 4, \quad \text{and} \\ y_2(t) y_1(t) &< 0 \quad \text{for } t \in [t_y, \infty). \end{aligned} \tag{5.4.8}$$

(vi) Let $\mathcal{N}_6 \subset \mathcal{N}$ be the set of those solutions $y \in \mathcal{N}$ for which there exists $t_y \in \mathbb{R}_+$ with the property

$$\begin{aligned} y_j(t)y_1(t) &< 0 \quad \text{for } j = 2, 4, \quad \text{and} \\ y_1(t)y_3(t) &> 0 \quad \text{for } t \in [t_y, \infty). \end{aligned}$$

(vii) Let $\mathcal{N}_7 \subset \mathcal{N}$ be the set of those solutions $y \in \mathcal{N}$ for which there exists $t_y \in \mathbb{R}_+$ with the property

$$\begin{aligned} y_j(t)y_1(t) &< 0 \quad \text{for } j = 2, 3, \quad \text{and} \\ y_1(t)y_4(t) &> 0 \quad \text{for } t \in [t_y, \infty). \end{aligned}$$

(viii) Let $\mathcal{N}_8 \subset \mathcal{N}$ be the set of those solutions $y \in \mathcal{N}$ for which there exists $t_y \in \mathbb{R}_+$ with the property

$$y_j(t)y_1(t) < 0 \quad \text{for } j = 2, 3, 4, \quad \text{and} \quad t \in [t_y, \infty).$$

The next two lemmas show that some of the sets in our classifications of solutions are empty.

Lemma 5.4.1. *The following statements hold.*

(i) $\mathcal{O} = \mathcal{O}_1 \cup \mathcal{O}_2$ *and* $\mathcal{N} = \mathcal{N}_0 \cup \mathcal{N}_1 \cup \mathcal{N}_3 \cup \mathcal{N}_4 \cup \mathcal{N}_5$.

(ii) *If* $\int_0^\infty a^{-\frac{1}{p}}(s)\,ds = \infty$, *then* $\mathcal{N}_4 = \emptyset = \mathcal{N}_5$, *so* $\mathcal{N} = \mathcal{N}_0 \cup \mathcal{N}_1 \cup \mathcal{N}_3$.

(iii) *If* $\int_0^\infty a^{-\frac{1}{p}}(s)\,ds = \infty$ *and* $\int_0^\infty r(s)\,ds = \infty$, *then* $\mathcal{N} = \mathcal{N}_0$.

Proof. (i) According to [6, Theorem 3], for any $y \in \mathcal{O}$ there exists a neighborhood of ∞ in which the zeros of y have no accumulation point. Hence, Theorem 2 in [6] implies y is strongly oscillatory, and so $\mathcal{O} = \mathcal{O}_1 \cup \mathcal{O}_2$ by [4, Lemma]. To show that $\mathcal{N}_2 = \emptyset = \mathcal{N}_6$, assume that $y(t) > 0$ for $t \geq t_y$ and integrate y_3' and y_4' to obtain

$$\begin{aligned} y_3(t) &= y_3(t_y) + \int_{t_y}^t \left[y_4(t_y) - \int_{t_y}^s r(\sigma)|y_1(\sigma)|^\lambda \, d\sigma \right] ds \\ &\leq y_3(t_y) + y_4(t_y) \int_{t_y}^t d\sigma \to -\infty \end{aligned}$$

as $t \to \infty$. This contradicts the fact that $y_3(t) > 0$ for both of these classes. Using y_1' and y_2' in a similar way, we can show that $\mathcal{N}_7 = \emptyset = \mathcal{N}_8$.

The proof of (ii) is similar to that of part (i) only using y_2' and y_3'. Part (iii) follows from [5, Theorem 1]. $\square$

For any solution y of (5.4.1), we define the function F by

$$F(t) = -y_4(t)y_1(t) + y_2(t)y_3(t)\,. \tag{5.4.9}$$

Lemma 5.4.2. (i) *Let y be a solution of (5.4.1) defined on $[0, \tau)$. Then*

$$F'(t) = r(t)|y(t)|^{\lambda+1} + a(t)|y''(t)|^{p+1}\,, \quad t \in [0, \tau)\,;$$

hence, F is a nondecreasing function for any solution y of (5.4.1).

(ii) *If $y \in \mathcal{O}_1$, then $\lim_{t\to\infty} F(t) \in (0, \infty]$.*

(iii) *If $y \in \mathcal{O}_2$, then $\lim_{t\to\infty} F(t) \in (-\infty, 0]$.*

(iv) *If $y \in \mathcal{N}_0 \cup \mathcal{N}_3 \cup \mathcal{N}_5$, then $\lim_{t\to\infty} F(t) \in (-\infty, 0]$.*

Proof. (i) Clearly, $F'(t) \geq 0$ by (5.4.2) and (5.4.3).

(ii) In view of (5.4.9) and (5.4.4), $F(t_k^1) = y_2(t_k^1)y_3(t_k^1) > 0$ for $k \in \{2, 3, \dots\}$, and the conclusion follows from case (i).

(iii) Similarly, (5.4.9) and (5.4.5) yield $F(t_k^1) = y_2(t_k^1)y_3(t_k^1) < 0$ for $k = 2, 3, 4, \dots$, and since F is nondecreasing, we have $\lim_{t\to\infty} F(t) \leq 0$.

(iv) If $y \in \mathcal{N}_3 \cup \mathcal{N}_5$, then (5.4.9), (5.4.7), and (5.4.8) yield $F(t) < 0$ for $t \geq t_y$, and the conclusion follows. If $y \in \mathcal{N}_0$, then $\lim_{t\to\infty} F(t) = 0$. □

The next lemma gives some preliminary results on the strong nonlinear limit-circle property.

Lemma 5.4.3. *Let $y \in \mathcal{S}$ satisfy $\lim_{t\to\infty} F(t) < \infty$.*

(i) *Then*

$$\int_0^\infty r(t)|y(t)|^{\lambda+1}\,dt < \infty \quad \textit{and} \quad \int_0^\infty a(t)|y''(t)|^{p+1}\,dt < \infty. \tag{5.4.10}$$

(ii) *If there exist positive constants r_0 and a_0 such that*

$$r_0 \leq r(t) \quad \textit{and} \quad a_0 \leq a(t) \quad \textit{on } \mathbb{R}_+, \tag{5.4.11}$$

then

$$\lim_{t\to\infty} y(t) = \lim_{t\to\infty} y'(t) = 0, \tag{5.4.12}$$

$$\int_0^\infty |y(t)|^{\lambda+1}\,dt < \infty, \quad \textit{and} \quad \int_0^\infty \frac{a(t)}{r(t)}|y''(t)|^{p+1}\,dt < \infty. \tag{5.4.13}$$

Proof. (i) The conclusion follows directly from Lemma 5.4.2 (i) by an integration on $\mathbb{R}_+$.

(ii) Property (5.4.13) follows from (5.4.10) and (5.4.11). Furthermore, (5.4.10) and (5.4.11) yield $y \in L_{\lambda+1}(\mathbb{R}_+)$ and $y'' \in L_{p+1}(\mathbb{R}_+)$. Hence, Lemma 1.5 in [132] yields (5.4.12). □

Our next lemma gives some additional properties of solutions of (5.4.1) that belong to the class $\mathcal{O}_1$.

Lemma 5.4.4. *Assume that $\lambda \le p$ and there are positive constants $\varepsilon \le p, a_0, r_0$, and M_0, and $t_0 \in \mathbb{R}_+$ such that*

$$a_0 \le a(t) \le M_0 t^{p-\varepsilon} \quad \text{and} \quad r_0 \le r(t) \quad \text{for } t \in [t_0, \infty). \tag{5.4.14}$$

Then $\lim_{t\to\infty} F(t) = \infty$ for any solution $y \in \mathcal{O}_1$.

Proof. Let $y \in \mathcal{O}_1$ and assume to the contrary that $\lim_{t\to\infty} F(t) = M \in (0, \infty)$. Then, (5.4.10) and (5.4.11) imply $y \in L_{\lambda+1}(\mathbb{R}_+)$ and $y'' \in L_{p+1}(\mathbb{R}_+)$ as before. Since $\lambda \le p$, Theorem 2 in [49, §V.3] implies $y' \in L_{p+1}(\mathbb{R}_+)$. Using Hölder's inequality, (5.4.10) and (5.4.14), we have

$$\begin{aligned}
\int_{t_0}^{t} a(s)|y'(s)|\,|y''(s)|^p &\le \left(\int_{t_0}^{t} |y'(s)|^{p+1}\,ds\right)^{\frac{1}{p+1}} \\
&\quad \times \left(\int_{t_0}^{t} a^{\frac{p+1}{p}}(s)|y''(s)|^{p+1}\,ds\right)^{\frac{p}{p+1}} \\
&\le \left(\int_{t_0}^{\infty} |y'(s)|^{p+1}\,ds\right)^{\frac{1}{p+1}} \\
&\quad \times \left(\int_{t_0}^{\infty} a(s)|y''(s)|^{p+1}\,ds\right)^{\frac{p}{p+1}} M_0^{\frac{1}{p}} t^{\frac{p-\varepsilon}{p}} \\
&\le C t^{1-\frac{\varepsilon}{p}}
\end{aligned} \tag{5.4.15}$$

for $t \in [t_0, \infty)$, where C is a suitable positive constant. Now define

$$Z(t) = -y_3(t)y_1(t) + 2\int_{t_1}^{t} a(s)y'(s)|y''(s)|^{p-1}y''(s)\,ds, \tag{5.4.16}$$

where $t_1 \in [t_0, \infty)$ satisfies

$$y_1(t_1) = 0 \quad \text{and} \quad F(t) \ge M/2 \quad \text{on } [t_1, \infty).$$

Since $Z'(t) = F(t)$ on $[t_1, \infty)$, if we take an increasing sequence $\{s_k\}_{k=1}^{\infty}$ of zeros of y_1 with $s_1 \ge t_1$, then (5.4.16) and (5.4.15) yield

$$\frac{M}{2}(s_k - t_1) \le \int_{t_1}^{s_k} F(s)\,ds = \int_{t_1}^{s_k} Z'(s)\,ds = Z(s_k) \le C s_k^{1-\frac{\varepsilon}{p}}.$$

This is a contradiction for large k and so the conclusion follows from Lemma 5.4.2(ii). □

Remark 5.4.1. If $a \equiv 1$, $p = 1$, and $y \in \mathcal{O}_1$, then $\lim_{t\to\infty} F(t) = \infty$ without any additional assumptions on r and λ (see [7, Lemma 3.4]).

Recall that we know from Lemma 5.2.6 that all solutions of equation (5.4.1) are continuable, i.e., they are defined on $\mathbb{R}_+$.

Our final lemma is actually a special case of Lemma 5.2.4 but we include it here because it actually gives explicit bounds on the solutions of the system (5.4.2).

Lemma 5.4.5. *Let $[a, b] \subset \mathbb{R}_+$ and a_0, a_1, r_0, and r_1 be positive constants. Let y be a solution of (5.4.2) such that y_i, $i = 1, 2, 3, 4$, have zeros on $[a, b]$, and let $\nu_i = \max_{a\le t\le b} |y_i(t)|$, $i = 1, 2, 3, 4$.*

(i) *If $a_0 \le a(t) \le a_1$ and $r(t) \le r_1$ on $[a, b]$, then*

$$\nu_2 \le K_0 \nu_1^{\delta}, \quad \nu_3 \le K_1 \nu_1^{\delta_1}, \quad \text{and} \quad \nu_4 \le K_2 \nu_1^{\delta_2}, \tag{5.4.17}$$

with

$$\delta = \frac{1}{2} + \frac{\lambda+1}{2(p+1)}, \quad \delta_1 = \frac{p(\lambda+1)}{p+1}, \quad \delta_2 = \frac{1}{2}\big(\lambda + \frac{p(\lambda+1)}{p+1}\big),$$

$$K_1 = \big(4a_0^{-\frac{1}{2p}} a_1^{\frac{1}{p}} r_1^{\frac{1}{2}}\big)^{2p/p+1}, \quad K_0 = \big(2a_0^{-\frac{1}{p}} K_1^{\frac{1}{p}}\big)^{1/2}, \quad \text{and}$$
$$K_2 = (2r_1 K_1)^{1/2}.$$

(ii) *If $a_0 \le a(t)$ and $r_0 \le r(t) \le r_1$ on $[a, b]$, then*

$$\nu_1 \le K_3 \nu_3^{\delta_3}, \quad \nu_2 \le K_4 \nu_3^{\delta_4}, \quad \text{and} \quad \nu_4 \le K_5 \nu_3^{\delta_5},$$

with

$$\delta_3 = \frac{p+1}{p(\lambda+1)}, \quad \delta_4 = \frac{\lambda+p+2}{2p(\lambda+1)}, \quad \delta_5 = \frac{1}{2}\left(1 + \frac{\lambda(p+1)}{p(\lambda+1)}\right),$$

$$K_3 = \big(4r_0^{-1} r_1^{\frac{1}{2}} a_0^{-\frac{1}{2p}}\big)^{\frac{2}{\lambda+1}}, \quad K_4 = \big(2a_0^{-\frac{1}{p}} K_3\big)^{\frac{1}{2}}, \quad \text{and} \quad K_5 = (2r_1 K_3^{\lambda})^{\frac{1}{2}}.$$

Proof. (i) We can choose intervals $J_i \subset [a,b]$, $i=1,2,3$ such that

$$\min_{t\in J_i} |y_i(t)| = 0 \quad \text{and} \quad \nu_i = \max_{t\in J_i} |y_i(t)|,$$

where ν_i occurs at one of the endpoints of the interval J_i and y_i does not change its sign on J_i, $i=1,2,3$. Then (5.4.2) and (5.4.3) yield

$$\begin{aligned} \nu_2^2 &\le 2\int_{J_1} |y_2(s)\, y_2'(s)|\, ds = 2\int_{J_1} a^{-\frac{1}{p}}(s)|y_2(s)|\, |y_3(s)|^{\frac{1}{p}}\, ds \\ &\le 2a_0^{-\frac{1}{p}} \nu_1 \nu_3^{\frac{1}{p}}, \qquad (5.4.18) \\ \nu_3^2 &\le 2\int_{J_2} |y_3(s)\, y_3'(s)\, ds| \le 2\int_{J_2} a^{\frac{1}{p}}(s)|y_2'(s)|\, |y_3(s)|^{\frac{p-1}{p}} |y_4(s)|\, ds \\ &\le 2a_1^{\frac{1}{p}} \nu_2 \nu_3^{\frac{p-1}{p}} \nu_4, \qquad (5.4.19) \end{aligned}$$

and

$$\begin{aligned} \nu_4^2 &\le 2\int_{J_3} |y_4(s)\, y_4'(s)|\, ds \le 2\int_{J_3} r(s)|y_1(s)|^{\lambda} |y_3'(s)|\, ds \\ &\le 2r_1 \nu_1^{\lambda} \nu_3. \qquad (5.4.20) \end{aligned}$$

From (5.4.18)–(5.4.20), we have

$$\nu_3^2 \le 2a^{\frac{1}{p}} \left(2a_0^{-\frac{1}{p}} \nu_1 \nu_3^{\frac{1}{p}}\right)^{\frac{1}{2}} (2r_1 \nu_1^{\lambda} \nu_3)^{\frac{1}{2}} \nu_3^{\frac{p-1}{p}}$$

and

$$\nu_3 \le K_1 \nu_1^{\frac{p(\lambda+1)}{p+1}}. \qquad (5.4.21)$$

Hence, the second inequality in (5.4.17) is proved. Furthermore, (5.4.18) and (5.4.21) yield $\nu_2 \le K_0 \nu_1^{\delta}$, so the first inequality in (5.4.17) holds. The last inequality in (5.4.17) follows from (5.4.20) and (5.4.21).

(ii) System (5.4.2) can be rewritten as the system $Z = (Z_1, \ldots, Z_4)$

$$\begin{aligned} Z_1' &= Z_2, & Z_3' &= Z_4, \\ Z_2' &= -r(t)|Z_3|^{\lambda} \operatorname{sgn} Z_3, & Z_4' &= a^{-\frac{1}{p}}(t)|Z_1|^{\frac{1}{p}} \operatorname{sgn} Z_1, \end{aligned} \qquad (5.4.22)$$

with $Z_1 = y_3$, $Z_2 = y_4$, $Z_3 = y_1$, and $Z_4 = y_2$. As the proof of (i) does not depend on the signs of the coefficients $a^{\frac{1}{p}}$ and $-r$ in (5.4.2), the results follow from part (i) applied to (5.4.22). □

5.5 Limit-Point/Limit-Circle and Related Properties of Solutions of Fourth-Order Equations

Our first theorem just concerns certain classes of solutions of (5.4.1)

Theorem 5.5.1.

(i) *If*

$$\int_0^\infty a^{-\frac{1}{p}}(\sigma)\,d\sigma = \infty \quad \textit{and} \quad \int_0^\infty r(\sigma)\,d\sigma = \infty,$$

then $\mathcal{N} = \mathcal{N}_0$.

(ii) *Assume there are positive constants* a_0 *and* r_0 *such that*

$$a_0 \le a(t) \quad \textit{and} \quad r_0 \le r(t) \quad \textit{on} \quad \mathbb{R}_+. \tag{5.5.1}$$

If $y \in \mathcal{O}_2 \cup \mathcal{N}_0$, *then*

$$\int_0^\infty |y(t)|^{\lambda+1}\,dt < \infty \quad \textit{and} \quad \int_0^\infty \frac{a(t)}{r(t)}|y''(t)|^{p+1}\,dt < \infty.$$

Proof. Part (i) follows from Lemma 5.4.1(ii). Part (ii) follows from Lemma 5.4.2 (iii) and Lemma 5.4.3 (ii) if $y \in \mathcal{O}_2$; if $y \in \mathcal{N}_0$, the result is obvious. □

The next theorem is concerned with the solutions of (5.4.1) that belong to the class $\mathcal{N}_1$.

Theorem 5.5.2. *Let* $y \in \mathcal{N}_1$. *Then* $\int_0^\infty |y(s)|^{\lambda+1}\,ds = \infty$. *Moreover,*

(i) *if* $\int_0^\infty a^{-\frac{1}{p}}(s) r^{-1}(s)\,ds = \infty$, *then* $\int\limits_0^\infty \frac{a(s)}{r(s)}|y''(s)|^{p+1}\,ds = \infty$;

(ii) *if*

$$\int_{t_0}^\infty r(s)\left(\int_{t_0}^s \int_{t_0}^\sigma a^{-\frac{1}{p}}(\sigma_1)\,d\sigma_1\,d\sigma\right)^\lambda ds = \infty \quad \textit{for} \quad t_0 \in \mathbb{R}_+, \tag{5.5.2}$$

then $\mathcal{N}_1 = \emptyset$.

Proof. Let $y \in \mathcal{N}_1$ and suppose for simplicity that $y_i(t) > 0$ for $i = 1, 2, 3, 4$ and $t \ge t_y$. In view of (5.4.2), $y_1(t) = y(t) \ge y_1(t_y) > 0$, so we have

$$\int_0^\infty |y(t)|^{\lambda+1}\,dt = \infty\,.$$

Moreover, since y_4 is positive and decreasing, we have

$$\begin{aligned} y_3(t_y) \le y_3(t) = y_3(t_y) + \int_{t_y}^t y_3'(s)\,ds \\ \le y_3(t_y) + y_4(t_y)(t-t_y) \le 2y_4(t_y)t \end{aligned} \tag{5.5.3}$$

for t large enough, say for $t \ge t_1 \ge t_y$. Hence,

$$Ca^{-\frac{1}{p}}(t) \le y''(t) = \big(y_3(t)a^{-1}(t)\big)^{\frac{1}{p}} \tag{5.5.4}$$

with $C = y_3^{\frac{1}{p}}(t_y)$. Part (i) then follows immediately.

To prove (ii), notice that the first inequality in (5.5.4) yields

$$y_1(t) \ge C\int_{t_y}^t \int_{t_y}^s a^{-\frac{1}{p}}(\sigma)\,d\sigma\,ds\,. \tag{5.5.5}$$

It follows from (5.4.2) that

$$y_4(t) = y_4(t_y) - \int_{t_y}^t r(s)\big(y_1(s)\big)^\lambda\,ds > 0 \quad \text{on } [t_y,\infty),$$

so (5.5.5) yields

$$C^\lambda \int_{t_y}^\infty r(s)\left[\int_{t_y}^s \int_{t_y}^\sigma a^{-\frac{1}{p}}(\sigma_1)\,d\sigma_1\,d\sigma\right]^\lambda ds \le \int_{t_y}^\infty r(s)y_1^\lambda(s)\,ds < \infty.$$

This contradicts (5.5.2) and completes the proof. □

Notation: Let $h \in C^0(\mathbb{R}_+)$. If $\int_t^\infty h(s)\,ds = \infty$ for $t \in \mathbb{R}_+$, then we set

$$\int_0^\infty \int_s^\infty h(\sigma)\,d\sigma\,ds = \infty.$$

Next, we turn our attention to solutions in the class $\mathcal{N}_3$.

Theorem 5.5.3. *Let* $y \in \mathcal{N}_3$.

(i) *Then* $\int_0^\infty |y(s)|^{\lambda+1}\,ds = \infty$.
(ii) *If*

$$\int_0^\infty a^{-\frac{1}{p}}(t)r^{-1}(t)\left(\int_t^\infty \int_s^\infty r(\sigma)\,d\sigma\,ds\right)^{\frac{p+1}{p}} dt = \infty,$$

then

$$\int_0^\infty \frac{a(s)}{r(s)} |y''(s)|^{p+1}\, ds = \infty\,.$$

Proof. Let $y \in \mathcal{N}_3$ be such that $y_1 > 0$, $y_2 > 0$, $y_3 < 0$, and $y_4 > 0$ on $[t_y, \infty)$. (If $y_1 < 0$, the proof is similar.) From (5.4.2), we have $|y_i|$ is decreasing for $i = 2, 3, 4$ and y_1 is increasing on $[t_y, \infty)$.

Part (i) follows from the fact that $y(t) \geq y(t_y) > 0$ on $[t_y, \infty)$. To prove (ii), note that using (5.4.2), we have

$$\begin{aligned} |y_3(t)| &\geq \int_t^\infty y_4(s)\, ds \geq \int_t^\infty \int_s^\infty r(\sigma) y^\lambda(\sigma)\, d\sigma\, ds \\ &\geq y^\lambda(t_y) \int_t^\infty \int_s^\infty r(\sigma)\, d\sigma\, ds\,, \end{aligned}$$

and thus

$$|y''(t)| \geq y^{\frac{\lambda}{p}}(t_y) a^{-\frac{1}{p}}(t) \left(\int_t^\infty \int_s^\infty r(\sigma)\, d\sigma\, ds \right)^{\frac{1}{p}}$$

for $t \geq t_y$. Hence,

$$\begin{aligned} &\int_{t_y}^\infty \frac{a(s)}{r(s)} |y''(s)|^{p+1}\, ds \\ &\quad \geq C \int_{t_y}^\infty a^{-\frac{1}{p}}(t) r^{-1}(t) \left(\int_t^\infty \int_s^\infty r(\sigma)\, d\sigma\, ds \right)^{\frac{p+1}{p}} dt = \infty \end{aligned}$$

with a suitable constant $C > 0$, and the statement holds. □

Applying Theorem 5.3.2 to equation (5.4.1) yields the following information about the solution of (5.4.1) in the class $\mathcal{O}_1$.

Theorem 5.5.4. *Let $\lambda \leq p$, let a_0, a_1, r_0, and r_1 be positive constants, and let $y \in \mathcal{O}_1$.*

(i) *If $a_0 \leq a(t) \leq a_1$ and $r(t) \leq r_1$ on $\mathbb{R}_+$, then*

$$\int_0^\infty |y(t)|^{\lambda+1} = \infty.$$

(ii) *If $a_0 \leq a(t) \leq a_1$ and $r_0 \leq r(t) \leq r_1$ on $\mathbb{R}_+$, then*

$$\int_0^\infty \frac{a(t)}{r(t)} |y''(t)|^{p+1}\, dt = \infty.$$

We can summarize the above results on solutions of (5.4.1) as follows.

Theorem 5.5.5. *Let $\lambda \le p$ and let a_0, a_1, r_0, and r_1 be positive constants such that*

$$a_0 \le a(t) \le a_1, \quad r_0 \le r(t) \le r_1 \quad on\ \mathbb{R}_+.$$

(i) *All solutions of* (5.4.1) *are continuable,* $\mathcal{S} = \mathcal{O}_1 \cup \mathcal{O}_2 \cup \mathcal{N}_0$, *and* $\mathcal{O}_1 \neq \emptyset$.

(ii) *For any solution y of* (5.4.1), *the following statements are equivalent:*

  (a) *y is of the nonlinear limit-circle type;*
  (b) *y is of the strong nonlinear limit-circle type;*
  (c) $y \in \mathcal{O}_2 \cup \mathcal{N}_0$.

(iii) *For any solution y of* (5.4.1), *the following statements are equivalent:*

  (a) *y is of the nonlinear limit-point type;*
  (b) *y is of the strong nonlinear limit-point type;*
  (c) $y \in \mathcal{O}_1$.

Proof. Parts (i) and (iii) of Lemma 5.4.1 together with Lemma 5.2.6 imply that all solutions of (5.4.1) are continuable and $\mathcal{S} = \mathcal{O}_1 \cup \mathcal{O}_2 \cup \mathcal{N}_0$. Now if y is a solution of (5.4.1) such that $F(0) > 0$, then parts (iii) and (iv) of Lemma 5.4.2 imply $y \notin \mathcal{O}_2 \cup \mathcal{N}_0$. Hence, $y \in \mathcal{O}_1$, and so part (i) holds. Parts (ii) and (iii) follow from Theorems 5.5.1 and 5.5.4. $\square$

It is reasonable to ask if it is possible to obtain results similar to those in Theorem 5.5.5 for $\lambda > p$. To show the difficulties involved, we consider the case where $p = 1$, $\lambda > 1$, and $a(t) \equiv 1$ in equation (5.4.1), i.e.,

$$y^{(4)} + r(t)|y|^{\lambda-1}y = 0. \tag{5.5.6}$$

We will need the following lemma.

Lemma 5.5.1. *Let $[T, T_1] \subset \mathbb{R}_+, T \le T_1 - 2, C_i \in \mathbb{R}$ for $i = 1, 2, 3, 4$, $\lambda > 1, \mu = (3\lambda^2 + 8\lambda + 5)/(3\lambda^2 + 6\lambda + 7)$, and*

$$0 < r_0 \le r(t) \le r_1 \quad on\ [T, T_1].$$

Then there exist constants $\delta > 0$ and $K > 0$, which do not depend on T, T_1, r_0, or r_1, such that a solution of the Cauchy initial value problem consisting of equation (5.4.1) *and the conditions*

$$y^{(i)} = C_{i+1}, \quad i = 0, 1, 2, 3, \tag{5.5.7}$$

is noncontinuable if

$$-C_1C_4 + C_2C_3 > K r_0^{-\frac{1}{\mu-1}} r_1^{\frac{3(\lambda+1)\mu}{(3\lambda+5)(\mu-1)}} (T_1 - T)^{-\frac{2\mu}{(\lambda+1)(\mu-1)}}$$

and

$$|C_1| + |C_2| + |C_3| + |C_4| > 4^{\frac{\mu+1}{\mu-1}} r_0^{-\frac{\lambda}{\lambda-1}} (\delta r_0 + r_1)(T_1 - T)^{-\frac{4}{\mu-1}}.$$

Proof. This follows from the proof of [61, Theorem 1]. Observe that $\mu > 1$. □

We then have the following result.

Theorem 5.5.6. *Assume that* $\lambda > 1, r_0 < r_1$, *and*

$$r_0 \le r(t) \le r_1 t^\sigma \quad \textit{for } t \in \mathbb{R}_+, \tag{5.5.8}$$

with $\sigma < \frac{2}{\lambda+1}\frac{3\lambda+5}{3\lambda+3}$. *Then equation* (5.5.6) *is of the strong nonlinear limit-circle type.*

Proof. By Lemma 5.4.1(ii), $\mathcal{N} = \mathcal{N}_0$. We will prove that $\mathcal{O}_1 = \emptyset$. Suppose $y \in \mathcal{O}_1 \neq \emptyset$. Then Lemma 5.4.2 (ii) implies $\lim_{t\to\infty} F(t) > 0$, and there exists $T \in \mathbb{R}$ such that

$$F(T) = -y(T)y'''(T) + y'(T)y''(T) > 0.$$

Choose $T_1 \in [T + 2, \infty)$ such that

$$F(T) > K r_0^{-\frac{1}{\mu-1}} (r_1 T_1)^{\frac{3(\lambda+1)\mu\sigma}{(3\lambda+5)(\mu-1)}} (T_1 - T)^{-\frac{2\mu}{(\lambda+1)(\mu-1)}}$$

and

$$\sum_{i=0}^{3} |y^{(i)}(T)| > 4^{\frac{\mu+1}{\mu-1}} r_0^{-\frac{\lambda}{\lambda-1}} (\delta r_0 + r_1 T_1^\sigma)(T_1 - T)^{-\frac{4}{\mu-1}},$$

where δ, K, and μ are given in Lemma 5.5.1; this choice is possible since $\mu > 1$,

$$\frac{2\mu}{(\lambda+1)(\mu-1)} > \frac{3(\lambda+1)\mu\sigma}{(3\lambda+5)(\mu-1)},$$

and $4/(\mu-1) > \sigma$. Now all the conditions of Lemma 5.5.1 hold with $r_0 = r_0$, $T = T$, $T_1 = T_1$, $r_1 = r_1 T_1^\sigma$, and so the solution y is noncontinuable. This contradiction proves that $\mathcal{O}_1 = \emptyset$. Thus, $\mathcal{O} = \mathcal{O}_2$ and the conclusion of the theorem follows from Theorem 5.5.1(ii). □

Remark 5.5.1. By [7, Lemma 3.4], for equation (5.5.6), we have $\lim_{t\to\infty} F(t) \in \{0, \infty\}$. If (5.5.8) holds, then we only have $F(t) \to 0$ as $t \to \infty$. Moreover, $F(t) \to 0$ as $t \to \infty$ if and only if y is a nonlinear limit-circle solution. It remains as an open problem whether Theorem 5.5.6 can be extended to equation (5.4.1).

To illustrate some of our results, we consider equation (5.4.1) with $a(t)$ and $r(t)$ being powers of t, namely, the equation

$$\left(t^{\alpha}|y''|^{p-1}y''\right)'' + t^{\beta}\,|y|^{\lambda-1}y = 0. \tag{5.5.9}$$

We see that, in view of Theorem 5.5.1(i), if

$$p \geq \alpha \quad \text{and} \quad \beta \geq -1,$$

then equation (5.5.9) has no proper nonoscillatory solutions. Now if

$$\beta \leq (p-\alpha)/p\,,$$

then any solution of (5.5.9) that belongs to $\mathcal{N}_1$ is of the strong nonlinear limit-point type by Theorem 5.5.2(i). By part (iii) of Theorem 5.5.2, if

$$\beta \geq -1,$$

then in fact $\mathcal{N}_1 = \emptyset$. By Theorem 5.5.3, if

$$\beta \geq -2 \quad \text{or} \quad -2 > \beta > \alpha - 3p - 2,$$

then any solution of equation (5.5.9) that belongs to $\mathcal{N}_3$ is of the strong nonlinear limit-point type.

It should be clear that the point-wise conditions on the coefficient functions a and r are less than desirable. This is especially true in the case of the function r; equation (5.4.1) and equations similar to it often appear in the literature with $a(t) \equiv 1$. However, by [122, Theorem 10.3],

$$\int_0^{\infty} s^3 r(s)\,ds < \infty \tag{5.5.10}$$

is a necessary and sufficient condition for proper nonoscillatory solutions of equation (5.5.6) to exist, that is, Theorem 5.5.1(i) is not true if (5.5.10) holds. In this same spirit, if (5.5.10) holds, then by [122, Theorem 16.9] there is a solution tending a nonzero constant so that equation (5.5.9) is of the nonlinear limit-point type contrary to Theorem 5.5.6. Now condition (5.5.10) implies $\liminf_{t\to\infty} r(t) = 0$, so in view of the above observations, the condition $r(t) \geq r_0$ may not be as restrictive as it first seems.

Hence, the form that results analogous to those here may take for equation (5.4.1) with $r(t) < 0$ remains as an open problem.

Chapter 6

Delay Equations I

6.1 Introduction

In this chapter, we examine the nonlinear limit-point and limit circle properties for delay differential equations of the form

$$\big(a(t)|y'|^{p-1}y'\big)' = r(t)f\big(y(\varphi(t))\big). \tag{6.1.1}$$

The first and very important issue that we must address is what is meant by saying a solution of equation (6.1.1) is of the nonlinear limit-circle or nonlinear limit-point type. This is not obvious even in the case of linear equations with a delay. This will be the focus of Section 6.2 in this chapter. In subsequent sections, we will discuss the nonlinear limit-point and nonlinear limit-circle properties as well as other asymptotic properties of solutions of (6.1.1). In particular, in Section 6.3, we present some preliminary lemmas and Section 6.4 deals primarily with nonoscillatory solutions. Oscillatory solutions are studied in Section 6.5. Section 6.6 contains some examples of our results.

6.2 Limit-Point/Limit-Circle and Other Properties of Solutions

Let us begin by considering the equation

$$\big(a(t)|y'|^{p-1}y'\big)' = r(t)f\big(y(\varphi(t))\big), \tag{6.2.1}$$

where a, r and φ are continuous functions defined on $\mathbb{R}_+ = [0,\infty)$, f is a continuous function defined on $\mathbb{R} = (-\infty,\infty)$, $a(t) > 0$, $r(t) > 0$, $\varphi(t) \le t$, $\lim_{t\to\infty}\varphi(t) = \infty$, $xf(x) > 0$ for $x \ne 0$, and p is a positive constant. We first define what we mean by a solution of equation (6.2.1).

Definition 6.2.1. Let $0 \le T_1 < T \le \infty, \sigma = \inf_{T_1 \le t < T} \varphi(t)$, $\phi \in C^0[\sigma, T_1]$, and $y_0' \in \mathbb{R}$. We say that a function y is a solution of (6.2.1) on $[T_1, T)$ (with the initial conditions (ϕ, y_0')) if $y \in C^0[\sigma, T)$, $y \in C^1[T_1, T)$, $a|y'|^{p-1}y' \in C^1[T_1, T)$, (6.2.1) holds on $[T_1, T)$, $y(t) = \phi(t)$ on $[\sigma, T_1]$, and $y'(T_1) = y_0'$.

Equation (6.2.1) is equivalent to the system of two differential equations

$$\begin{aligned} y_1' &= a^{-\frac{1}{p}}(t)|y_2|^{\frac{1}{p}} \operatorname{sgn} y_2, \\ y_2' &= r(t)\, f\big(y_1(\varphi)\big). \end{aligned} \tag{6.2.2}$$

The relationship between a solution y of (6.2.1) and a solution (y_1, y_2) of system (6.2.2) is

$$y_1(t) = y(t) \quad \text{and} \quad y_2(t) = a(t)\big|y'(t)\big|^{p-1} y'(t). \tag{6.2.3}$$

When discussing a solution y of (6.2.1), we will use (6.2.3) without mention.

Definition 6.2.2. A solution y of (6.2.1) is continuable if it is defined on $\mathbb{R}_+$. It is said to be noncontinuable if it is defined on $[0, T)$ with $T < \infty$, and it cannot be defined at $t = T$. A continuable solution y is called proper if it is nontrivial in any neighborhood of ∞. A proper solution y of (6.2.1) is called oscillatory if there exists a sequence of its zeros tending to ∞; otherwise it is called nonoscillatory.

Next we define what we will mean by nonlinear limit-point, nonlinear limit-circle, strong nonlinear limit-point, and strong nonlinear limit-circle types of solutions for delay equations.

Definition 6.2.3. A continuable solution y of (6.2.1) is said to be of the nonlinear limit-circle type if

$$\int_0^\infty \big|y(s)\, f\big(y(\varphi(s))\big)\big|\, ds < \infty \tag{NLC-D}$$

and it is said to be of the nonlinear limit-point type otherwise, i.e., if

$$\int_0^\infty \big|y(s)\, f\big(y(\varphi(s))\big)\big|\, ds = \infty. \tag{NLP-D}$$

Equation (6.2.1) will be said to be of the nonlinear limit-circle type if every solution y of (6.2.1) defined on $\mathbb{R}_+$ satisfies (NLC-D), and it will be said to be of the nonlinear limit-point type if there is at least one solution y for which (NLP-D) holds.

Definition 6.2.4. A continuable solution y of (6.2.1) is said to be of the strong nonlinear limit-point type if

$$\int_0^\infty \left|y(s)\,f\big(y(\varphi(s))\big)\right| ds = \infty \quad \text{and} \quad \int_0^\infty \frac{a(s)}{r(s)}\left|y'(s)\right|^{p+1} ds = \infty.$$

Equation (6.2.1) is said to be of the strong nonlinear limit-point type if equation (6.2.1) has proper solutions and every one of these is of the strong nonlinear limit-point type.

Definition 6.2.5. A continuable solution y of (6.2.1) is said to be of the strong nonlinear limit-circle type if

$$\int_0^\infty \left|y(s)\,f\big(y(\varphi(s))\big)\right| ds < \infty \quad \text{and} \quad \int_0^\infty \frac{a(s)}{r(s)}\left|y'(s)\right|^{p+1} ds < \infty.$$

Equation (6.2.1) is said to be of the strong nonlinear limit-circle type if every continuable solution is of the strong nonlinear limit-circle type.

When equation (6.2.1) is linear and $\varphi(t) \equiv t$, Definition 6.2.3 reduces to the (linear) limit-point and limit-circle definitions of Weyl. If $\varphi(t) \equiv t$, Definitions 6.2.3, 6.2.4, and 6.2.5 agree with their nonlinear versions for equations without delays.

As in Definition 3.11.1, for equations without delay, we say that a continuable solution y is of the nonlinear limit-circle type if

$$\int_0^\infty y(s)\,f(y(s))\,ds < \infty \tag{6.2.4}$$

and it is of the nonlinear limit-point type otherwise, i.e., if

$$\int_0^\infty y(s)\,f(y(s))\,ds = \infty. \tag{6.2.5}$$

Due to the fact that we ask that the sign condition, $xf(x) > 0$ if $x \neq 0$, holds, this is the same as Definition 6.2.3 above with $\varphi(t) \equiv t$. Of course one might wonder about using the definitions in (6.2.4) and (6.2.5) for delay equations, but this in a sense neutralizes the effect of the delay in the equation.

Another possibility would be to define the these properties in terms of whether the integral

$$\int_0^t y(s)\,f\big(y(\varphi(s))\big)\,ds$$

is bounded or not. Once again, this would agree with (6.2.4) and (6.2.5) in case $\varphi(t) \equiv t$. We should point out that all the results here remain true if this is taken as the definition.

It is our feeling that the form of the definitions appearing in Definition 6.2.3 above are the most natural and lend themselves to generalizations to higher order equations.

It is useful to note for computational purposes that

$$\int_0^\infty \frac{a(s)}{r(s)}|y'(s)|^{p+1}ds = \int_0^\infty a^{-\frac{1}{p}}(s)r^{-1}(s)|y_2(s)|^{\frac{p+1}{p}}ds. \tag{6.2.6}$$

For the results we will obtain next, it will be convenient to let S denote the set of all continuable solutions of (6.2.1) and we define the following classes:

$$\begin{aligned}
\mathcal{O} &= \{y \in S \,:\, y \text{ is oscillatory}\},\\
\mathcal{N} &= \{y \in S \,:\, y \text{ is nonoscillatory}\},\\
\mathcal{N}_0 &= \{y \in S \,:\, y \text{ is trivial in a neighborhood of infinity}\}.
\end{aligned}$$

We further subdivide $\mathcal{N}$ into the following classes:

$$\begin{aligned}
\mathcal{N}_1 &= \{y \in S :\ \text{for some } t_y \in \mathbb{R}_+,\ y(t)y'(t) > 0 \text{ for } t \in [t_y, \infty)\},\\
\mathcal{N}_2 &= \{y \in S :\ \text{for some } t_y \in \mathbb{R}_+,\ y(t)y'(t) < 0 \text{ for } t \in [t_y, \infty)\}.
\end{aligned}$$

Finally, we need the following subsets of $\mathcal{N}_2$:

$$\begin{aligned}
\mathcal{N}_{20} &= \left\{y \in \mathcal{N}_2 \,:\, \lim_{t\to\infty} y(t) = 0\right\},\\
\mathcal{N}_{21} &= \left\{y \in \mathcal{N}_2 \,:\, \lim_{t\to\infty} y(t) \neq 0\right\}.
\end{aligned}$$

6.3 Preliminary Lemmas

In this section, we give present some lemmas that are needed in the proofs of our main results. Our first lemma gives some basic information about the set S.

Lemma 6.3.1. *We have* $S = \mathcal{N}_0 \cup \mathcal{N}_1 \cup \mathcal{N}_2 \cup \mathcal{O}$.

Proof. It is easy to see that $S = \mathcal{N}_0 \cup \mathcal{N}_1 \cup \mathcal{N}_2 \cup \mathcal{N}_3 \cup \mathcal{O}$ where $\mathcal{N}_3 = \{y \in S : \text{for some } t_y \in \mathbb{R}_+,\ y(t) \neq 0 \text{ and } y' \text{ oscillates for } [t_y, \infty)\}$. But if $y \in \mathcal{N}_3$, then in view of (6.2.1), $a|y'|^{p-1}y'$ is either decreasing or increasing for large t. This contradicts y' being oscillatory and hence $\mathcal{N}_3 = \emptyset$. □

We will need the following lemma that describes properties of the equation

$$\left(a_0|z'|^{p-1}z'\right)' = r_0\left|z(\varphi_0)\right|^{\lambda-1}z(\varphi_0), \tag{6.3.1}$$

where a_0 and r_0 are positive constants, $\lambda > p$, $\varphi_0 \in C^0(\mathbb{R}_+)$ and $\varphi_0(t) \le t$ on $\mathbb{R}_+$.

Lemma 6.3.2. *Let $\lambda > p$, $T \in (0,\infty)$, $d \ge 1$, and $\varphi_0(t) = dt + T(1-d)$ on $(-\infty, T]$. Then* (6.3.1) *has a noncontinuable solution*

$$z(t) = M(T-t)^{-s} \quad on\ (-\infty, T), \tag{6.3.2}$$

where

$$s = \frac{p+1}{\lambda - p} \quad and \quad M = \left(\frac{a_0}{r_0}p\, s^p(s+1)\, d^{s\lambda}\right)^{\frac{1}{\lambda-p}}. \tag{6.3.3}$$

Proof. Note that $\varphi_0(T) = T$ and φ_0 is the delay on $(-\infty, T]$. The result can be obtained by the direct computation using the fact that $T - \varphi_0(t) = d(T-t)$. □

Lemma 6.3.3. *Let $0 \le \tau < T < \infty$, $a_0 > 0$, $r_0 > 0$, $\lambda > p$, and $\varphi_0 \in C^0(\mathbb{R}_+)$ be such that $\varphi(t) \ge \varphi_0(t)$ on $\mathbb{R}_+$. Assume that $\left|f(x)\right| \ge |x|^\lambda$ on $\mathbb{R}$ and that $a(t) \le a_0$ and $r(t) \ge r_0$ on $[\tau, T]$.*

Suppose z is a solution of (6.3.1) *defined on $[\tau, T)$ with the initial conditions $(\bar{\phi}, z_0')$ satisfying $\bar{\phi}(t) > 0$ on $[\bar{\sigma}, \tau]$ and $z_0' > 0$. Let y be a solution of* (6.2.1) *defined on $[\tau, T)$ with the initial conditions (ϕ, y_0') satisfying $\phi(t) > \max_{\bar{\sigma} \le s \le \tau} \bar{\phi}(s)$ for $t \in [\sigma, \tau]$ and $y_0' > z_0'$; here σ and $\bar{\sigma}$ are numbers from Definition* 6.2.1 *applied to* (6.2.1) *and* (6.3.1)*, respectively. Then,*

$$z(t) \le y(t) \quad for \quad t \in [\tau, T).$$

Proof. We can transform (6.3.1) into the system

$$z_1' = a_0^{\ -\frac{1}{p}}\left|z_2\right|^{\frac{1}{p}} \operatorname{sgn} z_2,$$

$$z_2' = r_0\left|z_1(\varphi_0)\right|^{\lambda-1} z_1(\varphi_0).$$

The result follows from a comparison theorem for this system and (6.2.2) that can be proved in the same way as well-known theorems on differential inequalities without delay (see, e.g., [112]). □

The following lemma gives a sufficient condition for the nonexistence of positive increasing solutions of (6.2.1) defined on $\mathbb{R}_+$.

Lemma 6.3.4. *Let $d \ge 1$, $\alpha \ge 0$, $\beta \le 0$, $\lambda > p$, and suppose there is an increasing sequence $\{T_k\}_{k=1}^{\infty}$ such that $T_1 > 0$, $\varphi(T_k) = T_k$, $k = 1, 2, \ldots,$*

$\lim_{k\to\infty} T_k = \infty$, *and*

$$\varphi(t) \geq dt + T_k(1-d) \quad \textit{for } t \in [T_{k-1}, T_k], \quad k = 2, 3, \ldots \tag{6.3.4}$$

Let $\alpha - \beta < p + 1$, $|f(x)| \geq |x|^\lambda$ for $x \in \mathbb{R}$, and assume that there are constants $M_1 > 0$ and $M_2 > 0$ such that

$$a(t) \leq M_1 t^\alpha \quad \textit{and } r(t) \geq M_2 t^\beta \quad \textit{for large} \quad t. \tag{6.3.5}$$

Then (6.2.1) has no solution y satisfying $y(t)y'(t) > 0$ for large t, i.e., $\mathcal{N}_1 = \emptyset$.

Proof. It follows from (6.3.4) that for $k \in \{1, 2, \ldots\}$ we have

$$t \geq \varphi(t) \geq dt + T_k(1-d) \quad \text{on } [T_1, T_k]. \tag{6.3.6}$$

Let y be a solution of (6.2.1) such that $y(t) > 0$ and $y'(t) > 0$ for large t; the case $y(t) < 0$ and $y' < 0$ for large t can be studied similarly. Then (6.3.5) and the fact that $\lim_{t\to\infty} \varphi(t) = \infty$ imply the existence of $\tau \in (0, \infty)$ such that $\varphi(t) \geq \varphi(\tau) \geq T_1$ and (6.3.5) hold for $t \geq \tau$, and

$$y(t) > 0 \quad \text{and} \quad y'(t) > 0 \quad \text{for} \quad t \geq \varphi(\tau).$$

Let $C = \min\{y'(\tau), \min_{\varphi(\tau)\leq s\leq\tau} y(s)\}$ and let $T \in \{T_k\}_{k=1}^\infty$ be such that $T > \tau + 1$, and

$$C \geq \max\{1, s\} \left(\frac{M_1}{M_2} p s^p (s+1)\, d^{s\lambda}\right)^{\frac{1}{\lambda - p}} T^{\frac{\alpha-\beta}{\lambda-p}} (T-\tau)^{-s}. \tag{6.3.7}$$

Together with (6.2.1), consider the equation

$$\left(a_0|z'|^{p-1}z'\right)' = r_0\left|z(\varphi^*)\right|^{\lambda-1} z(\varphi^*) \tag{6.3.8}$$

where $a_0 = M_1 T^\alpha$, $r_0 = M_2 T^\beta$, $\varphi^*(t) = dt + T(1-d)$ on $[0, T]$, and $\varphi^*(t) = t$ for $t > T$. Then,

$$a(t) \leq a_0 \quad \text{and} \quad r(t) \geq r_0 \quad \text{on } [\tau, T], \tag{6.3.9}$$

and (6.3.6) yields

$$\varphi(t) \geq \varphi^*(t) \quad \text{for } t \in [T_1, T]. \tag{6.3.10}$$

Let z be the solution of (6.3.8) given by (6.3.2) in Lemma 6.3.2 and let s and M be the constants given by (6.3.3); i.e.,

$$z(t) = M(T-t)^{-s} \quad \text{and} \quad z'(t) = sM(T-t)^{-s-1} \quad \text{for } t \in (-\infty, T).$$

From this, (6.3.6), and (6.3.7), we have

$$y(t) \geq C \geq \max_{-\infty < s \leq \tau} z(s) \tag{6.3.11}$$

and

$$y'(\tau) \geq C \geq \max_{-\infty < s \leq \tau} z'(s) \tag{6.3.12}$$

for $t \in [\varphi(\tau), \tau]$. Furthermore, (6.3.9)–(6.3.12) imply that the hypotheses of Lemma 6.3.3 (with $\varphi_0 = \varphi_*$) applied to (6.2.1) and (6.3.8) are satisfied, so

$$y(t) \geq z(t) \quad \text{on} \quad [\tau, T].$$

This contradicts the continuability of y and the noncontinuability of z. □

The conditions on the function φ in Lemma 6.3.4 may seem somewhat obscure, so we offer the following simple example of such a function.

Example 6.3.1. Let $\varphi(t) = t - (1-\sin t)/2$, $\{T_k\}_{k=0}^{\infty} = \{\pi/2 + 2k\pi\}$, and $d = 3/2$. Then we have $\varphi(t) = t$ if and only if $t = T_k$.

Remark 6.3.1. Lemma 6.3.4 generalizes the results of Lemma 3.13.5 which were proved for equation (6.2.1) with $\varphi(t) \equiv t$ on $\mathbb{R}_+$.

The following lemma sums up the needed properties of solutions in $\mathcal{N}_2$.

Lemma 6.3.5. (i) *Equation* (6.2.1) *has a solution* $y \in \mathcal{N}_{21}$ *if and only if*

$$\lim_{t\to\infty} \int_0^t a^{-\frac{1}{p}}(s) \left(\int_s^t r(\sigma)\, d\sigma \right)^{\frac{1}{p}} ds < \infty.$$

(ii) *Let* φ *be eventually nondecreasing and assume that there is a constant* $\varepsilon > 0$ *such that*

$$|f(x)| \geq |x|^p \quad \textit{for} \quad |x| \leq \varepsilon$$

and

$$\limsup_{t\to\infty} \int_{\varphi(t)}^t a^{-\frac{1}{p}}(s) \left(\int_s^t r(\sigma)\, d\sigma \right)^{\frac{1}{p}} ds > 1. \tag{6.3.13}$$

Then $\mathcal{N}_2 = \emptyset$.

(iii) *Let $y \in \mathcal{N}_{20}$, $\lambda > p$, and assume there exist $M > 0$ and $\varepsilon > 0$ such that*

$$a^{\frac{1}{p}}(t)r(t) \geq M \quad \text{on } \mathbb{R}_+, \qquad \int_0^\infty a^{-\frac{1}{p}}(s)\,ds = \infty, \tag{6.3.14}$$

and

$$|f(x)| \geq |x|^\lambda \quad \text{for } |x| \leq \varepsilon.$$

Then there is a constant $M_1 > 0$ such that

$$|y(t)| \leq M_1 \left(\int_0^t a^{-\frac{1}{p}}(s) \right)^{-\frac{p+1}{\lambda-p}} \qquad \text{for large } t.$$

Proof. (i) See [55, Theorem 3]

(ii) The conclusion $\mathcal{N}_2 = \emptyset$ is gained in [55, Theorem 2 and Corollary 1]in case $\varphi(t) < t$, eventually. But that assumption is not used in the proof.

(iii) Let $y \in \mathcal{N}_{20}$ and $y(t) > 0$ for large t; the case $y < 0$ for large t is similar. Since $\lim_{t\to\infty} \varphi(t) = \infty$, there exists $t_0 \in \mathbb{R}_+$ such that

$$\varepsilon \geq y(\varphi(t)) \geq y(t) > 0 \quad \text{and} \quad y'(t) < 0 \quad \text{on } [t_0, \infty).$$

From this, equation (6.2.1), and the fact that $\lim_{t\to\infty} y(t) = y(\infty) = 0$, we have

$$\begin{aligned}
|y_2(t)|^{\frac{p+1}{p}} &\geq |y_2(t)|^{\frac{p+1}{p}} - |y_2(\infty)|^{\frac{p+1}{p}} = -\frac{p+1}{p} \int_t^\infty |y_2(s)|^{\frac{1}{p}} |y_2(s)|' ds \\
&= -\frac{p+1}{p} \int_t^\infty a^{\frac{1}{p}}(s)\, r(s) f\big(y(\varphi(s))\big)\, y'(s)\,ds \\
&\geq -\frac{p+1}{p} \int_t^\infty a^{\frac{1}{p}}(s)\, r(s) y^\lambda(s) y'(s)\,ds \geq \frac{p+1}{p(\lambda+1)} M y^{\lambda+1}(t),
\end{aligned}$$

or

$$y'(t) \leq -C a^{-\frac{1}{p}}(t) y^{\frac{\lambda+1}{p+1}}(t)$$

for $t \geq t_0$, where $C = \left(\frac{p+1}{p(\lambda+1)} M\right)^{\frac{1}{p+1}}$. Integrating this inequality from t_0 to t yields

$$y^{-\frac{\lambda-p}{p+1}}(t) \geq y^{-\frac{\lambda-p}{p+1}}(t_0) + C\frac{\lambda-p}{p+1} \int_{t_0}^t a^{-\frac{1}{p}}(s)\,ds$$

for $t \geq t_0$. Now (6.3.14) implies the existence of $t_1 > t_0$ such that

$$y^{-\frac{\lambda-p}{p+1}}(t) \geq C_1 \int_0^t a^{-\frac{1}{p}}(s)\,ds \quad \text{for} \quad t \geq t_1$$

with $C_1 = C\frac{\lambda-p}{2(p+1)}$. Hence,

$$y(t) \le \left(C_1 \int_0^t a^{-\frac{1}{p}}(s)ds\right)^{-\frac{p+1}{\lambda-p}}, \quad t \ge t_1.$$

□

Lemma 6.3.6. *Assume that there are constants $r_1 \ge r_0 > 0$ such that $r_1 \ge r(t) \ge r_0$ on $\mathbb{R}_+$, and let $y \in \mathcal{O}$. If y is of the nonlinear limit-circle type, then y is of the strong nonlinear limit-circle type.*

Proof. Let $y \in \mathcal{O}$ and

$$\int_0^\infty \left|y(t)\, f(y(\varphi(t)))\right| dt < \infty. \tag{6.3.15}$$

Then,

$$\begin{aligned}(y_1(t)y_2(t))' &= y_1'(t)y_2(t) + y_1(t)y_2'(t)\\ &= a(t)\left|y_1'(t)\right|^{p+1} + r(t)\, y_1(t)\, f\big(y_1(\varphi(t))\big)\end{aligned}$$

on $\mathbb{R}_+$. Integrating, we have

$$\begin{aligned}y_1(t)\, y_2(t) - y_1(0)\, y_2(0) &= \int_0^t r(s)\left[\frac{a(s)}{r(s)}\left|y_1'(s)\right|^{p+1}\right.\\ &\quad \left. + y_1(s) f\big(y_1(\varphi(s))\big)\right] ds \ge r_0 \int_0^t \frac{a(s)}{r(s)}\left|y_1'(s)\right|^{p+1} ds\\ &\quad - r_1 \int_0^t \left|y_1(s)\, f\big(y_1(\varphi(s))\big)\right| ds. \end{aligned} \tag{6.3.16}$$

The conclusion follows from (6.3.16) applied on a sequence $\{t_k\}_{k=1}^\infty$ of zeros of y_1 tending to ∞ and from (6.3.15). □

Lemma 6.3.7. *Let either* (i) *$\varphi(t) < t$ on $\mathbb{R}_+$, or* (ii) *there is an $M \in \mathbb{R}_+$ with*

$$\left|f(x)\right| \le M + |x|^p \quad \text{on } \mathbb{R}. \tag{6.3.17}$$

Then every solution y of (6.2.1) *is defined on $\mathbb{R}_+$.*

Proof. (i) Let $\varphi(t) < t$ on $\mathbb{R}_+$ and to the contrary, let y be a noncontinuable solution of (6.2.1) defined on $[0, T)$, $T < \infty$. Then

$\limsup_{t\to T-} \big|y_2(t)\big| = \infty$. But since $\max_{0\le t\le T} \varphi(t) < T$, we have $\big|y_2'(t)\big| = \big|r(t)\, f\big(y(\varphi(t))\big)\big|$ is bounded on $[0, T)$. This contradiction proves (i).

(ii) Let (6.3.17) hold and let y be a noncontinuable solution of (6.2.1) defined on $[0, T)$, $T < \infty$, with the initial conditions (ϕ, y_0'). Then,

$$\limsup_{t\to T-} u(t) = \infty, \tag{6.3.18}$$

where

$$u(t) = \max_{0\le s\le t} \big|y_2(s)\big| + 1 \quad \text{for } t \in J \stackrel{\text{def}}{=} [0, T). \tag{6.3.19}$$

We have

$$\begin{aligned} \big|y\big(\varphi(t)\big)\big| &\le \max_{\sigma\le s\le 0} |\phi(s)| + |y(0)| + \int_0^t |y'(s)|\, ds \\ &\le C + \int_0^t a^{-\frac{1}{p}}(s)|y_2(s)|^{\frac{1}{p}}\, ds \le C + C_1 u^{\frac{1}{p}}(t) \le C_2 u^{\frac{1}{p}}(t) \end{aligned}$$

for $t \in J$ with $\sigma = \inf\{\varphi(t) : 0 \le t \le T\}$, $C = \max_{\sigma\le s\le 0} |\phi(s)| + |y(0)|$, $C_1 = \int_0^T a^{-\frac{1}{p}}(s)\, ds$, and $C_2 = C + C_1$. From this and (6.2.1) we have that

$$|y_2(t)| \le |y_2(0)| + \int_0^t r(s)\big|f\big(y(\varphi(s))\big)\big|\, ds \tag{6.3.20}$$

$$\le |y_2(0)| + r_2 \int_0^t \big(M + |y(\varphi(s))|^p\big)\, ds \tag{6.3.21}$$

$$\le C_3 + C_4 \int_0^t u(s)\, ds \tag{6.3.22}$$

for $t \in J$ with $r_2 = \max_{0\le t\le T} r(t)$, $C_3 = |y_2(0)| + r_2 TM$ and $C_4 = r_2 C_2^p$. Hence, (6.3.19) and (6.3.20) imply

$$u(t) \le C_3 + 1 + C_4 \int_0^t u(s)\, ds, \tag{6.3.23}$$

and applying Gronwall's inequality we see that u is bounded on J. This contradicts (6.3.18) and completes the proof. □

Remark 6.3.2. If (6.3.17) holds, then this result is proved in Theorem 1.4 in [47] under the restriction $t - C \le \varphi(t)$ where $C \in (0, \infty)$.

Our next two lemmas are somewhat of a technical nature but they will prove to be quite useful.

Lemma 6.3.8. *Let y be an oscillatory solution of* (6.2.1) *and let φ be nondecreasing on $\mathbb{R}_+$.*

(i) *Let $0 \le \tau_0 < \tau < \infty$ be such that $y(\tau_0) = 0$, $y(t) \ne 0$ on $(\tau_0, \tau]$, and $|y|$ have a local maximum at $t = \tau$. Then*

$$\tau - \varphi(\tau) > \tau - \tau_0.$$

(ii) *Let $0 \le \tau_0 < \tau < \infty$ be such that $y_2(\tau_0) = 0$, $y_2(t) \ne 0$ on $(\tau_0, \tau]$, and $|y_2|$ have a local maximum at $t = \tau$. Then*

$$\tau - \varphi(\tau) > \tau - \tau_0.$$

Proof. (i) Let $0 \le \tau_0 < \tau < \infty$ be such that $y(\tau_0) = 0$, and let $|y|$ have a local maximum at τ, and $y(t) \ne 0$ on $(\tau_0, \tau]$. If $\tau_0 \le \varphi(\tau)$, then φ being nondecreasing implies $y\big(\varphi(t)\big) \ne 0$ on $\big(\varphi(\tau), \tau\big)$, $y'(\tau) = y_2(\tau) = 0$, and (6.2.2) implies $y \ne 0$ on $[\tau, \infty)$. This contradicts y being oscillatory and proves $\varphi(\tau) < \tau_0$.

(ii) Let $0 \le \tau_0 < \tau < \infty$ be such that $y_2(\tau_0) = 0$, and let $|y_2|$ have a local maximum at τ, and $y_2(t) \ne 0$ on $(\tau_0, \tau]$. Suppose, for simplicity, that $y_2(\tau) < 0$; if $y_2(\tau) > 0$ the proof is similar. Since $y_2'(\tau) = 0$, (6.2.2) implies

$$y_1\big(\varphi(\tau)\big) = 0. \tag{6.3.24}$$

There are two possibilities: (a) $y_1(\tau) \le 0$ or (b) $y_1(\tau) > 0$.

Case (a): Let T_1 be a zero of y such that $T_1 \le \tau$ and $y_1(t) \le 0$ on $[T_1, \tau]$. If $\varphi(\tau) \ge T_1$, then $y_1(t) \le 0$ on $[\varphi(\tau), \tau]$, and since $y_2(\tau) < 0$ and φ is nondecreasing, (6.2.2) implies y is not oscillatory. Hence, $\varphi(\tau) < T_1$, and from (6.3.24) we have that there exists $T_0 < T_1$ such that $y_1(T_0) = 0$, T_0 and T_1 are consecutive zeros of y_1, and $\varphi(\tau) \le T_0$. It follows from the definition of τ_0 that $\varphi(\tau) \le T_0 < \tau_0$, and so the statement holds.

Case (b): Let $T_0 < T_1$ be two consecutive zeros of y_1 such that $\tau \in [T_0, T_1)$. These numbers exist due to (6.3.24) and the fact that y is oscillatory. Clearly, $T_0 \le \tau_0 < \tau$. Moreover, $y(t) > 0$ on $[\tau_0, \tau]$. Suppose $\varphi(\tau) \ge T_0$. Since φ is nondecreasing, $y(t) \ge 0$ on $[\varphi(\tau), \tau_0]$, and $y_2(\tau_0) = 0$, (6.2.2) implies y is nonoscillatory on $[\tau_0, \infty)$. This contradiction proves $\varphi(\tau) < T_0 < \tau_0$, and again the statement holds. □

Lemma 6.3.9. *Assume that φ is nondecreasing on $\mathbb{R}_+$ and y is an oscillatory solution of* (6.2.1).

(i) *Let y be bounded on $\mathbb{R}_+$ and let $\tau \in (0,\infty)$ be such that $|y_2|$ has a local maximum at $t = \tau$ and y_2 has a zero on $[0,\tau)$. Then there is a constant $K > 0$ such that K does not depend on τ and*

$$\big|y_2(\tau)\big| \le K \int_{\varphi(\tau)}^{\tau} r(t)\,dt. \tag{6.3.25}$$

(ii) *Let y_2 be bounded on $\mathbb{R}_+$ and let $\tau \in (0,\infty)$ be such that $|y_1|$ has a local maximum at $t = \tau$ and y_1 has a zero on $[0,\tau)$. Then there is a constant $K_1 > 0$ such that K_1 does not depend on τ and*

$$\big|y_1(\tau)\big| \le K_1 \int_{\varphi(\tau)}^{\tau} a^{-\frac{1}{p}}(t)\,dt.$$

Proof. (i) Let $\sup_{t\in[\omega,\infty)} |y_1(t)| = M < \infty$. Suppose $|y_2(\tau)| > 0$ for otherwise (6.3.25) would hold. If there is an increasing sequence $\{t_k\}_1^\infty$ of zeros of y_2 such that $t_k < \tau$ for $k = 1, 2, \ldots$ and $\lim_{k\to\infty} t_k = \tau$, then $y_2(\tau) = 0$ and (6.3.25) holds. Let τ_0 be the largest zero of y_2 smaller than τ. Then, according to Lemma 6.3.8(ii), $\varphi(\tau) < \tau_0$ and an integration of the second equality in (6.2.2) implies

$$\begin{aligned} \big|y_2(\tau)\big| &= \big|y_2(\tau) - y_2(\tau_0)\big| \le \int_{\tau_0}^{\tau} r(t)\,\big|f\big(y(\varphi(t))\big)\big|\,dt \\ &\le K \int_{\varphi(\tau)}^{\tau} r(t)\,dt \end{aligned}$$

with $K = \max_{|s|\le M} |f(s)| < \infty$. Hence, (6.3.25) holds.

(ii) The proof is similar if we use Lemma 6.3.8(i) and integrate the first inequality in (6.2.2). □

The following corollary is an immediate consequence of Lemma 6.3.9.

Corollary 6.3.1. *Let φ be nondecreasing and y be a bounded oscillatory solution of* (6.2.1).

(i) *If there is a positive constant K such that*

$$\int_{\varphi(t)}^{t} r(s)\,ds \le K \quad \textit{for } t \in \mathbb{R}_+, \tag{6.3.26}$$

then y_2 is bounded on $\mathbb{R}_+$.

(ii) *If $\lim_{t\to\infty} \int_{\varphi(t)}^{t} r(s)\,ds = 0$, then $\lim_{t\to\infty} y_2(t) = 0$.*

(iii) *If* (6.3.26) *holds and*

$$\lim_{t\to\infty} \int_{\varphi(t)}^{t} a^{-\frac{1}{p}}(s)\, ds = 0,$$

then $\lim_{t\to\infty} y_1(t) = 0$.

6.4 Nonoscillatory Solutions

The first theorem in this section concerns solutions that belong to the class $\mathcal{N}_2$.

Theorem 6.4.1.

(i) *If* $y \in \mathcal{N}_2$ *and there exists* $r_0 > 0$ *such that*

$$r(t) \geq r_0 \quad \textit{for large } t, \tag{6.4.1}$$

then y *is of the strong nonlinear limit-circle type.*

(ii) *Let* $y \in \mathcal{N}_{20}$ *and* $\lambda > p$. *Assume that* φ *is nondecreasing,* f *is nondecreasing on* $\mathbb{R}$, *and there exist* $\varepsilon > 0$, $M > 0$, *and* $\tau > 0$ *such that*

$$|f(x)| \geq |x|^{\lambda} \quad \textit{for } |x| \leq \varepsilon, \tag{6.4.2}$$

$a^{\frac{1}{p}}(t)\, r(t) \geq M$ *on* $\mathbb{R}_+$, *and* $\varphi(t) > 0$ *for* $t \geq \tau$. *If for every* $M_1 \in \mathbb{R}\backslash\{0\}$ *we have*

$$\int_{\tau}^{\infty} \left(\int_0^t a^{-\frac{1}{p}}(s)\, ds\right)^{-\frac{p+1}{\lambda-p}} \left| f\left(M_1 \left(\int_0^{\varphi(t)} a^{-\frac{1}{p}}(s)\, ds\right)^{-\frac{p+1}{\lambda-p}}\right)\right| dt < \infty, \tag{6.4.3}$$

then y *is of the nonlinear limit-circle type.*

(iii) *Let* $y \in \mathcal{N}_{21}$. *Then* y *is of the nonlinear limit-point type. If, in addition, we have*

$$\int_0^{\infty} a^{-\frac{1}{p}}(t)\, r^{-1}(t) \left(\int_t^{\infty} r(s)\, ds\right)^{\frac{p+1}{p}} dt = \infty, \tag{6.4.4}$$

then y *is of the strong nonlinear limit-point type.*

(iv) *Let* $y \in \mathcal{N}_{21}$ *and* $\int_0^{\infty} a^{-\frac{1}{p}}(t)\, dt = \infty$. *Then* $\int_0^{\infty} \frac{a(t)}{r(t)} |y'(t)|^{p+1}\, dt = \infty$ *if and only if* (6.4.4) *holds.*

Proof. Let $y \in \mathcal{N}_2$ and $t_0 \in \mathbb{R}_+$ be such that

$$y(t) > 0 \quad \text{on } [\sigma, \infty) \quad \text{and} \quad y_2(t) < 0 \quad \text{on } [t_0, \infty) \tag{6.4.5}$$

with $\sigma = \min_{t_0 \leq t < \infty} \varphi(t)$.

(i) From (6.2.2), and (6.4.5) we have

$$\infty > y_2(\infty) - y_2(t_0) = \int_{t_0}^{\infty} r(s)\, f\big(y(\varphi(s))\big)\, ds$$

$$\geq r_0 \int_{t_0}^{\infty} \frac{f\big(y(\varphi(s))\big)\, y(s)}{y(s)}\, ds$$

$$\geq \frac{r_0}{y(t_0)} \int_{t_0}^{\infty} \big|y(s)\, f\big(y(\varphi(s))\big)\big|\, ds;$$

hence, y is of the nonlinear limit-circle type.

Similarly, (6.2.2), (6.2.3), (6.2.6), and (6.4.1) imply $|y_2|$ is decreasing for $t \geq t_0$ and

$$\infty > y(t_0) - y(\infty) = -\int_{t_0}^{\infty} y'(s)\, ds = \int_{t_0}^{\infty} a^{-\frac{1}{p}}(s) |y_2(s)|^{\frac{p+1}{p}} |y_2(s)|^{-1}\, ds$$

$$\geq |y_2(t_0)|^{-1} \int_{t_0}^{\infty} \frac{a^{-\frac{1}{p}}(s)}{r(s)} |y_2(s)|^{\frac{p+1}{p}} r(s)\, ds$$

$$\geq r_0 |y_2(t_0)|^{-1} \int_{t_0}^{\infty} \frac{a(s)}{r(s)} |y'(s)|^{p+1} ds.$$

Thus, y is of the strong nonlinear limit-circle type.

(ii) Since y is decreasing to zero on $[t_0, \infty)$ and $f(x) \neq 0$ for $x \neq 0$, (6.4.2) implies the existence of $C > 0$ such that

$$f(x) \geq C x^{\lambda} \quad \text{for } 0 \leq x \leq y(t_0). \tag{6.4.6}$$

In view of (6.4.3) and (6.4.6), we must have $\int_0^{\infty} a^{-\frac{1}{p}}(s)\, ds = \infty$, and so the conclusion follows from Lemma 6.3.5(iii).

(iii) Now $\int_0^{\infty} \big|y(t)\, f\big(y(\varphi(t))\big)\big|\, dt = \infty$ follows from the fact that

$$\lim_{t \to \infty} y(t) = C_0 > 0 \quad \text{and} \quad \lim_{t \to \infty} \varphi(t) = \infty.$$

Furthermore, (6.2.2) and (6.4.5) imply

$$|y_2(t)| = |y_2(\infty)| + \int_t^{\infty} r(s) f\big(y_1(\varphi(s))\big)\, ds \geq C_1 \int_t^{\infty} r(s)\, ds \tag{6.4.7}$$

for $t \geq t_0$ with

$$C_1 = \min_{C_0 \leq x \leq C_2} f(x) > 0, \quad C_2 = \max_{\sigma \leq t < \infty} y_1(t) < \infty. \tag{6.4.8}$$

Hence, by (6.4.7)

$$\int_{t_0}^{\infty} \frac{a(t)}{r(t)} |y'(t)|^{p+1} dt = \int_{t_0}^{\infty} a^{-\frac{1}{p}}(t) r^{-1}(t) |y_2(t)|^{\frac{p+1}{p}} dt$$
$$\geq C_1^{\frac{p+1}{p}} \int_{t_0}^{\infty} a^{-\frac{1}{p}}(t) r^{-1}(t) \left(\int_t^{\infty} r(s)\, ds \right)^{\frac{p+1}{p}} dt$$

and the conclusion follows from (6.4.4).

(iv) Let $y \in \mathcal{N}_{21}$; then $\lim_{t\to\infty} y(t) = C_0 > 0$. It follows from (6.2.2) and (6.4.5) that $|y_2|$ is decreasing for $t \geq t_0$. Moreover, (6.2.2) implies

$$\infty > y(t_0) - y(\infty) = -\int_{t_0}^{\infty} y'(s)\, ds = \int_{t_0}^{\infty} a^{-\frac{1}{p}}(s) |y_2(s)|^{\frac{1}{p}} ds$$
$$\geq |y_2(\infty)|^{\frac{1}{p}} \int_{t_0}^{\infty} a^{-\frac{1}{p}}(s)\, ds; \tag{6.4.9}$$

hence, (6.4.9) implies $y_2(\infty) = 0$. Similarly to (6.4.7), there exists $t_1 > t_0$ such that

$$|y_2(t)| \leq C_3 \int_t^{\infty} r(s)\, ds$$

for $t \geq t_1$, where $C_3 = 2 \max_{C_0 \leq x \leq C_2} f(x) < \infty$ and C_2 is given by (6.4.8). From this and (6.2.6), we have

$$\int_{t_1}^{\infty} \frac{a(t)}{r(t)} |y'(t)|^{p+1}\, dt = \int_{t_1}^{\infty} a^{-\frac{1}{p}}(t) r^{-1}(t) |y_2(t)|^{\frac{p+1}{p}} dt$$
$$\leq C_3^{\frac{p+1}{p}} \int_{t_1}^{\infty} a^{-\frac{1}{p}}(t) r^{-1}(t) \left(\int_t^{\infty} r(s)\, ds \right)^{\frac{p+1}{p}} dt,$$

and the conclusion follows from this and case (iii). □

Example 6.4.1. Let $a(t) = t^p$, $r(t) = t^{-2}$, $\lambda > 0$ and $f(x) = |x|^{\lambda} \text{sgn}\, x$. Then according to Lemma 6.3.5(i), there exists $y \in \mathcal{N}_{21}$, so

$\int_0^\infty |y(t)|\,|y(\varphi(t))|^\lambda dt = \infty$. Theorem 6.4.1(iv) implies

$$\int_0^\infty \frac{a(t)}{r(t)} |y'(t)|^{p+1}\, dt = \infty \quad \text{if and only if} \quad p \geq 1.$$

Note that (6.4.1) does not hold here.

Remark 6.4.1. We noted above that the condition $r(t) \geq r_0 > 0$ did not hold in Example 6.4.1. This makes it seem reasonable that such a condition is needed in order to obtain a limit-circle type result as in Theorem 6.4.1(i).

Our next theorem concerns solutions in the class $\mathcal{N}_1$.

Theorem 6.4.2. *Let $y \in \mathcal{N}_1$.*

(i) *If*

$$\big|f(x)\big| \geq |x|^{-1} \quad \text{for large } |x|, \tag{6.4.10}$$

then y is of the nonlinear limit-point type.
Moreover, if

$$\int_0^\infty a^{-\frac{1}{p}}(t) r^{-1}(t)\, dt = \infty, \tag{6.4.11}$$

then y is of the strong nonlinear limit-point type.

(ii) *Assume that there is an $\varepsilon > 0$ such that*

$$\big|f(x)\big| \leq |x|^{-1-\varepsilon} \quad \text{for large } |x|, \tag{6.4.12}$$

and there is a $\tau > 0$ such that $\varphi(t) > 0$ for $t \geq \tau$. If

$$\int_\tau^\infty \int_0^t a^{-\frac{1}{p}}(s) \left(\int_0^s r(\sigma)\, d\sigma \right)^{\frac{1}{p}} ds \left(\int_0^{\varphi(t)} a^{-\frac{1}{p}}(\omega)\, d\omega \right)^{-1-\varepsilon} dt < \infty, \tag{6.4.13}$$

then y is of the nonlinear limit-circle type. If, in addition,

$$\int_0^\infty a^{-\frac{1}{p}}(s) r^{-1}(s) \left(\int_0^s r(\sigma)\, d\sigma \right)^{\frac{p+1}{p}} ds < \infty, \tag{6.4.14}$$

then y is of the strong nonlinear limit-circle type.

Proof. Let $y \in \mathcal{N}_1$, say $y(t) > 0$ for large t; the case $y < 0$ is similar. Since $\lim_{t\to\infty} \varphi(t) = \infty$, there exists $T \geq 0$ such that

$$y_1\big(\varphi(t)\big) > 0 \quad \text{and} \quad y_2(t) > 0 \quad \text{for} \quad t \geq T, \tag{6.4.15}$$

and (6.2.2) implies y_2 is increasing for $t \geq T$. We have

$$\begin{aligned} y(t) &= y(T) + \int_T^t y'(s)\, ds = y(T) + \int_T^t a^{-\frac{1}{p}}(s)\, y_2^{\frac{1}{p}}(s)\, ds \\ &\geq y(T) + y_2^{\frac{1}{p}}(T) \int_T^t a^{-\frac{1}{p}}(s)\, ds, \end{aligned}$$

and so $t_0 > T$ and $C_0 > 0$ exist so that

$$y(t) \geq C_0 \int_0^t a^{-\frac{1}{p}}(s)\, ds \quad \text{for } t \geq t_0. \tag{6.4.16}$$

(i) Since $f(x) \neq 0$ for $x \neq 0$, (6.4.10) implies there is a $C > 0$ such that

$$f(x) \geq Cx^{-1} \quad \text{for } x \geq \inf_{T \leq s < \infty} y\big(\varphi(s)\big) > 0.$$

From this and (6.4.15), we have

$$\int_0^\infty \big|y(t)\, f\big(y(\varphi(t))\big)\big|\, dt \geq C \int_T^\infty y(t) y^{-1}\big(\varphi(t)\big)\, dt \geq C \int_T^\infty dt = \infty,$$

and so y is of the nonlinear limit-point type. If (6.4.11) holds, then

$$\int_0^\infty a^{-\frac{1}{p}}(s) r^{-1}(s) y_2^{\frac{p+1}{p}}(s)\, ds \geq y_2^{\frac{p+1}{p}}(T) \int_T^\infty a^{-\frac{1}{p}}(s) r^{-1}(s)\, ds = \infty,$$

and (6.2.6) implies that y is of the strong nonlinear limit-point type.

(ii) It follows from (6.4.12) that there is a $C > 0$ such that

$$f(x) \leq Cx^{-1-\varepsilon} \quad \text{for } x \geq C_1 \stackrel{\text{def}}{=} \inf_{T \leq s < \infty} y\big(\varphi(s)\big) > 0. \tag{6.4.17}$$

We then have

$$\begin{aligned} y_2(t) &= y_2(T) + \int_T^t r(s) f\big(y(\varphi(s))\big)\, ds \leq y_2(T) + C \int_T^t r(s) y^{-1-\varepsilon}\big(\varphi(s)\big)\, ds \\ &\leq y_2(T) + C C_1^{-1-\varepsilon} \int_T^t r(s)\, ds \end{aligned}$$

for $t \geq T$. Hence,

$$y_2(t) \leq C_2 \int_0^t r(s)\,ds \tag{6.4.18}$$

for $t \geq t_1$ for some $t_1 > T$ and $C_2 > 0$. Thus,

$$y'(t) \leq C_2^{\frac{1}{p}} a^{-\frac{1}{p}}(t) \left(\int_0^t r(s)\,ds \right)^{\frac{1}{p}},$$

or

$$y(t) \leq y(t_1) + C_2^{\frac{1}{p}} \int_0^t a^{-\frac{1}{p}}(s) \left(\int_0^s r(\sigma)\,d\sigma \right)^{\frac{1}{p}} ds.$$

Hence, there is a $C_3 > 0$ such that

$$y(t) \leq C_3 \int_0^t a^{-\frac{1}{p}}(s) \left(\int_0^s r(\sigma)\,d\sigma \right)^{\frac{1}{p}} ds \quad \text{for} \quad t \geq t_1. \tag{6.4.19}$$

Let $t_2 = \max\{t_0, t_1\}$. Then (6.4.13), (6.4.16), (6.4.17) and (6.4.19) imply

$$\begin{aligned}
\int_{t_2}^{\infty} \left| y(t) f(y(\varphi(t))) \right| dt &\leq C \int_{t_2}^{\infty} y(t) y^{-1-\varepsilon}(\varphi(t))\,dt \\
&\leq C C_0^{-1-\varepsilon} C_3 \int_{t_2}^{\infty} \int_0^t a^{-\frac{1}{p}}(s) \left(\int_0^s r(\sigma) d\sigma \right)^{\frac{1}{p}} ds \\
&\quad \times \left(\int_0^{\varphi(t)} a^{-\frac{1}{p}}(s)\,ds \right)^{-1-\varepsilon} dt < \infty.
\end{aligned}$$

Thus, y is of the nonlinear limit-circle type. If (6.4.14) holds, then (6.4.18) implies

$$\begin{aligned}
&\int_{t_2}^{\infty} a^{-\frac{1}{p}}(s) r^{-1}(s) y_2^{\frac{p+1}{p}}(s)\,ds \\
&\quad \leq C_2^{\frac{p+1}{p}} \int_{t_2}^{\infty} a^{-\frac{1}{p}}(s) r^{-1}(s) \left(\int_0^s r(\sigma)\,d\sigma \right)^{\frac{p+1}{p}} ds < \infty.
\end{aligned}$$

From this and (6.2.6), we see that y is of the strong nonlinear limit-point type. □

The following theorem gives some easily verifiable criteria for equation (6.2.1) to be of the nonlinear limit-point type.

Theorem 6.4.3. *Let either*

(i) $\lim_{t\to\infty}\int_0^t a^{-\frac{1}{p}}(s)(\int_s^t r(\sigma)\,d\sigma)^{\frac{1}{p}}\,ds < \infty$,

(ii) $\varphi(t) < t$ *on* $\mathbb{R}_+$ *and*

$$\big|f(x)\big| \geq |x|^{-1} \quad \textit{for large } |x|, \quad \textit{or} \tag{6.4.20}$$

(iii) *condition* (6.4.20) *holds and there is a constant* $M \in \mathbb{R}_+$ *such that*

$$\big|f(x)\big| \leq M + |x|^p \quad \textit{on} \quad \mathbb{R}.$$

Then (6.2.1) *is of the nonlinear limit-point type.*

Proof. (i) This follows from Lemma 6.3.5(i) and Theorem 6.4.1(iii).

(ii) and (iii) Let y be a solution of (6.2.1) defined on its maximal interval $J \subset \mathbb{R}_+$ with the initial conditions (ϕ, y_0') such that

$$\phi(t) > 0 \quad \text{for} \quad \inf_{s\in\mathbb{R}_+}\varphi(s) \leq t \leq 0 \quad \text{and} \quad y_0 > 0.$$

Then (6.2.1) yields $y > 0$ and $y' > 0$ on J. Moreover, Lemma 6.3.7 implies $J = \mathbb{R}_+$ and $y \in \mathcal{N}_1$. Thus, the statement follows from Theorem 6.4.2 (i). □

The next theorem gives sufficient conditions for all nonoscillatory solutions to be of the strong nonlinear limit-point type.

Theorem 6.4.4. *Assume that* $\varphi(t) < t$ *on* $\mathbb{R}_+$, φ *is nondecreasing for large* t, *and there is an* $\varepsilon > 0$ *such that*

$$\big|f(x)\big| \geq |x|^p \quad \textit{for } |x| \leq \varepsilon \quad \textit{and} \quad \big|f(x)\big| \geq |x|^{-1} \quad \textit{for large } |x|.$$

If (6.4.11) *holds and*

$$\limsup_{t\to\infty}\int_{\varphi(t)}^t a^{-\frac{1}{p}}(s)\left(\int_s^t r(\sigma)\,d\sigma\right)^{\frac{1}{p}} ds > 1, \tag{6.4.21}$$

then any solution $y \in \mathcal{N}$ *of equation* (6.2.1) *is of the strong nonlinear limit-point type.*

Proof. By Lemma 6.3.5(ii), $\mathcal{N}_2 = \emptyset$. Theorem 6.4.2(i) implies $y \in \mathcal{N}_1$ is of the strong nonlinear limit-point type. Furthermore, according to Lemma 6.3.7, $\mathcal{N}_1 \neq \emptyset$. The statement follows from Definition 6.2.4. □

Next is a nonlinear limit-circle result for the solutions in the class $\mathcal{N}$

Theorem 6.4.5. *Assume that* $d \geq 1$, $\lambda > p$, *and there is an increasing sequence* $\{T_k\}_{k=1}^{\infty}$ *such that* $T_1 > 0$, $\varphi(T_k) = T_k$, $k = 1, 2, \ldots, \lim_{k\to\infty} T_k = \infty$,

$$\varphi(t) \geq dt + T_{k+1}(1-d) \quad \textit{for } t \in [T_k, T_{k+1}],\ k = 1, 2, \ldots,$$

φ *is nondecreasing for large* t, f *is nondecreasing on* $\mathbb{R}$, *and*

$$\left|f(x)\right| \geq |x|^{\lambda} \quad \textit{for } x \in \mathbb{R}.$$

In addition, assume that there are constants $M_2 > 0$, $M_3 > 0$, $M_4 > 0$, $\alpha \geq 0$, *and* $\beta \leq 0$, *with* $\alpha - \beta < p + 1$, *such that*

$$a(t) \leq M_2 t^{\alpha}, \quad r(t) \geq M_3 t^{\beta}, \quad \textit{and} \quad a^{\frac{1}{p}}(t)\, r(t) \geq M_4 \quad \textit{on } \mathbb{R}_+,$$

there is a $\tau > 0$ *such that* $\varphi(t) > 0$ *for* $t \geq \tau$, *and for every* $M_1 \in \mathbb{R} \setminus \{0\}$, (6.4.3) *holds. If*

$$\lim_{t\to\infty} \int_0^t a^{-\frac{1}{p}}(s) \left(\int_s^t r(\sigma)\, d\sigma \right)^{\frac{1}{p}} ds = \infty, \tag{6.4.22}$$

then any solution $y \in \mathcal{N}$ *of equation* (6.2.1) *is of the nonlinear limit-circle type.*

Proof. According to Theorem 6.4.1(ii) every solution $y \in \mathcal{N}_{20}$ is of the nonlinear limit-circle type; moreover, Lemma 6.3.5(i) implies $\mathcal{N}_{21} = \emptyset$. Furthermore, Lemma 6.3.4 yields $\mathcal{N}_1 = \emptyset$. Now, the statement follows from Lemma 6.3.1. □

Our last theorem in this section contains some additional limit-circle type results.

Theorem 6.4.6. *Assume that there exist* $\varepsilon > 0$ *and* $\tau > 0$ *such that* (6.4.12) *and* (6.4.13) *hold and* $\varphi(t) > 0$ *for* $t \geq \tau$.

(i) *Assume that there is an* $\varepsilon_1 > 0$ *such that*

$$\left|f(x)\right| \geq |x|^p \quad \textit{for } |x| \leq \varepsilon_1,$$

$\varphi(t)$ *is nondecreasing for large* t, *and* (6.4.21) *holds. Then equation* (6.2.1) *is of the nonlinear limit-circle type. If, moreover,* (6.4.14) *holds, then any solution* $y \in \mathcal{N}$ *of equation* (6.2.1) *is of the strong nonlinear limit-circle type.*

(ii) *Assume that* $\lambda > p$ *and there are constants* $\varepsilon > 0$, $\tau > 0$, *and* $M_5 > 0$ *such that* (6.4.2) *holds,* $\varphi(t) > 0$ *for* $t \geq \tau$,

$$a^{\frac{1}{p}}(t)\, r(t) \geq M_5 \quad \text{on } \mathbb{R}_+, \tag{6.4.23}$$

and for every $M_1 \in \mathbb{R} \setminus \{0\}$, (6.4.3) *holds. If* (6.4.22) *holds, then any solution* $y \in \mathcal{N}$ *of equation* (6.2.1) *is of the nonlinear limit-circle type.*

Proof. The hypotheses of Theorem 6.4.2(ii) are satisfied in both cases; hence, if $y \in \mathcal{N}_1$, then y is of the nonlinear limit-circle type, and if (6.4.14) holds, then it is of the strong nonlinear limit-circle type.

(i) Lemma 6.3.5(ii) implies $\mathcal{N}_2 = \emptyset$ and the statement follows from Lemma 6.3.1 and Definitions 6.2.3 and 6.2.5.

(ii) According to Theorem 6.4.1(ii), (6.4.23), and (6.4.3), $y \in \mathcal{N}_{20}$ is of the nonlinear limit-circle type. Moreover, Lemma 6.3.5(i) and (6.4.22) imply $\mathcal{N}_{21} = \emptyset$. The conclusion follows from Lemma 6.3.1. □

6.5 Oscillatory Solutions

We begin with a result that deals with the boundedness of oscillatory solutions.

Theorem 6.5.1. *Let* $0 < \lambda \leq p$, φ *be nondecreasing on* $\mathbb{R}_+$, *and assume that there are constants* $M > 0$ *and* $K_1 > 0$ *such that*

$$|f(x)| \leq |x|^{\lambda} \quad \text{for } |x| \geq M, \tag{6.5.1}$$

and

$$\int_{\varphi(t)}^{t} a^{-\frac{1}{p}}(s) \left(\int_s^t r(\sigma)\, d\sigma \right)^{\frac{1}{p}} ds \leq K_1 \quad \text{on } \mathbb{R}_+. \tag{6.5.2}$$

In addition, assume that $K_1 < 2^{-\frac{1}{p}}$ *if* $\lambda = p$. *Then every oscillatory solution* y *of* (6.2.1) *is bounded on* $\mathbb{R}_+$.

Proof. Set $M_1 = \max_{|x| \leq M} |f(x)|$ and let y be an oscillatory solution of (6.2.1). Suppose that y is unbounded. Then there exists an increasing sequence $\{\tau_k\}_{k=1}^{\infty}$ of local maxima of $|y_1|$ such that $\varphi(\tau_1) \geq 0$, y has a zero in $(0, \tau_1)$, $|y(\tau_k)| \geq (M_1)^{\frac{1}{\lambda}}$, $|y(t)| \leq |y(\tau_k)|$ for $t \in [\omega, \tau_k]$, $k = 1, 2, \ldots$, and

$\lim_{k\to\infty} \big|y(\tau_k)\big| = \infty$. Let

$$\tau \in \{\tau_k\}_{k=1}^{\infty} \tag{6.5.3}$$

and suppose, for simplicity, that $y(\tau) > 0$. Then there is a τ_0 such that $0 < \tau_0 < \tau$ and

$$y(\tau_0) = 0, \quad y(t) > 0 \quad \text{on } [\tau_0, \tau], \quad \text{and} \quad y'(\tau) = y_2(\tau) = 0. \tag{6.5.4}$$

Then Lemma 6.3.8(i) implies

$$\varphi(\tau) < \tau_0. \tag{6.5.5}$$

Also, we have

$$\big|y(\varphi(t))\big| \le A_0 = |y(\tau)| + 2$$

on $[0, \tau]$. Then an integration of the second equality in (6.2.2) and condition (6.5.1) imply

$$\big|y_2(t)\big| \le \int_t^{\tau} r(s)\{M_1 + \big|y(\varphi(s))\big|^{\lambda}\}\, ds \le 2A_0^{\lambda} \int_t^{\tau} r(\sigma)\, d\sigma \tag{6.5.6}$$

on $[\tau_0, \tau]$. Hence,

$$\big|y_1'(t)\big| \le 2^{\frac{1}{p}} A_0^{\frac{\lambda}{p}} a^{-\frac{1}{p}}(t) \left(\int_t^{\tau} r(\sigma)\, d\sigma\right)^{\frac{1}{p}}, \quad t \in [\tau_0, \tau], \tag{6.5.7}$$

and by (6.5.4) and (6.5.5),

$$\begin{aligned} \big|y_1(\tau)\big| &= \big|y_1(\tau) - y_1(\tau_0)\big| \le \int_{\tau_0}^{\tau} \big|y_1'(s)\big|\, ds \\ &\le 2^{\frac{1}{p}} A_0^{\frac{\lambda}{p}} \int_{\varphi(\tau)}^{\tau} a^{-\frac{1}{p}}(s) \left(\int_s^{\tau} r(\sigma)\, d\sigma\right)^{\frac{1}{p}} ds. \end{aligned} \tag{6.5.8}$$

If we set $D = 2^{\frac{1}{p}} \int_{\varphi(\tau)}^{\tau} a^{-\frac{1}{p}}(s)\big(\int_s^{\tau} r(\sigma)\, d\sigma\big)^{\frac{1}{p}}\, ds$, then

$$\big|y_1(t)\big| \le \big|y_1(\tau)\big| \le A_0^{\frac{\lambda}{p}} D \quad \text{for } t \in [\omega, \tau]. \tag{6.5.9}$$

Let $\lambda < p$. By repeating steps (6.5.6)–(6.5.9) with $A_0^{\frac{\lambda}{p}} D$ instead of A_0, then in place of (6.5.9) we obtain the estimation

$$|y_1(t)| \le \big(A_0^{\frac{\lambda}{p}} D\big)^{\frac{\lambda}{p}} D = A_0^{(\frac{\lambda}{p})^2} D^{1+\frac{\lambda}{p}}, \quad t \in [\omega, \tau].$$

Repeating this process n-times yields

$$|y_1(t)| \le A_0^{(\frac{\lambda}{p})^{n+1}} D^{1+\frac{\lambda}{p}+(\frac{\lambda}{p})^2+\cdots+(\frac{\lambda}{p})^n}, \tag{6.5.10}$$

and as $n \to \infty$, this implies

$$|y_1(t)| \le D^{\frac{p}{p-\lambda}} \le K_1^{\frac{p}{p-\lambda}} \quad \text{on } [\omega, \tau].$$

Letting $k \to \infty$ in (6.5.3) gives a contradiction to the fact that y is an unbounded solution.

Now let $\lambda = p$. Then, in a similar way, we obtain

$$|y_1(t)| \le A_0 D^{n+1}$$

instead of (6.5.10). Since $D < 1$ (see (6.5.2)), as $n \to \infty$ we have

$$|y_1(t)| \le 0 \quad \text{on } [\omega, \tau].$$

This contradicts $y_1(\tau) \ne 0$ and proves the statement in this case. □

The following theorem provides an answer to the limit-point/limit-circle question for bounded oscillatory solutions.

Theorem 6.5.2. *Let $\varphi \in C^1(\mathbb{R}_+)$ be increasing on $\mathbb{R}_+$ and assume there are positive constants a_0, r_0, K, and K_1 such that*

$$a_0 \le a(t), \quad r_0 \le r(t), \quad \int_{\varphi(t)}^t r(s)\,ds \le K, \quad \text{and} \quad t - \varphi(t) \le K_1 \quad \text{on } \mathbb{R}_+,$$

and

$$\int_0^\infty r(t)\big(t - \varphi(t)\big)\,dt < \infty. \tag{6.5.11}$$

If y is an oscillatory solution of (6.2.1) and either f is bounded on $\mathbb{R}$ or y is bounded on $\mathbb{R}_+$, then y is of the strong nonlinear limit-circle type.

Proof. Let y be oscillatory. Then there exists $M_1 > 0$ such that

$$M_1 = \sup_{t \in \mathbb{R}_+} \big|f\big(y(\varphi(t))\big)\big| < \infty.$$

Let τ be a zero of y and let $\tau \le t_0 < t_1 < t_2$ be such that

$$y(t_0) = y(t_2) = 0, \quad y(t) \ne 0 \quad \text{on } (t_0, t_2),$$

and $|y(t_1)| = \max_{t\in[t_0,t_2]} |y(t)|$. By Lemma 6.3.8(i) with $\tau_0 = t_0$ and $\tau = t_1$, we have

$$t_1 - \varphi(t_1) \ge t_1 - t_0. \tag{6.5.12}$$

Furthermore, $y_2(t_1) = 0$, (6.2.2) and (6.5.12) imply

$$|y_2(t)| \le \int_t^{t_1} r(s)\big|f\big(y(\varphi(s))\big)\big|\,ds \le M_1 \int_{\varphi(t_1)}^{t_1} r(s)\,ds.$$

Hence,

$$|y_1'(t)| \le a^{-\frac{1}{p}}(t) M_1^{\frac{1}{p}} \left(\int_{\varphi(t_1)}^{t_1} r(s)\,ds\right)^{\frac{1}{p}} \le a_0^{-\frac{1}{p}} M_1^{\frac{1}{p}} K^{\frac{1}{p}} \stackrel{\text{def}}{=} C \tag{6.5.13}$$

for $t \in [t_0, t_1]$. As C does not depend on t_0, t_1, or t_2, (6.5.13) holds for $t \ge \tau$. It follows from the boundedness of the derivatives, (6.5.13), and Lemma 6.3.9(ii) that y_1 and y_2 are bounded on $\mathbb{R}_+$.

We have

$$y\big(\varphi(t)\big) = y(t) - y'(\xi)\big(t - \varphi(t)\big) = y(t) + h(t) \tag{6.5.14}$$

with $h(t) = -y'(\xi(t))(t - \varphi(t))$ for $\xi \in [\varphi(t), t]$. From this and (6.5.13), we obtain

$$|h(t)| \le C\big(t - \varphi(t)\big), \quad t \ge \tau. \tag{6.5.15}$$

Then,

$$\begin{aligned}\big(y_1(t)\,y_2(t)\big)' &= y_1'(t)\,y_2(t) + y_1(t)\,y_2'(t)\\ &= a(t)\big|y_1'(t)\big|^{p+1} + r(t)\,y_1(t)\,f\big(y_1(\varphi(t))\big)\end{aligned}$$

on $\mathbb{R}_+$. Integrating, we have

$$\begin{aligned}y_1(t)\,y_2(t) = y_1(t)\,y_2(t) - y_1(\tau)\,y_2(\tau) &\ge r_0 \int_\tau^t \frac{a(s)}{r(s)}\big|y_1'(s)\big|^{p+1}\\ &+ \int_\tau^t r(s)\,y_1(s)\,f\big(y_1(\varphi(s))\big)\,ds.\end{aligned} \tag{6.5.16}$$

Making the transformation

$$v = \varphi(s), \quad s = \varphi^{-1}(v), \tag{6.5.17}$$

and using (6.5.14), we have

$$\begin{aligned}
&\int_\tau^t r(s)\, y(s)\, f\big(y(\varphi(s))\big)\, ds \\
&= \int_{\varphi(\tau)}^{\varphi(t)} r\big(\varphi^{-1}(v)\big)\, y\big(\varphi^{-1}(v)\big)\, f\big(y(v)\big) \frac{dv}{\varphi'\big(\varphi^{-1}(v)\big)} \\
&= \int_{\varphi(\tau)}^{\varphi(t)} r\big(\varphi^{-1}(v)\big)\, y(v)\, f\big(y(v)\big) \frac{dv}{\varphi'\big(\varphi^{-1}(v)\big)} \\
&\quad - \int_{\varphi(\tau)}^{\varphi(t)} r\big(\varphi^{-1}(v)\big)\, h\big(\varphi^{-1}(v)\big)\, f\big(y(v)\big) \frac{dv}{\varphi'\big(\varphi^{-1}(v)\big)}
\end{aligned}$$

for $t \geq \tau$, where in the second integral the relation $y(\varphi^{-1}(v)) = y(\varphi(\varphi^{-1}(v))) - h(\varphi^{-1}(v)) = y(v) - h(\varphi^{-1}(v))$ is used (see (6.5.14) with $t = \varphi^{-1}(v)$). Hence, using the transformation (6.5.17) and the fact that $f(x)x > 0$ for $x \neq 0$, we have

$$\int_\tau^t r(s)\, y(s)\, f\big(y(\varphi(s))\big)\, ds \geq r_0 \int_\tau^t y(\varphi(s))\, f\big(y(\varphi(s))\big)\, ds - J(t) \tag{6.5.18}$$

with

$$J(t) = \int_{\varphi(\tau)}^{\varphi(t)} r\big(\varphi^{-1}(v)\big) \left|h\big(\varphi^{-1}(v)\big)\right| \left|f\big(y(v)\big)\right| \frac{dv}{\varphi'\big(\varphi^{-1}(v)\big)}.$$

Furthermore, (6.5.11), (6.5.15), and (6.5.17) imply

$$\begin{aligned}
J(t) &\leq M_1 C \int_{\varphi(\tau)}^{\varphi(t)} r\big(\varphi^{-1}(v)\big) \big(\varphi^{-1}(v) - v\big) \frac{dv}{\varphi'\big(\varphi^{-1}(v)\big)} \\
&\leq M_1 C \int_\tau^\infty r(s)\, \big(s - \varphi(s)\big)\, ds < \infty.
\end{aligned}$$

Hence, J is bounded on $[\tau, \infty)$. Let $\{t_k\}_{k=1}^\infty$ be an increasing sequence of zeros of y tending to ∞. Then by setting $t = t_k$ in (6.5.16) and taking the limit as $k \to \infty$, we obtain

$$\int_\tau^\infty \frac{a(s)}{r(s)} |y_1'(s)|^{p+1}\, ds < \infty, \qquad \int_\tau^\infty y(\varphi(s))\, f\big(y(\varphi(s))\big)\, ds < \infty. \tag{6.5.19}$$

In fact, it follows from (6.5.16) and the boundedness of y_1 and y_2 that the integral on the left-hand side in (6.5.18) is bounded from above, and so the second inequality in (6.5.19) holds. Moreover, it follows from this, (6.5.18), and the boundedness of $J(t)$ that the integral on the left-hand side of (6.5.18) is bounded from below on $\mathbb{R}_+$. The first inequality in (6.5.19) follows from this, (6.5.16), and the boundedness of y_1 and y_2.

From (6.5.11), (6.5.14), (6.5.15), and (6.5.19), we have

$$\begin{aligned}\int_\tau^\infty |y(s)|\,\big|f\big(y(\varphi(s))\big)\big|\,ds &\le \int_\tau^\infty y(\varphi(s))\,f\big(y(\varphi(s))\big)\,ds\\ &\quad+\int_\tau^\infty |h(s)|\,\big|f\big(y(\varphi(s))\big)\big|\,ds\\ &\le \int_\tau^\infty y(\varphi(s))\,f\big(y(\varphi(s))\big)\,ds\\ &\quad+\frac{M_1C}{r_0}\int_0^\infty r(s)\,\big(s-\varphi(s)\big)\,ds<\infty.\end{aligned}$$

Hence, y is of the strong nonlinear limit-circle type. □

Our next theorem gives further properties of the oscillatory solutions of (6.2.1).

Theorem 6.5.3. *Assume that φ is nondecreasing on $\mathbb{R}_+$ and there are constants $M>0$ and $r_1>0$ such that*

$$r(t)\le r_1 \quad \text{and} \quad \int_{\varphi(t)}^t a^{-\frac{1}{p}}(s)\,ds\le M \quad \text{on } \mathbb{R}_+.$$

Then every oscillatory solution y of (6.2.1) satisfying

$$\limsup_{t\to\infty}\big|y(t)\big|>0 \tag{6.5.20}$$

is of the nonlinear limit-point type. Moreover, if

$$t-\varphi(t) \quad \text{is bounded on} \quad \mathbb{R}_+ \quad \text{and} \quad \lim_{t\to\infty}\int_{\varphi(t)}^t a^{-\frac{1}{p}}(s)\,ds=0, \tag{6.5.21}$$

then y is unbounded on $\mathbb{R}_+$.

Proof. Let y be an oscillatory solution of (6.2.1) and $\tau\ge 0$ be a zero of y_1. Suppose, to the contrary, that y is of the nonlinear limit-circle type, i.e.,

$$\int_0^\infty \big|y(t)\, f\big(y(\varphi(t))\big)\big|\, dt < \infty.$$

Then, $\int_0^\infty r(t)|y(t)\, f(y(\varphi(t)))|\, dt < \infty$ so

$$\lim_{t\to\infty} \int_0^t r(t)\, y(t)\, f\big(y(\varphi(t))\big)\, dt = K \in \mathbb{R}. \tag{6.5.22}$$

Since

$$\big(y_1(t)\, y_2(t)\big)' = a(t)\big|y_1'(t)\big|^{p+1} + r(t)\, y_1(t)\, f\big(y_1(\varphi(t))\big),$$

integrating on $[\tau, t]$ gives

$$y_1(t)\, y_2(t) = \int_\tau^t a(t)\big|y_1'(t)\big|^{p+1} + \int_\tau^t r(t)\, y_1(t)\, f\big(y_1(\varphi(t))\big)\, dt$$

for $t \geq \tau$, and thus (6.5.22) implies the existence of the limit of the right-hand side. Since y is oscillatory, we have

$$\lim_{t\to\infty} y_1(t)\, y_2(t) = 0. \tag{6.5.23}$$

In view of (6.5.20), there is an increasing sequence $\{t_k\}_{k=1}^\infty$ such that $|y_1|$ has a local maximum at $t = t_k$, $k \in \{1, 2, \ldots, \}$, y_1 has a zero on $[t_{k-1}, t_k)$, and

$$\big|y_1(t_k)\big| \geq \frac{1}{2} \limsup_{t\to\infty} \big|y_1(t)\big| \overset{\text{def}}{=} K_1 > 0, \quad k = 1, 2, \ldots.$$

Let $\{\tau_k\}_{k=1}^\infty$ be an increasing sequence such that $\tau_k < t_k$,

$$\big|y_1(\tau_k)\big| = \frac{K_1}{2} \quad \text{and} \quad \big|y_1(t)\big| \geq \frac{K_1}{2} \quad \text{for } t \in [\tau_k, t_k], \quad k = 1, 2, \ldots. \tag{6.5.24}$$

Note that, according to Lemma 6.3.8(i),

$$t_k - \tau_k < t_k - \varphi(t_k), \qquad k = 1, 2, \ldots. \tag{6.5.25}$$

Then an integration of the first equality in (6.2.2) together with (6.5.25) imply

$$\frac{K_1}{2} = \big|y_1(t_k) - y_1(\tau_k)\big| \leq \int_{\varphi(t_k)}^{t_k} a^{-\frac{1}{p}}(s)\big|y_2(s)\big|^{\frac{1}{p}}\, ds \leq M\big|y_2(s_k)\big| \tag{6.5.26}$$

for $k \in \{1, 2, \ldots\}$, where $\{s_k\}_{k=1}^{\infty}$ is a sequence such that $|y_2(s_k)| = \max_{\tau_k \le s \le t_k} |y_2(s)|$, $k = 1, 2, \ldots$. From this and from (6.5.24) and (6.5.26), we have

$$\left|y_1(s_k)\, y_2(s_k)\right| \ge \frac{K_1^2}{4M}, \quad k = 1, 2, \ldots$$

which contradicts (6.5.23). Hence, y is of the nonlinear limit-point type.

Now suppose that y is bounded on $\mathbb{R}_+$ and (6.5.21) holds. All assumptions of Corollary 6.3.1(iii) hold, so $\lim_{t\to\infty} y(t) = 0$. This contradicts (6.5.20) and proves that y is unbounded. □

Next, we give additional information about the behavior of oscillatory solutions of (6.2.1).

Theorem 6.5.4. *Let φ be nondecreasing and $\lim_{t\to\infty} \int_{\varphi(t)}^{t} a^{-\frac{1}{p}}(s)\, ds = 0$.*

(i) *Assume there are constants M and r_1 such that*

$$r(t) \le r_1 \quad \text{and} \quad t - \varphi(t) \le M \quad \text{on } \mathbb{R}_+.$$

Then for any oscillatory solution y of (6.2.1), either

$$\lim_{t\to\infty} y(t) = 0 \quad \text{or} \quad y \text{ is unbounded on } \mathbb{R}_+.$$

(ii) *Let $0 < \lambda \le p$ and $M_1 > 0$ and $K > 0$ be constants such that $|f(x)| \le |x|^\lambda$ for $|x| \ge M$, and $\int_{\varphi(t)}^{t} r(s)\, ds \le K$ on $\mathbb{R}_+$. Then any oscillatory solution y of (6.2.1) satisfies $\lim_{t\to\infty} y(t) = 0$.*

Proof. Part (i) follows from Lemma 6.5.3. Part (ii) follows from Theorem 6.5.1 and Corollary 6.3.1(iii) since (6.5.2) holds and $K < 2^{-\frac{1}{p}}$ for large t in case $\lambda = p$. □

The following two theorems address the limit-circle question for oscillatory solutions.

Theorem 6.5.5. *Let $p \ge \lambda > 0$, $\varphi \in C^1(\mathbb{R}_+)$ be increasing,*

$$\left|f(x)\right| \le |x|^\lambda \quad \text{for } |x| \ge M > 0, \qquad \int_0^\infty r(t)\left(t - \varphi(t)\right) dt < \infty,$$

and assume that there are positive constants a_0, r_0, K, K_1, and K_2 such that

$$a_0 \le a(t), \quad r_0 \le r(t), \quad \int_{\varphi(t)}^{t} r(s)\, ds \le K,$$

$$t - \varphi(t) \le K_1 \quad \text{on } \mathbb{R}_+,$$

and

$$\int_{\varphi(t)}^{t} a^{-\frac{1}{p}}(s) \int_s^t r(\sigma)\, d\sigma\, ds \le K_2 \quad \text{on } \mathbb{R}_+.$$

Moreover, let $K < 2^{-\frac{1}{p}}$ *in case* $\lambda = p$. *Then any oscillatory solution* y *of* (6.2.1) *is of the strong nonlinear limit-circle type.*

Proof. The conclusion follows from Theorem 6.5.1 and Lemma 6.5.2. □

Remark 6.5.1. Conditions like

$$\int_0^\infty r(t)\,\big(t - \varphi(t)\big)\,dt < \infty$$

or

$$t - \varphi(t) \le K_1 \quad \text{on } \mathbb{R}_+$$

in Theorem 6.5.5 and subsequent theorems on limit-circle results are not unreasonable. To see this, consider the equation

$$y'' = y(t - \pi)$$

which has the solution $y(t) = \sin t$ that is not of the limit circle type.

Theorem 6.5.6. *Let* $\varphi \in C^1(\mathbb{R}_+)$ *be increasing,* $\int_0^\infty (t - \varphi(t))\,dt < \infty$, *and assume there are positive constants* a_0, r_0, r_1, *and* K_1 *such that*

$$a_0 \le a(t), \quad r_0 \le r(t) \le r_1 \quad on \quad \mathbb{R}_+,$$

and

$$t - \varphi(t) \le K_1 \quad on\ \mathbb{R}_+.$$

Then any oscillatory solution y *of* (6.2.1) *is of the nonlinear limit-circle type if and only if* $\lim_{t\to\infty} y(t) = 0$.

Proof. The theorem follows from Lemmas 6.5.2 and 6.5.3. □

The next theorem gives sufficient conditions for equation (6.2.1) to be of the strong nonlinear limit-circle type.

Theorem 6.5.7. *Let* $\lambda > p$, $\varphi \in C^1(\mathbb{R}_+)$ *be increasing, and assume there are positive constants* a_0, r_0, K, k_1, ε, *and* ε_1 *such that*

$$a_0 \le a(t), \quad r_0 \le r(t), \int_{\varphi(t)}^t r(s)\,ds \le K \quad on\ \mathbb{R}_+,$$

$$|x|^\lambda \le |f(x)| \quad for\ |x| \le \varepsilon, \quad |f(x)| \le |x|^{-1-\varepsilon_1} \quad for\ large\ |x|,$$

and

$$t - \varphi(t) \le K_1 \quad on\ \mathbb{R}_+.$$

In addition, assume that

$$\int_0^\infty r(t)\left(t-\varphi(t)\right)dt < \infty, \quad \lim_{t\to\infty}\int_0^t a^{-\frac{1}{p}}(s)\left(\int_s^t r(\sigma)\,d\sigma\right)^{\frac{1}{p}} ds = \infty,$$

$$\int_1^\infty \left(\int_0^t a^{-\frac{1}{p}}(\sigma)\,d\sigma\right)^{-\frac{p+1}{\lambda-p}} dt < \infty, \tag{6.5.27}$$

and

$$\int_\tau^\infty \int_0^t a^{-\frac{1}{p}}(s)\left(\int_0^s r(\sigma)\,d\sigma\right)^{\frac{1}{p}} ds \left(\int_0^{\varphi(t)} a^{-\frac{1}{p}}(\omega)\,d\omega\right)^{-1-\varepsilon_1} dt < \infty,$$

for large τ. *Then equation* (6.2.1) *is of the strong nonlinear limit-circle type.*

Proof. If $y \in \mathcal{N}_0$, the statement is trivial. By Theorem 6.4.6, the conclusion holds for all nonoscillatory solutions of (6.2.1). Oscillatory solutions are of the strong nonlinear limit-circle type by Lemma 6.5.2 since f is bounded. □

Remark 6.5.2. We can relax condition (6.5.27) in Theorem 6.5.7 somewhat by imposing an additional condition on the function f. In particular, if in addition we ask that there exist $s > 0$ and $\varepsilon_2 > 0$ such that

$$|f(x)| \le |x|^s \quad \text{for } |x| \le \varepsilon_2,$$

then condition (6.5.27) can be replaced by

$$\int_1^\infty \left(\int_0^t a^{-\frac{1}{p}}(\sigma)\,d\sigma\right)^{-\frac{p+1}{\lambda-p}(1+s)} dt < \infty. \tag{6.5.28}$$

The last result in this section is for the case where f is linear, i.e., for the equation

$$\left(a(t)|y'|^{p-1}y'\right)' = r(t)y(\varphi(t)). \tag{6.5.29}$$

It gives conditions under which bounded oscillatory solutions are of the strong nonlinear limit-circle type.

Theorem 6.5.8. *Assume that* $t - \varphi(t)$ *is eventually nondecreasing and there are constants* $a_0 > 0$ *and* $r_0 > 0$ *such that* $a_0 \le a(t)$ *and* $r_0 \le r(t)$ *on* $\mathbb{R}_+$. *Let* $K > 0$ *be a constant and let* $\bar{r} \in C^0(\mathbb{R}_+)$ *be a nondecreasing function with* $r(t) \le \bar{r}(t)$,

$$\int_0^\infty r(t)\big(t-\varphi(t)\big)\,dt < \infty, \quad \text{and} \quad \bar{r}(t)\big(t-\varphi(t)\big) \le K \quad \text{on } \mathbb{R}_+. \tag{6.5.30}$$

Then every bounded oscillatory solution of (6.5.29) *is of the strong limit-circle type.*

Proof. Let $y \in \mathcal{O}$ be a bounded solution, say

$$\big|y(t)\big| \le B \quad \text{on} \quad [\sigma, \infty), \tag{6.5.31}$$

where $\sigma = \inf\{\varphi(t) : t \in \mathbb{R}_+\}$. Let $\tau \ge 0$ be a zero of y and let $\tau \le t_0 < t_1 < t_2$ be such that

$$y(t_0) = y(t_2) = 0, \quad y(t) \ne 0 \quad \text{on } (t_0, t_2), \quad \text{and} \quad \big|y(t_1)\big| = \max_{t \in [t_0, t_2]} \big|y(t)\big|. \tag{6.5.32}$$

To see that we must have

$$t_1 - \varphi(t_1) > t_1 - t_0, \tag{6.5.33}$$

notice that $y'(t_1) = 0$, (6.5.32), and (6.5.29) imply $y(t)\,y'(t) > 0$ for $t > t_1$, and this would contradict y being oscillatory. Furthermore, $y_2(t_1) = 0$, (6.5.31), and (6.5.33) imply

$$\big|y_2(t)\big| \le \int_t^{t_1} r(s)\,y(\varphi(s))\,ds \le B\bar{r}(t_1)(t_1 - t),$$

so

$$\big|y'(t)\big| \le a_0^{-\frac{1}{p}} B^{\frac{1}{p}} \bar{r}^{\frac{1}{p}}(t_1)\big(t_1 - \varphi(t_1)\big)^{\frac{1}{p}} \le C \quad \text{for } t \in [t_0, t_1], \tag{6.5.34}$$

with $C = a_0^{-\frac{1}{p}} B^{\frac{1}{p}} K^{\frac{1}{p}}$. Since C does not depend on t_0, t_1, or t_2, (6.5.34) holds for all $t \ge \tau$. Hence, we have

$$y\big(\varphi(t)\big) = y(t) - y'(\xi)\big(t - \varphi(t)\big) = y(t) + h(t)$$

with $h(t) = -y'\big(\xi(t)\big)\big(t - \varphi(t)\big)$ and $\xi \in [\varphi(t), t]$. From this and (6.5.34), we see that

$$\big|h(t)\big| \le C\big(t - \varphi(t)\big) \quad \text{for } t \ge \tau. \tag{6.5.35}$$

Thus,

$$\big(y_1(t)\, y_2(t)\big)' = y_1'(t)\, y_2(t) + y_1(t)\, y_2'(t) = a(t)\big|y_1'(t)\big|^{p+1} + r(t)\, y_1(t)\, y_1\big(\varphi(t)\big)$$

on $\mathbb{R}_+$. Integrating, we have

$$\begin{aligned} y_1(t)\, y_2(t) - y_1(\tau)\, y_2(\tau) \ge r_0 \int_\tau^t \frac{a(s)}{r(s)}\big|y_1'(s)\big|^{p+1} ds + r_0 \int_\tau^t y_1^2(s)\, ds \\ + \int_\tau^t r(s)\, y_1(s)\, h(s)\, ds \end{aligned} \tag{6.5.36}$$

for $t \ge \tau$. But (6.5.35) and (6.5.30) imply

$$\int_\tau^\infty r(s)\big|y_1(s)\, h(s)\big|\, dt \le CB \int_\tau^\infty r(t)\big(t - \varphi(t)\big)\, dt < \infty. \tag{6.5.37}$$

Evaluating (6.5.36) along a sequence $\{t_k\}_{k=1}^\infty \to \infty$ of zeros of y_1 gives

$$\int_0^\infty \frac{a(t)}{r(t)}\big|y_1'(s)\big|^{p+1} ds < \infty \quad \text{and} \quad \int_0^\infty y^2(s)\, ds < \infty. \tag{6.5.38}$$

In view of (6.5.37) and (6.5.38),

$$\begin{aligned} \int_\tau^\infty \big|y(t)\, y(\varphi(t))\big|\, dt &\le \int_\tau^\infty |y(t)|\big(|y(t)| + |h(t)|\big)\, dt \\ &\le \int_\tau^\infty y^2(t)\, dt + \int_\tau^\infty |y(t)|\, |h(t)|\, dt \\ &\le \int_\tau^\infty y^2(t)\, dt + \frac{CB}{r_0} \int_\tau^\infty r(t)\big(t - \varphi(t)\big)\, dt \\ &< \infty, \end{aligned}$$

so the conclusion of the theorem follows. □

6.6 Examples

The following examples illustrate some of our results in this chapter.

Example 6.6.1. Consider the equation

$$\big(t^\alpha |y'|^{p-1} y'\big)' = t^\beta |y(\varphi(t))|^\lambda \operatorname{sgn} y(\varphi(t)), \quad t \ge 1. \tag{6.6.1}$$

If either (i) $\beta < \min\{-1, \alpha - p - 1\}$, (ii) $\varphi(t) < t$ and $\lambda > 0$, or (iii) $\lambda \le p$, then equation (6.6.1) is of the nonlinear limit-point type by Theorem 6.4.3.

Example 6.6.2. Consider the equation

$$(t^\alpha y')' = t^\beta |y|^\lambda y(t/2) \operatorname{sgn} y(\mathrm{t}/2), \quad \mathrm{t} \geq 1. \tag{6.6.2}$$

If $\lambda \leq 1$, $\alpha < 1$, $\beta > -1$, and $\beta - \alpha + 2 > 0$, then any nonoscillatory solution of (6.6.2) is of the strong nonlinear limit-point type by Theorem 6.4.4.

Example 6.6.3. Let $\varphi(t) = t - (1 - \sin t)/2$ (see Example 6.3.1). If $0 \leq -\beta \leq \alpha < \min\{\frac{\lambda+3}{2(\lambda+1)}, \beta + 2\}$, $\beta \neq -1$, and $\lambda > 1$, then any nonoscillatory solution of the equation

$$(t^\alpha y')' = t^\beta |y|^\lambda y(\varphi(t)) \operatorname{sgn} y(\varphi(\mathrm{t})), \quad \mathrm{t} \geq 1, \tag{6.6.3}$$

is of the nonlinear limit-circle type by Theorem 6.4.5. It proved to be convenient to use the estimate $\varphi(t) \geq t/2$ for large t in order to verify condition (6.4.3).

Example 6.6.4. Consider the equation

$$\left(t^\alpha |y'|^{p-1} y'\right)' = t^\beta y(\varphi(t)), \quad t \geq 1, \tag{6.6.4}$$

where $\alpha > 0$, $\beta > 0$, and $\varphi(t) = (t^{\beta+3} - 1)/(t^{\beta+2})$. The hypotheses of Theorem 6.5.8 are satisfied with $\bar{r}(t) = t^{\beta+1}$ and $K = 1$, so all bounded oscillatory solutions of (6.6.4) are of the strong limit-circle type.

Example 6.6.5. Consider the equation

$$\left(t^\alpha |y'|^{p-1} y'\right)' = t^\beta |y(\varphi(t))|^\lambda \operatorname{sgn} y(\varphi(\mathrm{t})), \quad \mathrm{t} > 1, \tag{6.6.5}$$

where $0 < \lambda \leq p$, $\varphi \in C^1(\mathbb{R}_+)$ is nondecreasing on $\mathbb{R}_+$, $t^\gamma \leq \varphi(t) \leq t$ for large t, and $0 < \gamma \leq 1$. If either (i) $\alpha > p$, and $\alpha \geq \frac{\beta+1}{\gamma} + p$, or (ii) $\alpha \leq p$ and $\beta + 1 + p \leq \alpha$, then, by Theorem 6.5.1, every oscillatory solution of (6.6.5) is bounded.

Example 6.6.6. Consider the equation

$$\left(t^\alpha |y'|^{p-1} y'\right)' = t^\beta |y(\varphi(t))|^\lambda \operatorname{sgn} y(\varphi(\mathrm{t})), \quad \mathrm{t} \geq 1, \tag{6.6.6}$$

where $\alpha > p$ and φ is nondecreasing on $\mathbb{R}_+$. By Theorem 6.5.4(i), if $\beta \leq 0$ and there is a constant $M > 0$ such that $t - \varphi(t) \leq M$, then any oscillatory solution y of (6.6.6) is either unbounded or satisfies $\lim_{t\to\infty} y(t) = 0$. On the other hand, By Theorem 6.5.4(ii) if $\lambda = p$, and $\beta < -1$, then any oscillatory solution y of (6.6.6) satisfies $\lim_{t\to\infty} y(t) = 0$.

Example 6.6.7. Consider the equation

$$\left(t^{\alpha}|y'|^{p-1}y'\right)' = t^{\beta}|y(\varphi(t))|^{\lambda}\text{sgn } y(\varphi(\text{t})), \quad \text{t} \geq 1, \tag{6.6.7}$$

where $p \geq \lambda > 0$ and $\varphi \in C^1(\mathbb{R}_+)$ is increasing. Assume that $\alpha \geq 0$, $\beta \geq 0$, and $t - ct^{-s} \leq \varphi(t) \leq t$ with $s > \beta + 1$. Then, by Theorem 6.5.5, any oscillatory solution y of (6.6.7) is of the strong nonlinear limit-circle type.

Example 6.6.8. Consider the equation

$$\left(t^{\alpha}|y'|^{p-1}y'\right)' = r_0|y(\varphi(t))|^{\lambda}\text{sgn } y(\varphi(\text{t})), \quad \text{t} \geq 1, \tag{6.6.8}$$

where $r_0 > 0$ is a constant, $\varphi \in C^1(\mathbb{R}_+)$ is increasing, $\alpha \geq 0$, and $t - ct^{-s} \leq \varphi(t) \leq t$ with $s > 1$. By Theorem 6.5.6, any oscillatory solution y of (6.6.8) will be of the nonlinear limit-circle type if and only if $\lim_{t\to\infty} y(t) = 0$.

Example 6.6.9. Consider the equation

$$y'' = r_0 f(y(\varphi(t))), \quad t \geq 1,$$

where $r_0 > 0$, there are constants $\varepsilon > 0$ and $\varepsilon_1 > 0$ such that

$$f(x) = |x|^{-3-\varepsilon}\text{sgn } x \quad \text{for } |x| \geq 1,$$
$$f(x) = x^3 \quad \text{for } |x| < 1,$$

and $\varphi \in C^1(\mathbb{R}_+)$ is increasing and satisfies $t - \frac{1}{t^{1+\varepsilon_1}} \leq \varphi(t) \leq t$ on $(1, \infty)$. Then all assumptions of Theorem 6.5.7 are satisfied (with $p = 1$, $3 \leq \lambda$, $a \equiv 1$, $r(t) \equiv r_0$, $\varepsilon = 1$, $\varepsilon_1 = 2 + \varepsilon$), and hence the above equation is of the strong nonlinear limit-circle type. Note that Remark 6.5.2 holds if $0 < s \leq 3$ and $3 \leq \lambda < 2s + 3$.

Example 6.6.10. Consider the equation

$$y'' = t^{\beta} f(y(t/2)), \quad t \geq 1, \tag{6.6.9}$$

where

$$f(y) = \begin{cases} |y|^{\lambda} \text{ sgn } y, & |y| \leq 1, \\ |y|^{-1-\epsilon} \text{ sgn } y, & |y| > 1, \end{cases}$$

and $\epsilon > \beta + 2 > 0$. If either (i) $\lambda \leq 1$ or (ii) $\beta \geq 0$, then equation (6.6.9) is of the nonlinear limit circle type by Theorem 6.4.6.

Chapter 7

Delay Equations II

7.1 Introduction

In this chapter, we will examine the nonlinear limit-point and limit-circle properties of solutions of the second-order nonlinear delay differential equation

$$\left(a(t)|y'|^{p-1}y'\right)' + r(t)\,|y\,(\varphi(t))|^{\lambda}\,\mathrm{sgn}\,y\,(\varphi(t)) = 0\,. \tag{7.1.1}$$

As is the case of equations without delays, if $p = \lambda$, this is known as the *half-linear* equation, and we say that equation (7.1.1) is of the *super-half-linear* type if $\lambda \geq p$, and that it is of the *sub-half-linear* type if $\lambda \leq p$.

The definitions of nonlinear limit-point, nonlinear limit-circle, strong nonlinear limit-point, and strong nonlinear limit-circle solutions are the same in this chapter as they are in Chapter 6 with $f(u)$ replaced with $|u|^{\lambda}\mathrm{sgn}\,u$.

Although many of the results in this chapter on asymptotic properties of solutions other than limit-point and limit-circle properties are independent of the value of λ, essentially all the limit-point and limit-circle results are for sub-half-linear equations. However, we may have $\lambda \leq p \leq 1$ or $1 \leq \lambda \leq p$.

In Section 7.2 we give some preliminary notions need for our discussion in the remainder of the chapter. Section 7.3 is concerned with nonoscillatory solutions and Section 7.4 deals with oscillatory solutions as well as sufficient conditions for all solutions of equation (7.1.1) to satisfy one of the limit-point/limit-circle properties. Examples to illustrate our results are contained in the last section of the chapter.

7.2 Preliminaries

Consider the second-order nonlinear delay differential equation

$$\left(a(t)|y'|^{p-1}y'\right)' + r(t)\,|y\,(\varphi(t))|^{\lambda}\,\mathrm{sgn}\,y\,(\varphi(t)) = 0\,, \tag{7.2.1}$$

where $a \in C^1(\mathbb{R}_+)$, $r \in C^1(\mathbb{R}_+)$, $a(t) > 0$, $r(t) > 0$, $\varphi : C^1(\mathbb{R}_+) \to \mathbb{R}_+$, $\varphi(0) = 0$, $\varphi(t) \le t$ on $\mathbb{R}_+$, and $\lim_{t\to\infty}\varphi(t) = \infty$. We will write equation (7.2.1) as the equivalent system

$$\begin{aligned} y_1' &= a^{-\frac{1}{p}}(t)|y_2|^{\frac{1}{p}}\,\mathrm{sgn}\,y_2, \\ y_2' &= -r(t)\,\,|y\,(\varphi(t))|^{\lambda}\,\mathrm{sgn}\,y\,(\varphi(t))\,, \end{aligned} \tag{7.2.2}$$

where the relationship between a solution y of (7.2.1) and a solution (y_1, y_2) of the system (7.2.2) is given by

$$y_1(t) = y(t) \quad \text{and} \quad y_2(t) = a(t)\,|y'(t)|^{p-1}\,y'(t).$$

Definition 7.2.1. A solution y is said to be continuable if it is defined on $\mathbb{R}_+$. It is proper if it is continuable and nontrivial in any neighborhood of ∞. A proper solution y of (7.2.1) is oscillatory if there exists a sequence of its zeros tending to ∞, and it is nonoscillatory otherwise.

The definitions of nonlinear limit-point, nonlinear limit-circle, strong nonlinear limit-point, and strong nonlinear limit-circle solutions are given in Section 6.2 of Chapter 6.

We will make use of the following constants some of which were defined earlier but are included here for convenience:

$$\alpha = \frac{p+1}{(\lambda+2)p+1}, \qquad \beta = \frac{(\lambda+1)p}{(\lambda+2)p+1}, \qquad \gamma = \frac{p+1}{p(\lambda+1)},$$

$$\gamma_1 = \alpha\gamma^{-\frac{1}{\lambda+1}}, \qquad \delta = \frac{p+1}{p}, \qquad \hat{\omega} = -\omega = \frac{(\lambda+1)(p+1)}{p-\lambda} \quad \text{for } \lambda \ne p,$$

$$\beta_1 = 7\frac{(\lambda+2)p+1}{(\lambda+1)(p+1)}, \qquad \beta_2 = \frac{p}{(\lambda+2)p+1}.$$

Notice that $\alpha = 1 - \beta$. We define the functions $R,\ g : \mathbb{R}_+ \to \mathbb{R}$ and $\bar{a} : \mathbb{R}_+ \to \mathbb{R}_+$ by

$$R(t) = a^{\frac{1}{p}}(t)\,r(t), \quad g(t) = -\frac{a^{\frac{1}{p}}(t)\,R'(t)}{R^{\alpha+1}(t)}$$

and

$$\bar{a}(t) = \min\{a(s) : \varphi(t) \le s \le t, s \ge 0\}, \quad t \in \mathbb{R}_+,$$

and for any continuous function $h : \mathbb{R}_+ \to \mathbb{R}_+$, we let $h_+(t) = \max\{h(t), 0\}$ and $h_-(t) = \max\{-h(t), 0\}$, so that $h(t) = h_+(t) - h_-(t)$. For any solution of (7.2.1), we let

$$\begin{aligned} F(t) &= R^{\beta}(t)\left[\frac{a(t)}{r(t)}\left|y'(t)\right|^{p+1} + \gamma\left|y(t)\right|^{\lambda+1}\right] \\ &= R^{\beta}(t)(R^{-1}(t)\left|y_2(t)\right|^{\frac{p+1}{p}} + \gamma\left|y(t)\right|^{\lambda+1}). \end{aligned}$$

Notice that $F \ge 0$ for every solution of (7.2.1). In certain results we will need to assume that

$$\lim_{t\to\infty} g(t) = 0 \quad \text{and} \quad \int_0^{\infty} |g'(\sigma)|\, d\sigma < \infty. \tag{7.2.3}$$

It will be convenient to note that (see (6.2.6))

$$\int_0^{\infty} \frac{a(s)}{r(s)}\left|y'(s)\right|^{p+1} ds = \int_0^{\infty} R^{-1}(s)\left|y_2(s)\right|^{\frac{p+1}{p}} ds. \tag{7.2.4}$$

Let $\mathcal{N}$ be the set of all nonoscillatory solutions of (7.2.1). For our results in this chapter, it will be convenient to divide $\mathcal{N}$ into the subsets (see [57]) $\mathcal{N} = \mathcal{N}_1 \cup \mathcal{N}_2 \cup \mathcal{N}_3$, $\mathcal{N}_i = \mathcal{N}_{i0} \cup \mathcal{N}_{i1}$, $i = 1, 2, 3$ where

$$\mathcal{N}_1 = \left\{y \in \mathcal{N} : \lim_{t\to\infty} |y(t)| = \infty\right\},$$

$$\mathcal{N}_2 = \left\{y \in \mathcal{N} : \lim_{t\to\infty} y(t) = C_y,\ |C_y| \in (0,\infty)\right\},$$

$$\mathcal{N}_3 = \left\{y \in \mathcal{N} : \lim_{t\to\infty} y(t) = 0\right\},$$

$$\mathcal{N}_{10} = \left\{y \in \mathcal{N}_1 : \lim_{t\to\infty} y_2(t) = 0\right\},$$

$$\mathcal{N}_{11} = \left\{y \in \mathcal{N}_1 : \lim_{t\to\infty} y_2(t) = d_y,\ |d_y| \in (0,\infty)\right\},$$

$$\mathcal{N}_{20} = \left\{y \in \mathcal{N}_2 : \lim_{t\to\infty} y_2(t) = 0\right\},$$

$$\mathcal{N}_{21} = \left\{y \in \mathcal{N}_2 : \lim_{t\to\infty} |y_2(t)| = \infty\right\},$$

$$\mathcal{N}_{30} = \left\{y \in \mathcal{N}_3 : \lim_{t\to\infty} y_2(t) = d_y,\ |d_y| \in (0,\infty)\right\},$$

$$\mathcal{N}_{31} = \left\{y \in \mathcal{N}_3 : \lim_{t\to\infty} |y_2(t)| = \infty\right\}.$$

The following lemma is based on Theorems 4.1, 4.2, and 5.1 in [57] and gives useful information about some of the subsets defined above.

Lemma 7.2.1. (i) *If* $\int_0^\infty a^{-\frac{1}{p}}(t)\,dt = \infty$, $\int_0^\infty r(t)\,dt < \infty$, *and* $\int_0^\infty r(t) (\int_0^{\varphi(t)} a^{-\frac{1}{p}}(\sigma)\,d\sigma)^\lambda dt < \infty$, *then* $\mathcal{N}_{11} \neq \emptyset$.

(ii) *If* $\int_0^\infty a^{-\frac{1}{p}}(t)\,dt = \infty$, $\int_0^\infty r(t)\,dt < \infty$, *and* $\int_0^\infty a^{-\frac{1}{p}}(t)(\int_t^\infty r(\sigma)\,d\sigma)^{\frac{1}{p}} dt < \infty$, *then* $\mathcal{N}_{20} \neq \emptyset$.

(iii) *If* $\int_0^\infty a^{-\frac{1}{p}}(t)\,dt < \infty$, $\int_0^\infty r(t)\,dt = \infty$, $\int_0^\infty r(t)(\int_{\varphi(t)}^t a^{-\frac{1}{p}}(\sigma)\,d\sigma)^{\frac{1}{p}} dt = \infty$, *and* $\int_0^\infty a^{-\frac{1}{p}}(t)(\int_0^t r(\sigma)\,d\sigma)^{\frac{1}{p}} dt < \infty$, *then* $\mathcal{N}_{31} \neq \emptyset$.

We will also make use of the auxiliary equation

$$\left(r^{-\frac{1}{\lambda}}(t)|Z'|^{\frac{1}{\lambda}}\operatorname{sgn} Z'\right)' + a^{-\frac{1}{p}}(\varphi(t))\varphi'(t)|Z(\varphi)|^{\frac{1}{p}}\operatorname{sgn} Z(\varphi) = 0\,. \tag{7.2.5}$$

We let φ^{-1} denote the inverse function of φ.

Lemma 7.2.2. *Let* φ *be increasing,* Z *be a solution of* (7.2.5) *defined on* $\mathbb{R}_+$, *and*

$$Z_2(t) = r^{-\frac{1}{\lambda}}(t)|Z'(t)|^{\frac{1}{\lambda}}\operatorname{sgn} Z'(t).$$

Then equation (7.2.1) *has a solution* $y(t) = Z_2(\varphi^{-1}(t))$ *on* $\mathbb{R}_+$ *and* $y_2(t) = -Z(t)$.

Proof. Let y be a solution of (7.2.1) defined on $\mathbb{R}_+$. From (7.2.2) we have

$$y_1(\varphi(t)) = -(r^{-1}(t)|y_2'(t)|)^{\frac{1}{\lambda}}\operatorname{sgn} y_2'(t).$$

Hence,

$$\begin{aligned}
[r^{-\frac{1}{\lambda}}(t)|y_2'(t)|^{\frac{1}{\lambda}}\operatorname{sgn} y_2'(t)]' &= -y_1'(\varphi(t))\varphi'(t)\\
&= -a^{-\frac{1}{p}}(\varphi(t))\varphi'(t)|y_2(\varphi(t))|^{\frac{1}{p}}\operatorname{sgn} y_2(\varphi(t))
\end{aligned}$$

and so $Z(t) = y_2(t)$ is a solution of (7.2.5). Moreover,

$$\begin{aligned}
Z_2(t) &= r^{-\frac{1}{\lambda}}(t)|Z'(t)|^{\frac{1}{\lambda}}\operatorname{sgn} Z'(t) = r^{-\frac{1}{\lambda}}(t)|y_2'(t)|^{\frac{1}{\lambda}}\operatorname{sgn} y_2'(t)\\
&= r^{-\frac{1}{\lambda}}(t)|-r(t)|y(\varphi(t))|^\lambda \operatorname{sgn} y(\varphi(t))|^{\frac{1}{\lambda}}(-\operatorname{sgn} y(\varphi(t))) = -y(\varphi(t)).
\end{aligned}$$

Thus, if Z is a solution of (7.2.5), then $y(t) = Z_2(\varphi^{-1}(t))$ is a solution of (7.2.1) with $y_2(t) = -Z(t)$. □

Next we have some information about the behavior of nonlinear limit-circle type solutions of (7.2.1).

Lemma 7.2.3. *Assume there exists $r_1 > 0$ such that*

$$r(t) \le r_1 \quad on\ \mathbb{R}_+ \tag{7.2.6}$$

and let y be an oscillatory nonlinear limit-circle type solution of (7.2.1). *Then*

$$\lim_{t\to\infty} y_1(t)\, y_2(t) = 0$$

and there are positive constants K and K_1 such that

$$|y_1(t)| \le K \left(\int_0^t a^{-\frac{1}{p}}(s)\, ds \right)^{\frac{p}{p+1}} \tag{7.2.7}$$

and

$$|y_2(t)| \le K_1 \int_0^t \left(\int_0^s a^{-\frac{1}{p}}(\sigma)\, d\sigma \right)^{\frac{\lambda p}{p+1}} ds \tag{7.2.8}$$

for large t. If, in addition, there is an $r_0 > 0$ such that $r_0 \le r(t)$ on $\mathbb{R}_+$, then y is of the strong nonlinear limit-circle type.

Proof. Let y be a nonlinear limit-circle type oscillatory solution of (7.2.1). Then (NLC-D) holds and (7.2.6) implies

$$\int_0^\infty r(t)\, |y(t)|\, |y(\varphi(t))|^\lambda\, dt < \infty;$$

hence,

$$\lim_{t\to\infty} \int_0^t r(t)\, y(t)\, |y(\varphi(t))|^\lambda \operatorname{sgn} y(\varphi(t))\, dt \in \mathbb{R}. \tag{7.2.9}$$

Now

$$(y_1(t)\, y_2(t))' = a(t)\, |y_1'(t)|^{p+1} - r(t)\, y_1(t)\, |y_1(\varphi(t))|^\lambda \operatorname{sgn} y_1(\varphi(t))$$

so we have

$$y_1(t)\, y_2(t) = C + \int_0^t a(s)\, |y_1'(s)|^{p+1}\, ds \\ - \int_0^t r(s)\, y_1(s)\, |y_1(\varphi(s))|^\lambda \operatorname{sgn} y_1(\varphi(s))\, ds \tag{7.2.10}$$

with $C = y_1(0)\, y_2(0)$. From this and (7.2.9), we see that $\lim_{t\to\infty} y_1(t)y_2(t)$ exists, and since y_1 is oscillatory, we must have

$$\lim_{t\to\infty} y_1(t)\, y_2(t) = 0\,. \tag{7.2.11}$$

From (7.2.9)–(7.2.11) we have

$$\int_0^\infty a(s)\ |y_1'(s)|^{p+1}\ ds < \infty. \tag{7.2.12}$$

Next, we derive an estimate on the solution y. In view of (7.2.11), there is a constant $C_1 > 0$ such that $|y_1(t)\, y_2(t)| \le C_1$ on $\mathbb{R}_+$ and so

$$|\,|y(t)|^{\frac{1}{p}} y'(t)| \le (C_1)^{\frac{1}{p}} a^{-\frac{1}{p}}(t) \tag{7.2.13}$$

on $\mathbb{R}_+$. Let $t_0 \in \mathbb{R}_+$ be a zero of y. An integration of (7.2.13) yields

$$\begin{aligned}\frac{p}{p+1}\, |y(t)|^{\frac{p+1}{p}} &= \left|\frac{p}{p+1}(|y(t)|^{\frac{p+1}{p}} \operatorname{sgn} y(t) - |y(t_0)|^{\frac{p+1}{p}} \operatorname{sgn} y(t_0))\right| \\ &= \left|\int_{t_0}^t \left| y(s)\right|^{\frac{1}{p}} y'(s)\, ds\right| \le (C_1)^{\frac{1}{p}} \int_{t_0}^t a^{-\frac{1}{p}}(s)\, ds\end{aligned}$$

for $t \ge t_0$. From this we obtain

$$|y(t)| \le C_2 \left(\int_{t_0}^t a^{-\frac{1}{p}}(s)\, ds\right)^{\frac{p}{p+1}}$$

for $t \ge t_0$ with $C_2 = (\frac{p+1}{p}(C_1)^{\frac{1}{p}})^{\frac{p}{p+1}}$ and so (7.2.7) holds. Let t_1 be a zero of y_2. Then (7.2.1) and (7.2.7) imply

$$\begin{aligned}|y_2(t)| = |y_2(t) - y_2(t_1)| &\le \int_{t_1}^t r(s)\, |y\,(\varphi(s))|^\lambda\ ds \\ &\le r_1 K^\lambda \int_{t_1}^t \left(\int_0^s a^{-\frac{1}{p}}(\sigma)\, d\sigma\right)^{\frac{\lambda p}{p+1}} ds\end{aligned}$$

for $t \ge t_1$, so (7.2.8) holds.

The remainder of the statement follows from (7.2.12) and the fact that r is bounded from below by r_0. □

Our next lemma gives growth estimates on solutions of equation (7.2.1).

Lemma 7.2.4. *Let y be a solution of (7.2.1). Then there exist positive constants C and C_1 such that for all large t:*

(i) *if $p > \lambda$, then*

$$|y_1(t)| \le C \left[\int_0^t a^{-\frac{1}{p}}(s) \left(\int_0^s r(\sigma)\, d\sigma \right)^{\frac{1}{p}} ds \right]^{\frac{p}{p-\lambda}} \tag{7.2.14}$$

and

$$|y_2(t)| \le C_1 \left[\int_0^t r(s) \left(\int_0^s a^{-\frac{1}{p}}(\sigma)\, d\sigma \right)^{\lambda} ds \right]^{\frac{p}{p-\lambda}}; \tag{7.2.15}$$

(ii) *if $p = \lambda$, then*

$$|y_1(t)| \le C \exp\left\{ \int_0^t k(u)\, du \right\} \tag{7.2.16}$$

and

$$|y_2(t)| \le C_1 \exp\left\{ \int_0^t k_1(u)\, du \right\}, \tag{7.2.17}$$

where $k(t) = 2^{\frac{1}{p}} a^{-\frac{1}{p}}(t) \left(\int_0^t r(s)\, ds \right)^{\frac{1}{p}}$ and $k_1(t) = 2^p r(t) \left(\int_0^t a^{-\frac{1}{p}}(\sigma)\, d\sigma \right)^p$.

Proof. Let y_1 be a solution of (7.2.1) with the initial conditions (ϕ, y_0'). First, we prove that inequality (7.2.15) holds. If y_2 is bounded on $\mathbb{R}_+$, then (7.2.15) holds, so suppose that y_2 is unbounded on $\mathbb{R}_+$ and set $v(t) = \max_{0 \le s \le t} |y_2(s)|$. Integrating (7.2.2), we see that there exists $t_0 \ge 0$ such that

$$\begin{aligned} |y_1(t)| &\le |y_1(t_0)| + \int_0^t a^{-\frac{1}{p}}(s)\, |y_2(s)|^{\frac{1}{p}}\, ds \\ &\le 2 \int_0^t a^{-\frac{1}{p}}(s)\, v^{\frac{1}{p}}(s)\, ds \end{aligned} \tag{7.2.18}$$

for $t \ge t_0$. If $t_1 \ge t_0$ is such that $\varphi(t) \ge t_0$ for $t \ge t_1$, then (7.2.2) and (7.2.18) imply

$$\begin{aligned} |y_2(t)| &\le |y_2(t_0)| + \int_{t_0}^t r(s)\, |y_1(\varphi(s))|^{\lambda}\, ds \\ &\le |y_2(t_0)| + 2^{\lambda} \int_{t_0}^t r(s) v^{\frac{\lambda}{p}}(s) \left(\int_0^s a^{-\frac{1}{p}}(\sigma)\, d\sigma \right)^{\lambda} ds \end{aligned} \tag{7.2.19}$$

for $t \geq t_1$. Since $p > \lambda$, v is nondecreasing, and $\lim_{t\to\infty} v(t) = \infty$, (7.2.19) implies the existence of $t_2 \geq t_1$ and a constant $C_2 > 0$ such that

$$|y_2(t)| \leq C_2 v^{\frac{\lambda}{p}}(t) \int_0^t r(s) \left(\int_0^s a^{-\frac{1}{p}}(\sigma)\, d\sigma \right)^{\lambda} ds\,, \qquad t \geq t_2\,.$$

Hence,

$$v(t) \leq C_2 v^{\frac{\lambda}{p}}(t) \int_0^t r(s) \left(\int_0^s a^{-\frac{1}{p}}(\sigma)\, d\sigma \right)^{\lambda} ds\,, \quad t \geq t_2,$$

and (7.2.15) follows from this and the fact that $|y_2(t)| \leq v(t)$. Inequality (7.2.14) can be proved similarly by setting $v(t) = \max_{-\sigma<s<t}|y_1(s)|$ in the first inequality in (7.2.19) to obtain

$$|y_2(t)| \leq 2v^{\lambda}(t) \int_0^t r(s)\, ds$$

for $t \geq t_3$ for some $t_3 \geq t_0$. Substituting this into the first inequality in (7.2.18) leads to (7.2.14).

Let $p = \lambda$. Then (7.2.18) and (7.2.19) imply

$$|y_2(t)| \leq |y_2(t_0)| + \int_{t_0}^t k_1(s)\, v(s)\, ds\,, \quad t \geq t_0,$$

or

$$|v(t)| \leq |y_2(t_0)| + \int_{t_0}^t k_1(s)\, v(s)\, ds\,, \quad t \geq t_0.$$

Hence, Gronwall's inequality implies (7.2.17). Inequality (7.2.16) can be proved in the same way as (7.2.14) in the case $\lambda < p$. □

If R is nondecreasing, we can get better estimates.

Lemma 7.2.5. *Let $p \geq \lambda \geq 1$, $R'(t) \geq 0$ on $\mathbb{R}_+$, and*

$$\int_0^{\infty} \bar{a}^{-\frac{2}{p}}(s) R^{\frac{2}{p+1}}(s)\, (s - \varphi(s))\, ds < \infty. \tag{7.2.20}$$

Then every solution y of (7.2.1) is bounded on $\mathbb{R}_+$ and there exists a constant $K = K(y) > 0$ such that

$$|y'(t)| \leq K a^{-\frac{1}{p}}(t) R^{\frac{1}{p+1}}(t)$$

on $\mathbb{R}_+$.

Proof. Let y be a solution of (7.2.1) and let $t_0 \geq 0$ be such that $\varphi(t) \geq 0$ on $[t_0, \infty)$ and

$$\lambda\delta\gamma^{-\frac{\lambda-1}{\lambda+1}} \int_{t_0}^{\infty} \bar{a}^{-\frac{2}{p}}(s) R^{\frac{2}{p+1}}(s)\,(s-\varphi(s))\, ds \leq \frac{1}{2}\,. \tag{7.2.21}$$

If

$$E(t) = R^{-1}(t)\,|y_2(t)|^{\delta} + \gamma\,|y(t)|^{\lambda+1} \geq 0\,, \quad t \geq 0, \tag{7.2.22}$$

then

$$\begin{aligned} E'(t) &= -\frac{R'(t)}{R^2(t)}\,|y_2(t)|^{\delta} - \frac{\delta}{R(t)}\,|y_2(t)|^{\frac{1}{p}} \operatorname{sgn} y_2(t)\, r(t)\,|y\,(\varphi(t))|^{\lambda} \operatorname{sgn} y(\varphi(t)) \\ &\quad + \delta\,|y(t)|^{\lambda}\, y'(t) \operatorname{sgn} y(t) \\ &\leq -\frac{R'(t)}{R^2(t)}\,|y_2(t)|^{\delta} + \delta\,|y'(t)|\,\left|\,|\,y(t)\,|^{\lambda} \operatorname{sgn} y(t) - |\,y\,(\varphi(t))\,|^{\lambda} \operatorname{sgn} y\,(\varphi(t))\right| \\ &\leq -\frac{R'(t)}{R^2(t)}\,|y_2(t)|^{\delta} + \lambda\delta\,|y'(t)|\,|y(\xi)|^{\lambda-1}\,|y'(\xi)|\,(t-\varphi(t))\,, \end{aligned} \tag{7.2.23}$$

where $\xi \in [\varphi(t), t]$, $t \geq t_0$. Define Z by

$$Z(t) = \max\nolimits_{0 \leq s \leq t} E(s) + 1.$$

Then (7.2.22) implies

$$|y_1(t)| \leq \left(\frac{Z(t)}{\gamma}\right)^{\frac{1}{\lambda+1}}, \quad |y_2(t)| \leq (R(t)\,Z(t))^{\frac{p}{p+1}}, \tag{7.2.24}$$

and

$$|y'(t)| \leq a^{-\frac{1}{p}}(t)\,(R(t)\,Z(t))^{\frac{1}{p+1}} \tag{7.2.25}$$

for $t \geq t_0$. From this, (7.2.24), and (7.2.23), we have

$$\begin{aligned} E'(t) &\leq \lambda\delta a^{-\frac{1}{p}}(t)\,\left(R(t)\,Z(t)\right)^{\frac{1}{p+1}} \left(\frac{Z(\xi)}{\gamma}\right)^{\frac{\lambda-1}{\lambda+1}} \\ &\quad \times a^{-\frac{1}{p}}(\xi)\,\left(R(\xi)\,Z(\xi)\right)^{\frac{1}{p+1}}\,(t-\varphi(t)) \\ &\leq C_1 \bar{a}^{-\frac{2}{p}}(t)\,R^{\frac{2}{p+1}}(t)\,(t-\varphi(t))\;Z(t) \end{aligned}$$

for $t \geq t_0$ with $C_1 = \lambda\delta\gamma^{-\frac{\lambda-1}{\lambda+1}}$. Hence, applying (7.2.21), we obtain

$$E(t) \leq E(t_0) + C_1 Z(t) \int_{t_0}^{t} \bar{a}^{-\frac{2}{p}}(s)\,R^{\frac{2}{p+1}}(s)\;(s-\varphi(s))\;ds$$

and so $Z(t) \le E(t_0) + 1 + Z(t)/2$ for $t \ge t_0$. This implies $Z(t) \le 2E(t_0) + 2$ so Z is bounded on $\mathbb{R}_+$. The conclusions of the lemma follow from this and (7.2.25). □

Our next lemma reveals information related to the nonlinear limit-point and limit-circle properties of solutions.

Lemma 7.2.6. *Let y be a solution of (7.2.1).*

(i) *If there exists $\varepsilon > 0$ such that*

$$\varphi'(t) \ge \varepsilon \quad \text{for large } t \tag{7.2.26}$$

and

$$\int_0^\infty |y(t)|^{\lambda+1} dt < \infty,$$

then

$$\int_0^\infty |y(t)|\, |y(\varphi(t))|^\lambda \, dt < \infty. \tag{7.2.27}$$

(ii) *Let A and B be positive functions, B be nondecreasing, and $M > 0$ be a constant such that*

$$0 < \varphi'(t) \le M \quad \text{for large } t, \tag{7.2.28}$$

$$|y(t)| \le A(t) \quad \text{and} \quad |y'(t)| \le B(t) \quad \text{on} \quad \mathbb{R}_+, \tag{7.2.29}$$

$$\textstyle\int_0^\infty A^\lambda(\varphi(t))\, B(t)\, (t - \varphi(t))\, dt < \infty, \tag{7.2.30}$$

and

$$\int_0^\infty |y(t)|^{\lambda+1} dt = \infty.$$

Then

$$\int_0^\infty |y(t)|\, |y(\varphi(t))|^\lambda \, dt = \infty. \tag{7.2.31}$$

Proof. Let y be a solution of (7.2.1) and $\tau \ge 0$ be such that $\varphi'(t) \ge \varepsilon$ and $\varphi'(t) \le M$ hold for $t \ge \tau$ in cases (i) and (ii), respectively.

(i) By Hölder's inequality

$$\int_\tau^\infty |y(t)|\, |y(\varphi(t))|^\lambda \, dt \le \left(\int_\tau^\infty |y(\varphi(t))|^{\lambda+1} \, dt \right)^{\frac{\lambda}{\lambda+1}} \times \left(\int_\tau^\infty |y(t)|^{\lambda+1} \, dt \right)^{\frac{1}{\lambda+1}}$$

$$\leq \varepsilon^{-\frac{\lambda}{\lambda+1}} \left(\int_\tau^\infty |y(\varphi(t))|^{\lambda+1} \varphi'(t)\, dt \right)^{\frac{\lambda}{\lambda+1}}$$

$$\times \left(\int_\tau^\infty |y(t)|^{\lambda+1}\, dt \right)^{\frac{1}{\lambda+1}}$$

$$\leq \varepsilon^{-\frac{\lambda}{\lambda+1}} \int_{\varphi(\tau)}^\infty |y(t)|^{\lambda+1}\, dt < \infty.$$

Thus, (7.2.27) holds.

(ii) First note that

$$y(t) = y(\varphi(t)) + y'(\xi)(t - \varphi(t)) = y(\varphi(t)) + h(t)$$

for $t \geq \tau$ where $\xi \in [\varphi(t), t]$ and $h(t) = y'(\xi)(t - \varphi(t))$. Then (7.2.29) implies

$$|h(t)| \leq B(t)(t - \varphi(t)) .$$

Since $\lim_{t\to\infty} \varphi(t) = \infty$, we see that

$$\int_\tau^t |y(s)| \left| y(\varphi(s)) \right|^\lambda ds \geq \int_\tau^t \left| y(\varphi(s)) \right|^{\lambda+1} ds - J(t)$$

where, by (7.2.30),

$$J(t) = \int_\tau^t |h(s)|\, |y(\varphi(s))|^\lambda\, ds \leq \int_\tau^\infty A^\lambda(\varphi(s))\, B(s)(s - \varphi(s))\, ds < \infty.$$

Hence, (7.2.31) holds. □

The following lemma gives sufficient conditions for all solutions of (7.2.1) to oscillate.

Lemma 7.2.7 (**[1, Theorem 3.1]**). *Let* $\lambda < p$,

$$\int_0^\infty a^{-\frac{1}{p}}(t)\, dt = \infty \quad \textit{and} \quad \int_0^\infty \left(\int_0^{\varphi(t)} a^{-\frac{1}{p}}(s)\, ds \right)^\lambda r(t)\, dt = \infty.$$

Then every proper solution of (7.2.1) *is oscillatory.*

Our next lemma is concerned with an equation of the same form as equation (7.2.1) but without a delay, i.e.,

$$\left(a(t)|Z'|^{p-1}Z'\right)' + r(t)\,|Z|^\lambda \operatorname{sgn} Z = e(t)\,, \tag{7.2.32}$$

where $e \in C^0(\mathbb{R}_+)$. At times, we will need the assumption

$$\int_t^\infty R^{-\beta_2}(\sigma)\,|e(\sigma)|\,d\sigma \le K \int_t^\infty |g'(\sigma)|\,d\sigma \tag{7.2.33}$$

for large t, where $K > 0$ is a constant.

Lemma 7.2.8. (See Theorem 3.9.4). *Let* (7.2.3) *and* (7.2.33) *hold and either* (i) $\lambda = p$, *or* (ii) $\lambda < p$,

$$\int_0^\infty \frac{|e(\sigma)|}{a(\sigma)}\,d\sigma < \infty\,, \quad \int_0^\infty \frac{|e(\sigma)|}{r(\sigma)} < \infty, \tag{7.2.34}$$

and

$$\liminf_{t\to\infty} R^\beta(t) \left(\int_t^\infty |g'(s)|\,ds\right)^{\hat{\omega}} \exp\left\{\int_0^t \left(R^{-1}(\sigma)\right)'_+ R(\sigma)\,d\sigma\right\} = 0\,. \tag{7.2.35}$$

If

$$\int_0^\infty R^{-\beta}(\sigma)\,d\sigma < \infty,$$

then for every solution Z of (7.2.32),

$$\int_0^\infty |Z(s)|^{\lambda+1}\,ds < \infty \quad \textit{and} \quad \int_0^\infty \frac{a(s)}{r(s)}\,|Z'(s)|^{p+1}\,ds < \infty.$$

For any solution Z of (7.2.32), let

$$Z_2(t) = a(t)\,|Z'(t)|^{p-1}\,Z'(t) \quad \text{and} \quad G(t) = R^\beta \left[\frac{|Z_2(t)|^\delta}{R(t)} + \gamma |Z(t)|^{\lambda+1}\right].$$

Also note that if y is a solution of (7.2.1), then y is also a solution of equation (7.2.32) with

$$e(t) = r(t)\left[|y(t)|^\lambda \operatorname{sgn} y(t) - |y(\varphi(t))|^\lambda \operatorname{sgn} y(\varphi(t))\right], \quad t \in \mathbb{R}_+. \tag{7.2.36}$$

Lemma 7.2.9. *Let* $1 \le \lambda < p$, (7.2.3) *hold, and let A and B be nondecreasing positive functions such that, for any solution y of* (7.2.1), *we have*

$$|y(t)| \le CA(t) \quad \textit{and} \quad |y'(t)| \le C_1 B(t) \tag{7.2.37}$$

on $\mathbb{R}_+$ with constants C and C_1 depending on y. If

$$\int_0^\infty R^{-\beta_2}(t)\, r(t)\, A^{\lambda-1}(t)\, B(t)\, (t-\varphi(t))\, dt < \infty, \tag{7.2.38}$$

then equation (7.2.1) has a solution y for which F is bounded from below by a positive constant on $\mathbb{R}_+$.

Proof. Let $K_1 = \delta + \gamma_1 \sup_{t\in\mathbb{R}_+} |g(t)|$, $N = \frac{1}{12}\left[\left(\frac{3}{2}\right)^{\beta_1}\gamma_1 + K_1\right]^{-1}$, and choose $T \in (0,\infty)$ such that

$$\int_T^\infty |g'(\sigma)|\, d\sigma \le N\,, \quad \int_T^\infty R^{-\beta_2}(t)\, r(t)\, A^{\lambda-1}(t)\, B(t)\, (t-\varphi(t))\, dt \le N,$$

and

$$|g(t)| \le N \quad \text{for } t \ge T.$$

Let y be a solution of (7.2.1) satisfying

$$y(t) \equiv D \quad \text{on } [\sigma, 0] \quad \text{and} \quad y'(0) = 0,$$

where D is defined by $C = \int_0^T a^{-\frac{1}{p}}(s)\left(\int_0^s r(\sigma)\, d\sigma\right)^{\frac{1}{p}} ds$,

$$D - C(2D)^{\frac{\lambda}{p}} > \left[\gamma^{-1} \min_{0\le s\le T} R^{-\beta}(s)\right]^{\frac{1}{\lambda+1}}, \quad \text{and} \quad D \ge \frac{1}{2}(2C)^{\frac{p}{p-\lambda}}. \tag{7.2.39}$$

Set $v(t) = \max_{0\le\sigma\le t} |y(\sigma)|$. Then, from (7.2.2) we obtain

$$|y_2(t)| \le \int_0^t r(s)\, |y(\varphi(s))|^\lambda\, ds \le v^\lambda(t) \int_0^t r(s)\, ds\,,$$

and

$$|y(t) - D| \le \int_0^t a^{-\frac{1}{p}}(s)\, |y_2(s)|^{\frac{1}{p}}\, ds \le C v^{\frac{\lambda}{p}}(t)$$

on $[0,T]$. This implies

$$D - C v^{\frac{\lambda}{p}}(t) \le y(t) \le D + C v^{\frac{\lambda}{p}}(t)\,, \quad t \in [0,T], \tag{7.2.40}$$

or

$$v(t) \le D + C v^{\frac{\lambda}{p}}(t)\,, \quad t \in [0,T]\,.$$

From this and (7.2.39), we have $v(t) \le \max\left(2D, (2C)^{\frac{p}{p-\lambda}}\right) = 2D$. Hence, the first inequality in (7.2.39) and (7.2.40) imply

$$y(t) \ge D - C(2D)^{\frac{\lambda}{p}} > \left[\gamma^{-1} \min_{0\le s\le T} R^{-\beta}(s)\right]^{\frac{1}{\lambda+1}}$$

on $[0,T]$. Thus, $F(t) > 1$ on $[0,T]$, and either

$$F(t) > 1 \quad \text{on } \mathbb{R}_+ , \tag{7.2.41}$$

or there exists $t_0 \in (T,\infty)$ such that

$$F(t_0) = 1 \quad \text{and} \quad F(t) > 1 \quad \text{on } [T,t_0) . \tag{7.2.42}$$

If (7.2.41) holds, we are done, so suppose (7.2.42) holds. If e is given by (7.2.36), then y is a solution of (7.2.32) with $G(t_0) = 1$, and from (7.2.37), we have

$$\begin{aligned} |e(t)| &\le \lambda r(t)\, |y(\xi)|^{\lambda-1}\, |y'(\xi)|\, (t - \varphi(t)) \\ &\le \lambda C^{\lambda-1} C_1 r(t)\, A^{\lambda-1}(t)\, B(t)\, (t-\varphi(t)) \,. \end{aligned}$$

From this and (7.2.38) we have

$$\int_0^\infty R^{-\beta_2}(t)\, |e(t)|\, dt < \infty. \tag{7.2.43}$$

Furthermore, it follows from Lemma 3.9.3 and its proof (with $t_0 = t_0$ and $N = N$) that

$$F(t) = G(t) \ge \tfrac{1}{2} \quad \text{for } t \ge t_0 ; \tag{7.2.44}$$

notice that all the assumptions on t_0 and N are satisfied, (3.9.7) holds by (7.2.43), and (3.9.8) is not needed to prove (7.2.44) (it is used to show $\lim_{t\to\infty} F(t) \in (0,\infty)$). The conclusion of the lemma now follows from (7.2.41), (7.2.42), and (7.2.44). □

Our next three lemmas contain results concerning the auxiliary equation (7.2.32).

Lemma 7.2.10. *Let the hypotheses of Lemma* 7.2.9 *hold,* $R'(t) \ge 0$ *on* $\mathbb{R}_+$,

$$\int_0^\infty \left[\frac{r(t)}{a(t)} + 1\right] A^{\lambda-1}(t) B(t) (t-\varphi(t))\, dt < \infty,$$

and

$$\int_0^\infty \frac{|r'(t)| a^{\frac{1}{p+1}}(t) dt}{r^{\frac{p+2}{p+1}}(t)} < \infty. \tag{7.2.45}$$

If

$$\int_0^\infty R^{-\beta}(t)\, dt = \infty \tag{7.2.46}$$

and e is given by (7.2.36), *then there exists a solution Z of equation* (7.2.32) *satisfying* $\int_0^\infty |Z(t)|^{\lambda+1}\,dt = \infty$.

Proof. Let y be a solution of (7.2.1) as given in Lemma 7.2.9. Then $Z = y$ is a solution of (7.2.32) and G is bounded from below by a positive constant. Condition (7.2.46) then implies

$$\int_0^\infty \frac{|Z_2(t)|^\delta}{R(t)}\,dt + \gamma \int_0^\infty |Z(t)|^{\lambda+1}\,dt = \int_0^\infty \frac{G(t)}{R^\beta(t)}\,dt = \infty.$$

If Z is oscillatory, then the result follows from this and Theorem 2.6.4 since

$$\int_0^\infty |Z(t)|^{\lambda+1}\,dt = \infty \quad \text{if and only if} \quad \int_0^\infty |Z_2(t)|^\delta R^{-1}(t)\,dt = \infty\,.$$

If $Z(t)\,Z'(t) > 0$ eventually, then we are done, so assume that $Z(t)\,Z'(t) < 0$ for large t. The case

$$\int_0^\infty |Z(t)|^{\lambda+1}\,dt < \infty \quad \text{and} \quad \int_0^\infty \frac{|Z_2(t)|^\delta}{R(t)}\,dt = \infty$$

is impossible due to (2.6.6) and (2.6.7) in the proof of Theorem 2.6.4. □

We will need the condition

$$\int_0^\infty R^{-\beta_2}(\sigma)\,|e(t)|\,d\sigma$$

$$\le M \stackrel{\text{def}}{=} \begin{cases} \left[24\left(\frac{3}{2}\right)^{\beta_1}\gamma_1 N\right]^{\frac{p(\lambda+1)}{\lambda-p}} [8K_1N_1]^{-1} & \text{if } p > \lambda, \\ [8K_1N_1]^{-1} & \text{if } p = \lambda, \end{cases} \tag{7.2.47}$$

where $N_1 = \left(\frac{3}{2}\right)^{\frac{1}{p+1}} + \left(\frac{3}{2}\right)^{\frac{1}{\lambda+1}}$, $K_1 = \delta + \gamma_1 N$, and N is defined by

$$\sup_{t\in\mathbb{R}_+} |y(t)| \le N\,, \quad \int_0^\infty |y'(\sigma)|\,d\sigma$$

$$\le N, \quad \text{and} \quad \begin{cases} 24N\gamma_1\left(\frac{3}{2}\right)^{\beta_1} \le 1 & \text{if } p > \lambda, \\ N \le [36\gamma_1]^{-1} & \text{if } p = \lambda. \end{cases}$$

Lemma 7.2.11. *Let* (7.2.3) *and* (7.2.47) *hold. Then for any solution y of* (7.2.32) *satisfying*

$$G(0) = C \stackrel{\text{def}}{=} \begin{cases} \left[24N\gamma_1\left(\frac{3}{2}\right)^{\beta_1}\right]^{\frac{(p+1)(\lambda+1)}{\lambda-p}} & \textit{if } p > \lambda \ge 1, \\ 1 & \textit{if } p = \lambda \ge 1, \end{cases} \tag{7.2.48}$$

the inequality

$$\tfrac{3}{4}C \le G(t) \le \tfrac{3}{2}C \tag{7.2.49}$$

holds on $\mathbb{R}_+$.

Proof. Let y be a solution of (7.2.32) satisfying $G(0) = C$. Then, by Lemma 2.5.1 and (2.5.11), we have

$$|y(t)| \le \gamma^{-\frac{1}{\lambda+1}} R^{-\beta_2}(t) G^{\frac{1}{\lambda+1}}(t), \quad |y_2(t)| \le R^{\beta_2}(t) G^{\frac{p}{p+1}}(t) \tag{7.2.50}$$

on $\mathbb{R}_+$, and

$$\begin{aligned} G(t) = G(\tau) &- \alpha g(\tau)\, y(\tau)\, y_2(\tau) + \alpha g(t)\, y(t)\, y_2(t) \\ &- \alpha \int_\tau^t g'(s)\, y(s)\, y_2(s)\, ds + D(t,\tau) \end{aligned} \tag{7.2.51}$$

for $0 \le \tau < t$, where $D(t,\tau) \le K_1 \int_\tau^t R^{-\beta_2}(s) \left(G^{\frac{1}{p+1}}(s) + G^{\frac{1}{\lambda+1}}(s) \right)$ $|e(s)|\, ds$. First, we will show that

$$G(t) \le \tfrac{3}{2}C \quad \text{on } \mathbb{R}_+ . \tag{7.2.52}$$

Suppose that (7.2.52) does not hold. Then there exist $t_2 > t_1 \ge t_0$ such that

$$G(t_2) = \tfrac{3}{2}C, \quad G(t_1) = C, \quad \text{and} \quad C < G(t) < \tfrac{3}{2}C \quad \text{for } t \in (t_1, t_2).$$

Then (7.2.50) and (7.2.51) with $\tau = t_1$ and $t = t_2$ imply

$$\begin{aligned} \frac{C}{2} &\le 3\alpha N \max_{t_1 \le \sigma \le t_2} |y(\sigma)\, y_2(\sigma)| + K_1 N_1 C^{\frac{1}{p+1}} \int_{t_1}^{t_2} R^{-\beta_2}(\sigma)\, |e(\sigma)|\, d\sigma \\ &\le 3N\gamma_1 \left(\frac{3}{2}\right)^{\beta_1} C^{\beta_1} + M K_1 N_1 C^{\frac{1}{p+1}} \le \frac{C}{8} + \frac{C}{8} = \frac{C}{4}. \end{aligned}$$

This contradiction proves that (7.2.52) holds.

Now from (7.2.52), (7.2.50), and (7.2.51) with $t = t$ and $\tau = 0$ we obtain

$$\begin{aligned} |G(t) - C| &\le 3\alpha N \sup_{\sigma \in \mathbb{R}_+} |y(\sigma)\, y_2(\sigma)| + K_1 N_1 C^{\frac{1}{p+1}} \int_0^t R^{-\beta_2}(\sigma)\, |e(\sigma)|\, d\sigma \\ &\le 3\gamma_1 N \left(\frac{3}{2}\right)^{\beta_1} C^{\beta_1} + M K_1 N_1 C^{\frac{1}{p+1}} \le \frac{C}{8} + \frac{C}{8} = \frac{C}{4}. \end{aligned}$$

Hence, $\tfrac{3}{4}C \le G(t) \le \tfrac{5}{4} < \tfrac{3}{2}C$ on $\mathbb{R}_+$. □

Lemma 7.2.12. *Let $p \geq \lambda \geq 1$, (7.2.3) and (7.2.47) hold,*

$$\lim_{t\to\infty} \frac{a'(t)}{a^{1-\beta/p}(t) r^{\alpha}(t)} = 0, \tag{7.2.53}$$

and

$$\lim_{t\to\infty} \frac{e(t)}{r(t)} R^{\lambda\beta_2}(t) = 0. \tag{7.2.54}$$

If

$$\int_0^{\infty} R^{-\beta}(t)\,dt = \infty, \tag{7.2.55}$$

then any solution Z of (7.2.32) satisfying (7.2.48) is proper and

$$\int_0^{\infty} |Z(t)|^{\lambda+1}\,dt = \infty. \tag{7.2.56}$$

If, in addition,

$$r(t) \geq r_0 > 0 \quad \text{for } t \in \mathbb{R}_+,$$

then

$$\int_0^{\infty} \frac{a(t)}{r(t)} |Z'(t)|^{p+1}\,dt = \infty\,. \tag{7.2.57}$$

Proof. Note that the hypotheses of Lemma 7.2.11 hold. Let y be a solution of (7.2.1) satisfying (7.2.48). Then (7.2.49) holds, y is proper, and properties (7.2.56) and (7.2.57) follow from Theorem 3.9.5. Note that condition (3.9.8) in Theorem 3.9.5 is only used for the existence of a solution y satisfying (7.2.49). But here (7.2.49) is proved without using it. □

Remark 7.2.1. Note that (7.2.48) does not depend on the function e.

7.3 Behavior of Nonoscillatory Solutions

In this section, we discuss limit-point and limit-circle properties for various classes of nonoscillatory solutions of (7.2.1). First we consider solutions in the class $\mathcal{N}_2$.

Theorem 7.3.1. *Let $y \in \mathcal{N}_2$. Then y is of the nonlinear limit-point type. In addition,*

(i) *if $y \in \mathcal{N}_{20}$, then y is of the strong nonlinear limit-point type if and only if*

$$\int_0^{\infty} R^{-1}(t) \left(\int_t^{\infty} r(s)\,ds \right)^{\frac{p+1}{p}} dt = \infty; \tag{7.3.1}$$

(ii) *if* $y \in \mathcal{N}_{21}$, *then* y *is of the strong nonlinear limit-point type if and only if*

$$\int_0^\infty R^{-1}(t)\left(\int_0^t r(s)\,ds\right)^{\frac{p+1}{p}} dt = \infty\,.$$

Proof. Let $y \in \mathcal{N}_2$ be such that

$$\lim_{t\to\infty} y(t) = C > 0;$$

case $C < 0$ can be handled similarly. Then y is of the nonlinear limit-point type. Let $\tau \geq 0$ be such that

$$\frac{C}{2} \leq y\left(\varphi(t)\right) \leq 2C \quad \text{for } t \geq \tau.$$

If $y \in \mathcal{N}_{20}$, then an integration of (7.2.1) implies

$$y_2(t) = \int_t^\infty r(s)\, y^\lambda\left(\varphi(s)\right)\, ds,$$

and so

$$\left(\frac{C}{2}\right)^\lambda \int_t^\infty r(s)\,ds \leq y_2(t) \leq (2C)^\lambda \int_t^\infty r(s)\,ds$$

for $t \geq \tau$. The conclusion then follows from (7.3.1).

Now let $y \in \mathcal{N}_{21}$. Suppose, without loss of generality, that $y_2(t) \leq 0$ for $t \geq \tau$. Then an integration of (7.2.1) yields

$$|y_2(t)| = |y_2(\tau)| + \int_\tau^t r(s)\, y^\lambda\left(\varphi(s)\right)\, ds,$$

and there exists $T \geq \tau$ such that

$$\begin{aligned} 2^{-1-\lambda} C^\lambda \int_T^t r(s)\,ds &\leq \tfrac{1}{2}\int_T^t r(s)\, y^\lambda\left(\varphi(s)\right)\, ds \leq |y_2(t)| \\ &\leq 2\int_T^t r(s) y^\lambda\left(\varphi(s)\right)\, ds \leq 2^{\lambda+1} C^\lambda \int_T^t r(s)\,ds \end{aligned}$$

for $t \geq T$. The conclusion follows from this and Definitions 6.2.4 and 6.2.5. □

Next, we examine solutions in the class $\mathcal{N}_1$.

Theorem 7.3.2. *Let $y \in \mathcal{N}_1$. Then y is of the nonlinear limit-point type. In addition,*

(i) *if $y \in \mathcal{N}_{11}$, then y is of the strong nonlinear limit-point type if and only if*

$$\int_0^\infty R^{-1}(t)\,dt = \infty\,;$$

(ii) *if $y \in \mathcal{N}_{10}$ and*

$$\int_0^\infty R^{-1}(s)\left(\int_s^\infty r(\sigma)\,d\sigma\right)^{\frac{p+1}{p}} ds = \infty, \tag{7.3.2}$$

then y is of the strong nonlinear limit-point type. If

$$\int_0^\infty R^{-1}(s)\left[\int_s^\infty r(\sigma)\left(\int_0^{\varphi(\sigma)} a^{-\frac{1}{p}}(v)\,dv\right)^{\lambda} d\sigma\right]^{\frac{p+1}{p}} ds < \infty \tag{7.3.3}$$

then y is not of the strong nonlinear limit-point type.

Proof. Let $y \in \mathcal{N}_1$ with $\lim_{t\to\infty} y(t) = \infty$; the case $\lim_{t\to\infty} y(t) = -\infty$ can be handled similarly. Then y is of the nonlinear limit-point type. If $y \in \mathcal{N}_{11}$, then $\lim_{t\to\infty} y_2(t) = d_y$, $|d_y| > 0$, and the statement follows from Definitions 6.2.4 and 6.2.5 and expression (7.2.4).

Let $y \in \mathcal{N}_{10}$; then $\lim_{t\to\infty} y_2(t) = 0$. In view of (7.2.1), y_2 is positive and decreasing for large t. Thus, there exist positive constants T, C_0, and C_1 such that

$$0 < y'(t) \le \frac{C_1}{2} a^{-\frac{1}{p}}(t)$$

and

$$C_0 \le y\left(\varphi(t)\right) \le C_1 \int_0^{\varphi(t)} a^{-\frac{1}{p}}(s)\,ds \tag{7.3.4}$$

for $t \ge T$. An integration of (7.2.1) gives

$$y_2(t) = \int_t^\infty r(s) y^{\lambda}\left(\varphi(s)\right)\,ds \ge C_0^{\lambda} \int_t^\infty r(s)\,ds. \tag{7.3.5}$$

Hence, $\int_0^\infty r(s)\,ds < \infty$, and if (7.3.2) holds, then

$$\int_T^\infty R^{-1}(s) y_2^{\delta}(s)\,ds \ge C_0^{\lambda\delta} \int_T^\infty R^{-1}(s)\left(\int_s^\infty r(\sigma)\,d\sigma\right)^{\delta} ds = \infty.$$

That is, y is of the strong nonlinear limit-point type.

Now let (7.3.3) hold. Then (7.3.3)–(7.3.5) imply

$$y_2(t) \le C_1^\lambda \int_t^\infty r(s) \left(\int_0^{\varphi(s)} a^{-\frac{1}{p}}(\sigma)\, d\sigma \right)^\lambda ds,$$

or

$$\int_T^\infty R^{-1}(s) y_2^\delta(s)\, ds$$

$$\le C_1^{\lambda\delta} \int_T^\infty R^{-1}(s) \left[\int_s^\infty r(\sigma) \left(\int_0^{\varphi(\sigma)} a^{-\frac{1}{p}}(v)\, dv \right)^\lambda d\sigma \right]^\delta ds < \infty$$

for $t \ge T$. That is, y is of the nonlinear limit-point type but not of the strong nonlinear limit-point type. □

Our final theorem in this section concerns the solutions in the class $\mathcal{N}_3$.

Theorem 7.3.3. *Let $y \in \mathcal{N}_{30}$. Then y is of the nonlinear limit-circle type if and only if*

$$\int_0^\infty \int_t^\infty a^{-\frac{1}{p}}(s)\, ds \left(\int_{\varphi(t)}^\infty a^{-\frac{1}{p}}(s)\, ds \right)^\lambda dt < \infty. \tag{7.3.6}$$

Moreover,

$$\int_0^\infty R^{-1}(t) |y_2(t)|^\delta\, dt < \infty \quad \textit{if and only if} \quad \int_0^\infty R^{-1}(t)\, dt < \infty. \tag{7.3.7}$$

Proof. From [57], if $y \in \mathcal{N}_3$, then

$$\int_0^\infty a^{-\frac{1}{p}}(t)\, dt < \infty\,.$$

Let $y \in \mathcal{N}_{30}$, and for simplicity, assume $y(\varphi(t)) > 0$ for $t \ge t_0$. Then,

$$\lim_{t\to\infty} |y_2(t)| = d > 0,$$

from which (7.3.7) follows immediately. Furthermore, there exists $t_1 \ge t_0$ such that

$$\frac{d}{2} \le |y_2(t)| \le 2d \quad \text{for } t \ge t_1\,.$$

From this and an integration of the first equality in (7.2.2), we have

$$y(t) = -\int_t^\infty y'(s)\, ds = \int_t^\infty a^{-\frac{1}{p}}(s) |y_2(s)|^{\frac{1}{p}}\, ds,$$

or

$$\left(\frac{d}{2}\right)^{\frac{1}{p}} \int_t^\infty a^{-\frac{1}{p}}(s)\,ds \le y(t) \le (2d)^{\frac{1}{p}} \int_t^\infty a^{-\frac{1}{p}}(s)\,ds,$$

and (7.3.6) follows from this and Definitions 6.2.3 and 6.2.4. □

Remark 7.3.1. It is not difficult to show that under somewhat mild conditions, the set $\mathcal{N}_{31}$ is empty. For example, it is known that taken together

$$\int_0^\infty a^{-\frac{1}{p}}(t)\,dt < \infty \quad \text{and} \quad \int_0^\infty r(t)\,dt = \infty$$

forms a necessary condition for the class $\mathcal{N}_{31}$ to be nonempty. However, if in addition, $R(t) = a^{\frac{1}{p}}(t)r(t)$ has a finite limit R_1, then l'Hôpital's rule shows

$$0 = \lim_{t\to\infty} \frac{\int_0^t a^{-\frac{1}{p}}(s)\,ds}{\int_0^t r(s)\,ds} = \lim_{t\to\infty} \frac{1}{R(t)} = \frac{1}{R_1},$$

which is a contradiction. It remains an open problem as to what conditions would imply that solutions in the class $\mathcal{N}_{31}$ are nonlinear limit-point.

Corollary 7.3.1. *If*

$$\int_0^\infty a^{-\frac{1}{p}}(t)\,dt = \infty\,, \quad \int_0^\infty r(t)\,dt < \infty\,, \quad \textit{and}$$

$$\int_0^\infty a^{-\frac{1}{p}}(t) \left(\int_t^\infty r(\sigma)\,d\sigma\right)^{\frac{1}{p}} dt < \infty\,,$$

then equation (7.2.1) *is of the nonlinear limit-point type. There is a strong nonlinear limit-point type solution if, in addition,*

$$\int_0^\infty R^{-1}(t) \left(\int_t^\infty r(s)\,ds\right)^{\frac{p+1}{p}} dt = \infty\,.$$

Proof. This result follows from Lemma 7.2.1(ii) and Theorem 7.3.1. □

Corollary 7.3.2. *Let*

$$\int_0^\infty a^{-\frac{1}{p}}(t)\,dt = \infty\,, \quad \int_0^\infty r(t)\,dt < \infty\,, \quad \textit{and}$$

$$\int_0^\infty r(t) \left(\int_0^{\varphi(t)} a^{-\frac{1}{p}}(\sigma)\,d\sigma\right)^{\lambda} dt < \infty\,.$$

Then equation (7.2.1) *is of the nonlinear limit-point type and* $\mathcal{N}_{11} \neq \emptyset$. *Moreover,* $y \in \mathcal{N}_{11}$ *is of the strong nonlinear limit-point type if and only if* $\int_0^\infty R^{-1}(t)dt = \infty$.

Proof. This follows from Lemma 7.2.1(i) and Theorem 7.3.2. □

7.4 Behavior of Oscillatory Solutions

Now we turn our attention to the oscillatory solutions of equation (7.2.1). Recall the auxiliary equation (see (7.2.32))

$$\left(a(t)|Z'|^{p-1}Z'\right) + r(t)|Z|^{\lambda}\operatorname{sgn} Z = e(t), \tag{7.4.1}$$

where $e \in C^0(\mathbb{R}_+)$.

Theorem 7.4.1. *Let* $1 \leq \lambda < p$, (7.2.3) *and* (7.2.28) *hold, and assume there are constants* r_0 *and* r_1 *such that* $0 < r_0 \leq r(t) \leq r_1$ *on* $\mathbb{R}_+$,

$$\int_0^\infty a^{-\frac{1}{p}}(t)\, dt = \infty\,, \quad \textit{and} \quad \int_0^\infty a^{-\frac{\beta}{p}}(t)\, dt = \infty. \tag{7.4.2}$$

Set

$$A(t) = \left(\int_0^t a^{-\frac{1}{p}}(s)\, ds\right)^{\frac{p}{p+1}}$$

and

$$B(t) = a^{-\frac{1}{p}}(t)\left[\int_0^t \left(\int_0^s a^{-\frac{1}{p}}(\sigma)\, d\sigma\right)^{\frac{p\lambda}{p+1}} ds\right]^{\frac{1}{p}}.$$

Then equation (7.2.1) *is of the nonlinear limit-point type if*

$$\int_0^\infty A^{\lambda}(t)\, B(t)\, (t - \varphi(t))\, dt < \infty \tag{7.4.3}$$

and

$$\int_0^\infty a^{-\frac{1}{(\lambda+2)p+1}}(t) A^{\lambda-1}(t)\, B(t)\, (t - \varphi(t))\, dt < \infty. \tag{7.4.4}$$

Proof. In view of Lemmas 7.2 and 7.2.9 and (7.4.4), there is an oscillatory solution y of (7.2.1) such that $F(t) \geq C > 0$ for a suitable constant C. To prove that y is of the nonlinear limit-point type, suppose that it is of the nonlinear limit-circle type. Then Lemma 7.2.3 implies

$$\lim_{t\to\infty} y(t)y_2(t) = 0 \quad \text{and} \quad \int_0^\infty R^{-1}(t)|y_2(t)|^\delta \, dt < \infty . \tag{7.4.5}$$

Furthermore, condition (7.4.2) and the boundedness of r imply

$$\int_0^\infty R^{-\beta}(t)\, dt = \infty. \tag{7.4.6}$$

Hence,

$$0 < C \le F(t) = R^\beta(t) \left[R^{-1}(t)\, |y_2(t)|^\delta + \gamma\, |y(t)|^{\lambda+1}\right]$$

for large t, say $t \ge t_0$. From this, (7.4.6), and an integration, we obtain

$$\int_{t_0}^\infty R^{-1}(t)\, |y_2(t)|^\delta \; dt + \gamma \int_{t_0}^\infty |y(t)|^{\lambda+1} \; dt = \infty.$$

In view of (7.4.5), we have

$$\int_{t_0}^\infty |y(t)|^{\lambda+1} \; dt = \infty\, .$$

Note that the assumptions of Lemma 7.2.6(i) hold and (7.2.31) contradicts y being a nonlinear limit-circle type solution. Hence, y is of the nonlinear limit-point type. □

Remark 7.4.1. From [1, Theorem 3.1], condition (7.4.2), and the boundedness of r away from zero, all proper solutions of (7.2.1) are oscillatory.

Theorem 7.4.2. *In addition to the hypotheses of Theorem 7.4.1, let*

$$\int_0^\infty \frac{R'_-(t)}{R(t)}\, dt < \infty \quad \textit{and} \quad \int_0^\infty \frac{|r'(t)|a^{\frac{1}{p+1}}(t)}{r^{\frac{p+2}{p+1}}(t)}\, dt < \infty\, . \tag{7.4.7}$$

Then equation (7.2.1) *has a strong nonlinear limit-point type solution.*

Proof. By Theorem 7.4.1 and its proof, there is an oscillatory solution y of equation (7.2.1) that is of the nonlinear limit-point type and

$$\int_0^\infty |y(t)|^{\lambda+1}\, dt = \infty\, . \tag{7.4.8}$$

Also, y is a solution of equation (7.4.1) with e given by

$$e(t) = r(t)\left[|y(t)|^{\lambda} \operatorname{sgn} y(t) - |y(\varphi(t))|^{\lambda} \operatorname{sgn} y(\varphi(t))\right].$$

Then,

$$|e(t)| = \lambda r(t)\, |y(\xi)|^{\lambda-1}\, |y'(\xi)|\, (t - \varphi(t))$$

for some $\xi \in [\varphi(t), t]$. From (7.2.7) and (7.2.8), we have

$$|e(t)| \le C_0 A^{\lambda-1}(t)\, B(t)\, (t - \varphi(t)) \tag{7.4.9}$$

for large t and some positive constant C_0. Now the estimate (7.4.9) and condition (7.4.3) imply

$$\begin{aligned}\int_0^\infty |e(t)|\, dt &\le C_0 \int_0^\infty A^{\lambda-1}(t)\, B(t)\, (t - \varphi(t))\, dt\\ &\le C_1 \int_0^\infty A^{\lambda}(t)\, B(t)\, (t - \varphi(t))\, dt < \infty\end{aligned}$$

for some $C_1 > 0$. From this and (7.4.7), the assumptions of Theorem 2.6.5 are satisfied and so together with (7.4.8) this implies

$$\int_0^\infty R^{-1}(t)\, |y_2(t)|^{\delta}\, dt = \infty.$$

This completes the proof of the theorem. □

Next, we define on $\mathbb{R}_+$ the functions

$$\hat{R}(t) = r^{-1}(t) a^{-\frac{1}{p}}(\varphi(t))\, \varphi'(t) \quad \text{and} \quad \hat{g}(t) = -r^{-1}(t)\hat{R}'(t)\hat{R}^{-1-\beta}(t).$$

Theorem 7.4.3. *Let $\lambda < p \le 1$, $\varphi \in C^2(\mathbb{R}_+)$, and assume there are positive constants ε, M, a_0, and a_1 such that*

$$0 < \varepsilon \le \varphi'(t) \le M \quad \textit{and} \quad a_0 \le a(t) \le a_1. \tag{7.4.10}$$

Let

$$\int_0^\infty r(t)\, dt = \infty, \quad \int_0^\infty r^{\alpha}(t)\, dt = \infty, \quad \int_0^\infty \frac{\hat{R}'_-(t)}{\hat{R}(t)}\, dt < \infty,$$

$$\lim_{t\to\infty} \hat{g}(t) = 0, \quad \int_0^\infty |\hat{g}'(t)|\, dt < \infty,$$

$$\int_0^\infty r^{-\frac{1}{\lambda+1}}(t)\left|(a^{-\frac{1}{p}}(\varphi(t))\varphi'(t))'\right| dt < \infty,$$

$$\int_0^\infty r(t)\left(\int_0^t r(s)\,ds\right)^{\frac{1}{p(\lambda+1)}}\left[\int_0^t\left(\int_0^s r(\sigma)\,d\sigma\right)^{\frac{1}{p(\lambda+1)}} ds\right]^\lambda (t-\varphi(t))\,dt < \infty,$$

$$\int_0^\infty r^{\frac{(\lambda+3)p+1}{(\lambda+2)p+1}}(t)\left[\int_0^t r(s)\,ds\right]^{\frac{1-p}{p(\lambda+1)}}$$
$$\times\left[\int_0^t\left(\int_0^s r(\sigma)\,d\sigma\right)^{\frac{1}{p(\lambda+1)}} ds\right]^\lambda (t-\varphi(t))dt < \infty,$$

and

$$\int_0^\infty t^{\lambda+\frac{1}{p}}(t-\varphi(t))\,dt < \infty\,. \tag{7.4.11}$$

Then (7.2.1) *has a nonlinear limit-point type solution. If, moreover,*

$$r(t) \le r_1 \quad on \quad \mathbb{R}_+, \tag{7.4.12}$$

then this solution is of the strong nonlinear limit-point type.

Proof. Note, that $1 \le \frac{1}{p} < \frac{1}{\lambda}$ and we can apply Theorems 7.4.1 and 7.4.2 to equation (7.2.5). In equation (7.2.5) the quantities p, λ, β, a, r, R, g have, respectively, the values

$$\frac{1}{\lambda},\quad \frac{1}{p},\quad \alpha,\quad r^{-\frac{1}{\lambda}},\quad a^{-\frac{1}{p}}(\varphi)\varphi',\quad \hat{R},\quad \text{and}\quad \hat{g}\,.$$

From this and from our assumptions, all hypotheses of Theorems 7.4.1 and 7.4.2 are satisfied. Hence, equation (7.2.5) has a strong nonlinear limit-point type solution Z, i.e.,

$$\int_0^\infty |Z(t)|\,|Z(\varphi(t))|^{\frac{1}{p}}\,dt = \infty \quad \text{and} \quad \int_0^\infty \hat{R}^{-1}(t)|Z_2(t)|^{\lambda+1}\,dt = \infty. \tag{7.4.13}$$

Observe that the condition $\int_0^\infty \frac{\hat{R}'_-(t)}{\hat{R}(t)}\,dt < \infty$ implies $\hat{R}(t) \ge \hat{R}_0$ for some $\hat{R}_0 > 0$. From this and the second integral in (7.4.13), we have

$$\infty = \int_0^\infty \hat{R}^{-1}(t)|Z_2(t)|^{\lambda+1}\,dt \le \hat{R}_0^{-1}\int_0^\infty |Z_2(t)|^{\lambda+1}\,dt\,. \tag{7.4.14}$$

The first integral in (7.4.13), (7.4.14), and Hölder's inequality imply

$$\infty = \int_0^\infty |Z(t)|\,|Z(\varphi(t))|^{\frac{1}{p}}\,dt \tag{7.4.15}$$

$$\leq \left(\int_0^\infty |Z(t)|^{\frac{p+1}{p}}\,dt\right)^{\frac{p}{p+1}} \left(\int_0^\infty |Z(\varphi(t))|^{\frac{p+1}{p}}\,dt\right)^{\frac{1}{p+1}}$$

$$\leq \varepsilon^{-1}\int_0^\infty |Z(t)|^{\frac{p+1}{p}}\,dt\,. \tag{7.4.16}$$

According to Lemma 7.2.2, equation (7.2.1) has a solution $y(t) = Z_2(\varphi^{-1}(t))$ on $\mathbb{R}_+$ and $y_2(t) = -Z(t)$ holds. From this and from (7.4.14),

$$\int_0^\infty |y(t)|^{\lambda+1}\,dt = \int_0^\infty |Z_2(\varphi^{-1}(t))|^{\lambda+1}\,dt \geq \varepsilon \int_0^\infty |Z_2(s)|^{\lambda+1}\,ds = \infty\,. \tag{7.4.17}$$

It follows from Lemma 7.2.3 and our assumptions that there are positive constants K and K_1 such that

$$|y(t)| \leq K t^{\frac{p}{p+1}} \quad \text{and} \quad |y_2(t)| \leq K_1 t^{\frac{\lambda p}{p+1}+1}$$

for large t. From this and (7.4.11) we see that the hypotheses of Lemma 7.2.6(ii) are satisfied, so (7.4.17) and (7.2.30) imply $\int_0^\infty |y(t)|\,|y(\varphi(t))|^\lambda\,dt = \infty$, i.e., y is of the nonlinear limit-point-type.

Let (7.4.12) hold. Then (7.4.16), (7.4.10), and (7.4.14) imply

$$\int_0^\infty R^{-1}(t)|y_2(t)|^\delta\,dt = \int_0^\infty R^{-1}(t)|Z(t)|^\delta\,dt \geq a_1^{-\frac{1}{p}} r_1^{-1}\int_0^\infty |Z(t)|^\delta\,dt = \infty,$$

and so y is of the strong nonlinear limit-point type. □

Since the hypotheses of the previous theorems are rather complicated, we will apply our results to the case $a \equiv 1$, i.e., to the equation

$$\left(|y'|^{p-1}y'\right)' + r(t)\,|y\,(\varphi(t))|^\lambda \operatorname{sgn} y\,(\varphi(t)) = 0. \tag{7.4.18}$$

Corollary 7.4.1. *Let $1 \leq \lambda < p$ and (7.2.28) hold, and assume there exist constants r_0 and r_1 such that $0 < r_0 \leq r(t) \leq r_1$ on $\mathbb{R}_+$,*

$$\int_0^\infty \left(|r''(s)| + r'^2(s)\right)\,ds < \infty \quad \textit{and} \quad \int_0^\infty t^{\lambda+\frac{1}{p}}\,(t-\varphi(t))\,dt < \infty.$$

Then there is an oscillatory solution of (7.4.18) that is of the nonlinear limit-point type. Moreover, if $r' \in L^1(\mathbb{R}_+)$, then equation (7.4.18) has a strong nonlinear limit-point type solution.

Proof. The conclusion follows from Theorems 7.4.1 and 7.4.2. Note that since $r'' \in L_1(\mathbb{R}_+)$ and $r' \in L_2(\mathbb{R}_+)$, Theorem 1 in [49, §5.2,] implies $\lim_{t\to\infty} r'(t) = 0$. □

7.5 Limit-Point/Limit-Circle Properties of Solutions

Our first theorem gives a sufficient condition for equation (7.2.1) to be of the strong nonlinear limit-circle type.

Theorem 7.5.1. *Assume that* (7.2.3), (7.2.26), *and* (7.2.33) *hold. In addition, assume that one of the following conditions holds:*

(i) $1 \le \lambda < p$, (7.2.35) *holds,*

$$\int_0^\infty \frac{e(\sigma)}{a(\sigma)}\,d\sigma < \infty, \quad \text{and} \quad \int_0^\infty \frac{e(\sigma)}{r(\sigma)} < \infty,$$

where

$$e(t) = \bar{a}^{-\frac{1}{p}}(t)\,r(t)\left[\int_0^t a^{-\frac{1}{p}}(s)\left(\int_0^s r(\sigma)\,d\sigma\right)^{\frac{1}{p}} ds\right]^{\frac{(\lambda-1)p}{p-\lambda}}$$

$$\times\left[\int_0^t r(s)\left(\int_0^s a^{-\frac{1}{p}}(\sigma)\,d\sigma\right)^{\lambda} ds\right]^{\frac{1}{p-\lambda}}(t-\varphi(t))\,; \qquad (7.5.1)$$

(ii) $1 \le \lambda = p$ *and*

$$e(t) = \bar{a}^{-\frac{1}{p}}(t)r(t)\exp\left\{(\lambda-1)\int_0^t k(u)\,du\right\}$$

$$\times \exp\left\{\frac{1}{p}\int_0^t k_1(u)\,du\right\}(t-\varphi(t))\,, \qquad (7.5.2)$$

where

$$k(t) = 2^{\frac{1}{p}}a^{-\frac{1}{p}}(t)\left(\int_0^t r(s)\,ds\right)^{\frac{1}{p}}$$

and

$$k_1(t) = 2^p r(t)\left(\int_0^t a^{-\frac{1}{p}}(\sigma)\,d\sigma\right)^p.$$

If

$$\int_0^\infty R^{-\beta}(\sigma)\,d\sigma < \infty,$$

then (7.2.1) *is of the strong nonlinear limit-circle type.*

Proof. Let y be a solution of (7.2.1). Then, as we noted earlier, y is also a solution of equation (7.2.32) with $e(t)$ given by

$$e(t) = r(t)\left[|y(t)|^{\lambda}\operatorname{sgn} y(t) - |y(\varphi(t))|^{\lambda}\operatorname{sgn} y(\varphi(t))\right], \quad t \in \mathbb{R}_+. \tag{7.5.3}$$

Thus,

$$e(t) = \lambda r(t)\,|y(\xi)|^{\lambda-1}\,y'(\xi)\,(t-\varphi(t)) \tag{7.5.4}$$

for some $\xi \in [\varphi(t), t]$.

For $t \geq 1$, define

$$A(t) \stackrel{\text{def}}{=} \left[\int_0^t a^{-\frac{1}{p}}(s)\left(\int_0^s r(\sigma)\,d\sigma\right)^{\frac{1}{p}} ds\right]^{\frac{p}{p-\lambda}},$$

$$B(t) \stackrel{\text{def}}{=} \bar{a}^{-\frac{1}{p}}(t)\left[\int_0^t r(s)\left(\int_0^s a^{-\frac{1}{p}}(\sigma)\,d\sigma\right)^{\lambda} ds\right]^{\frac{1}{p-\lambda}}$$

if $\lambda < p$, and

$$A(t) \stackrel{\text{def}}{=} \exp\left\{\int_0^t k(u)\,du\right\},$$

$$B(t) \stackrel{\text{def}}{=} \bar{a}^{-\frac{1}{p}}(t)\exp\left\{\frac{1}{p}\int_0^t k_1(u)\,du\right\}$$

if $\lambda = p$.

Since the hypotheses of Lemma 7.2.4 are satisfied, the estimates (7.2.14)–(7.2.17) and (7.5.4) imply

$$|y(t)| \leq CA(t), \quad |y_1'(t)| \leq C_1 B(t), \tag{7.5.5}$$

and

$$|e(t)| \leq C_2 A^{\lambda-1}(t)\,B(t)\,r(t)\,(t-\varphi(t)) \tag{7.5.6}$$

for some positive constants C, C_1, and C_2. By Lemma 7.2.8, we have

$$\int_0^{\infty} |y(s)|^{\lambda+1}\,ds < \infty \quad \text{and} \quad \int_0^{\infty} \frac{a(s)}{r(s)}\,|y'(s)|^{p+1}\,ds < \infty.$$

The hypotheses of Lemma 7.2.6(i) hold, so the conclusion follows from (7.2.27). □

Our next strong nonlinear limit-circle result allows for a larger class of delays but requires $R'(t) \geq 0$.

Theorem 7.5.2. *Assume that* (7.2.3) *and* (7.2.26) *hold,* $\lambda \geq 1$, $R'(t) \geq 0$ *on* $\mathbb{R}_+$,

$$\int_0^\infty \bar{a}^{-\frac{2}{p}}(s) R^{\frac{2}{p+1}}(s)\,(s - \varphi(s))\, ds < \infty,$$

and (7.2.33) *holds with*

$$e(t) = \bar{a}^{-\frac{1}{p}}(t)\, r(t)\, R^{\frac{1}{p+1}}(t)\,(t - \varphi(t))\,.$$

In addition, assume that either (i) $\lambda = p$, *or* (ii) $\lambda < p$, (7.2.34) *holds, and*

$$\liminf_{t\to\infty} R^\beta(t) \left(\int_t^\infty |g'(s)|\, ds \right)^{\hat{\omega}} = 0\,.$$

Then if

$$\int_0^\infty R^{-\beta}(t)\, dt < \infty,$$

equation (7.2.1) *is of the strong nonlinear limit-circle type.*

Proof. The proof is similar to that of Theorem 7.5.1 except that we apply Lemma 7.2.5 instead of Lemma 7.2.4. Note that

$$|y_1(t)| \leq CA(t) = \text{const.} \quad \text{and} \quad |y_1'(t)| \leq C_1 B(t) = C_1 \bar{a}^{-\frac{1}{p}}(t)\, R^{\frac{1}{p+1}}(t)$$

for some positive constants C and C_1. $\square$

Theorem 7.5.3. *Let conditions* (7.2.3), (7.2.28), (7.2.47), *and* (7.2.53)–(7.2.55) *hold, and let*

$$k(t) = a^{-\frac{1}{p}}(t) \left(\int_0^t r(s)\, ds \right)^{\frac{1}{p}}.$$

(i) *For* $1 \leq \lambda < p$, *let* e *be given by* (7.5.1) *and* $A(t) = \left[\int_0^t k(s)\, ds \right]^{\frac{p}{p-\lambda}}$.

(ii) *For* $1 \leq \lambda = p$, *let* e *be given by* (7.5.2) *and* $A(t) = \exp\left\{ \int_0^t k(s)\, ds \right\}$.

Then, if

$$\int_0^\infty \frac{e(t)\, A(t)}{r(t)}\, dt < \infty, \tag{7.5.7}$$

then equation (7.2.1) *has a solution* y *that is of the nonlinear limit-point type, i.e.,* (7.2.1) *is of the nonlinear limit-point type. If, moreover, there is a constant* $r_0 > 0$ *such that*

$$r(t) \geq r_0 > 0 \quad \text{for } t \in \mathbb{R}_+ \,, \tag{7.5.8}$$

then y *is of the strong nonlinear limit-point type.*

Proof. Let y be a solution of (7.2.1) satisfying $F(0) = C$ where C is given by (7.2.48). Then, in a manner similar to the proof of Theorem 7.5.1, only now using Lemma 7.2.12 instead of Lemma 7.2.8, we can show that

$$\int_0^\infty |y(s)|^{\lambda+1}\, ds = \infty \,. \tag{7.5.9}$$

(If (7.5.8) holds, we can show that $\int_0^\infty \frac{a(s)}{r(s)} |y'(s)|^{p+1}\, ds = \infty$ as well.) Now, part (ii) of Lemma 7.2.6 holds since (7.2.30) follows from (7.5.5), (7.5.6) and (7.5.7). The conclusion then follows from this and (7.5.9). □

Theorem 7.5.4. *Let* $1 \leq \lambda < p$, *conditions* (7.2.3), (7.2.20), (7.2.28), *and* (7.2.45) *hold,* $R'(t) \geq 0$ *on* $\mathbb{R}_+$, *and*

$$\int_0^\infty \left[1 + r(t) + \frac{r(t)}{a(t)}\right] a^{-\frac{1}{p}}(t) R^{\frac{1}{p+1}}(t)(t - \varphi(t))\, dt < \infty.$$

If

$$\int_0^\infty R^{-\beta}(t)\, dt = \infty \,, \tag{7.5.10}$$

then equation (7.2.1) *is of the nonlinear limit-point type.*

Proof. In Lemma 7.2.9, we take the estimates $A(t) = \text{const.}$ and $B(t) = a^{-\frac{1}{p}}(t) R^{\frac{1}{p+1}}(t)$ from Lemma 7.2.5, and we let y be a solution of (7.2.1) as given in Lemma 7.2.9. Then we see that the hypotheses of Lemma 7.2.10 hold. As in the proof of Theorem 7.5.1, if we set

$$e(t) = r(t) \left[|y(t)|^\lambda \operatorname{sgn} y(t) - |y(\varphi(t))|^\lambda \operatorname{sgn} y(\varphi(t))\right], \quad t \in \mathbb{R}_+,$$

then y is the solution of (7.2.32) for which G is bounded from below by a positive constant. The conclusion can be proved similarly to the proofs of Theorems 7.5.1 and 7.5.2 using Lemmas 7.2.5 and 7.2.10 instead of Lemmas 7.2.4 and 7.2.8, respectively. □

Since the hypotheses of the previous theorems are rather complicated, we will apply our results to the case $a \equiv 1$, i.e., to the equation

$$\left(|y'|^{p-1}y'\right)' + r(t)\,|y\,(\varphi(t))|^{\lambda}\,\mathrm{sgn}\,y\,(\varphi(t)) = 0\,. \tag{7.5.11}$$

Corollary 7.5.1. *Let $1 \le \lambda < p$, $r'(t) \ge 0$ on $\mathbb{R}_+$, condition* (7.2.28) *hold,*

$$\lim_{t\to\infty} \frac{r'(t)}{r^{1+\alpha}(t)} = 0, \quad \int_0^\infty \frac{|r'(t)|}{r^{\frac{p+2}{p+1}}(t)} < \infty, \quad \int_0^\infty \left|\left(\frac{r'(t)}{r^{1+\alpha}(t)}\right)'\right| dt < \infty,$$

$$\int_0^\infty r^{\frac{p+2}{p+1}}(t)(t-\varphi(t))\,dt < \infty\,, \quad \textit{and} \quad \int_0^\infty r^{-\beta}(t)\,dt = \infty\,.$$

Then equation (7.5.11) *is of the nonlinear limit-point type.*

Proof. This is a special case of Theorem 7.5.4. □

7.6 Examples

We end this chapter with some examples to illustrate some of our results.

Example 7.6.1. Consider the equation

$$\left(|y'|^{p-1}y'\right)' + (1+\sigma t^{v})\,|y\,(t-t^{m})|^{\lambda}\,\mathrm{sgn}\,y\,(\varphi(t)) = 0\,, \quad t \ge 1, \tag{7.6.1}$$

where $1 \le \lambda < p$, $\sigma \in \{-1, 1\}$, $v \le 0$, and $m < -1 - \lambda - \frac{1}{p}$. Then equation (7.6.1) has an oscillatory solution that is of the strong nonlinear limit-point type by Corollary 7.4.1.

Example 7.6.2. Consider the equation

$$\left(|y'|^{p-1}y'\right)' + t^{s}\,|y\,(t-t^{m})|^{\lambda}\,\mathrm{sgn}\,y\,(t-t^{m}) = 0\,, \quad t \ge 1, \tag{7.6.2}$$

where $\lambda < p \le 1$, $s \in (-1, 0]$, and $m < -1 - \lambda - \frac{1}{p}$. Then equation (7.6.2) has a strong nonlinear limit-point type solution by Theorem 7.4.3.

Example 7.6.3. Consider the equation

$$\left(|y'|^{p-1}y'\right)' + t^{s}\,|y\,(\varphi(t))|^{\lambda}\,\mathrm{sgn}\,y\,(\varphi(t)) = 0, \quad t \ge 1, \tag{7.6.3}$$

where $1 \le \lambda < p$, $\varphi' > 0$, and $t - \frac{1}{t^{\nu}} \le \varphi(t) \le t$ for large t. If $0 \le s \le \frac{1}{\beta}$ and $\nu \ge 1 + \frac{p+2}{p+1}s$, then (7.6.3) is of the nonlinear limit-point type by Theorem 7.5.4. On the other hand, if $\lambda \le p$, $\varphi' \ge \epsilon > 0$, $\frac{1}{\beta} < s$, and

$\nu \geq 2 + [\frac{p+2}{p+1} + \frac{1}{(\lambda+2)p+1}]s$, then Theorem 7.5.2 implies equation (7.6.3) is of the strong nonlinear limit-circle type.

Example 7.6.4. Consider the equation

$$\left(|y'|^{p-1}y'\right)' + Ct^s \left|y\left(\varphi(t)\right)\right|^\lambda \operatorname{sgn} y\left(\varphi(t)\right) = 0, \quad t \geq 1, \tag{7.6.4}$$

where $C \geq [24\gamma_1 \left(\frac{3}{2}\right)^{\beta_1}]^{\frac{1}{\alpha}}$, $1 \leq \lambda < p$, $-\frac{1}{\alpha} \leq s < 0$, $0 < \varphi' \leq M$, and $t - \frac{1}{t^\nu} \leq \varphi(t) \leq t$ for large t. If either $s \in (-1, 0]$ and $\nu > 1 + \frac{(s+p+2)\lambda+s+1}{p-\lambda}$, or $s \in [-\frac{1}{\alpha}, -1)$ and $\nu > 1 + \frac{\lambda p}{p-\lambda} + \max\{0, \frac{s+\lambda+1}{p-\lambda}\}$, then (7.6.4) is of the nonlinear limit-point type by Theorem 7.5.3(i).

Chapter 8

Transformations of the Basic Equation

8.1 Introduction

The use of transformations of the independent variable "t" in the study of the (linear) limit-point/limit-circle problem goes back to the work of Dunford and Schwartz [72, p. 1410] who considered the linear equation

$$(a(t)y')' + r(t)y = 0. \tag{8.1.1}$$

Setting

$$s = \int_0^t \left(\frac{r(\tau)}{a(\tau)}\right)^{\frac{1}{2}} d\tau \quad \text{and} \quad y(t) = x(s), \tag{T_1}$$

equation (8.1.1) becomes

$$\ddot{x}(s) + \frac{(a(t)r(t))'}{2a^{1/2}(t)r^{3/2}(t)}\dot{x}(s) + x(s) = 0,$$

where "$\cdot$" $= \frac{d}{ds}$. We should point out that if

$$\int_0^\infty \left(\frac{r(\tau)}{a(\tau)}\right)^{\frac{1}{2}} d\tau < \infty,$$

then the interval $0 \le t \le \infty$ is mapped into an interval of the form $0 \le s \le c$ for some $0 < c < \infty$, i.e., a finite interval.

Burton and Patula [53], in studying the linear equation

$$y'' + r(t)y = 0, \tag{8.1.2}$$

took

$$s = \int_0^t r^{\frac{1}{2}}(\tau)d\tau \quad \text{and} \quad y(t) = x(s), \tag{T_2}$$

in which case equation (8.1.2) becomes

$$\ddot{x} + 2\theta(t)\dot{x} + x = 0,$$

where $\theta(t) = \frac{r'(t)}{4r^{3/2}(t)}$. They in turn wrote this as the system

$$\dot{x} = z - \theta(t)x,$$
$$\dot{z} = -\theta(t)z + [\theta^2(t) + \dot{\theta}(t) - 1]x.$$

Transformations were also used early in the development of the nonlinear limit-point and nonlinear limit-circle properties as can be seen in the papers of Graef [98, 99], Spikes [155, 156], and Graef and Spikes [104]. For example, a special case of the equation studied in [98] is

$$(a(t)y')' + r(t)y^{2k-1} = 0, \tag{8.1.3}$$

where $k > 1$ is a positive integer. Defining $\alpha = 1/2(k+1)$ and $\beta = (2k+1)/2(k+1)$, under the transformation

$$s = \int_0^t [r^\alpha(u)/a^\beta(u)]du, \quad x(s) = y(t), \tag{T_3}$$

equation (8.1.3) becomes

$$\ddot{x} + \alpha p(t)\dot{x} + P(t)x^{2k-1} = 0, \tag{8.1.4}$$

where

$$p(t) = (a(t)r(t))'/a^\alpha(t)r^{\alpha+1}(t) \quad \text{and} \quad P(t) = (a(t)r(t))^{\beta-\alpha}.$$

Note that

$$\beta - \alpha = 2\beta - 1 = k/(k+1).$$

The form of this transformation is motivated by the shape of the nonlinear term in equation (8.1.3). When $k = 1$, it does not reduce to the transformations (T_1) or (T_2) above. Equation (8.1.4) was then written as the system

$$\begin{aligned} \dot{x} &= z - \alpha p(t)x, \\ \dot{z} &= \alpha\dot{p}(t)x - P(t)x^{2k-1}. \end{aligned} \tag{8.1.5}$$

In the next section, we discuss the use of transformations in obtaining nonlinear limit-point and nonlinear limit-circle solutions.

8.2 Preliminaries

Here we will consider the Emden–Fowler equation

$$(a(t)y')' + r(t)|y|^{\lambda}\mathrm{sgn}\, y = 0, \tag{8.2.1}$$

where $\lambda > 0$, $\lambda \neq 1$, $a(t) > 0$, $r(t) > 0$, and a, $r \in AC^1_{\mathrm{loc}}(\mathbb{R}_+)$. As we will see, since a and r belong to $AC^1_{\mathrm{loc}}(\mathbb{R}_+)$, all nontrivial solutions of (8.2.1) are proper, that is, there are no singular solutions (see Lemma 8.2.1(i) below or Lemma 2.3.1 above).

It is well known that oscillatory and other asymptotic properties of solutions of equation (8.2.1) may differ depending on whether

$$\int_0^{\infty} \frac{d\sigma}{a(\sigma)} = \infty \tag{8.2.2}$$

or

$$\int_0^{\infty} \frac{d\sigma}{a(\sigma)} < \infty. \tag{8.2.3}$$

We will try to take a unified approach by first observing that if (8.2.2) holds, then the transformation

$$s = \int_0^{t} \frac{d\sigma}{a(\sigma)}, \quad y(t) = w(s) \tag{T_4}$$

applied to equation (8.2.1) yields

$$\ddot{w}(s) + a(t)r(t)|w^{\lambda}|\mathrm{sgn}\, w = 0, \quad t = t(s), \quad s \geq s_0 = 0,$$

where again "$\cdot$" denotes d/ds. In this case

$$\int_0^{\infty} |y(t)|^{\lambda+1} dt = \int_{s_0}^{\infty} |w(s)|^{\lambda+1} a(t(s)) ds. \tag{8.2.4}$$

Now if (8.2.3) holds, then the transformation

$$s = \left(\int_t^{\infty} \frac{d\sigma}{a(\sigma)}\right)^{-1}, \quad sy(t) = v(s) \tag{T_5}$$

applied to equation (8.2.1) gives

$$\ddot{v}(s) + a(t)r(t)\left(\int_t^{\infty} \frac{d\sigma}{a(\sigma)}\right)^{3+\lambda} |v|^{\lambda}\mathrm{sgn}\, z = 0$$

for

$$s \geq s_0 = \left(\int_0^\infty \frac{d\sigma}{a(\sigma)} \right)^{-1} > 0.$$

We then have

$$\int_0^\infty |y^{\lambda+1}(t)| dt = \int_{s_0}^\infty \frac{|v(s)|^{\lambda+1}}{s^{\lambda+1}} a(t(s)) \left(\int_{t(s)}^\infty \frac{d\sigma}{a(\sigma)} \right)^2 ds$$

$$= \int_{s_0}^\infty |v(s)|^{\lambda+1} a(t(s)) \left(\int_{t(s)}^\infty \frac{d\sigma}{a(\sigma)} \right)^{\lambda+3} ds. \quad (8.2.5)$$

Hence, in both cases we need to study conditions under which the equation

$$z'' + Q(s)|z|^\lambda \operatorname{sgn} z = 0, \quad Q > 0, \quad \lambda > 0, \quad \lambda \neq 1, \quad (8.2.6)$$

has solutions satisfying

$$\int_{s_0}^\infty |z(s)|^{\lambda+1} A(s) ds < \infty \quad (\text{or } = \infty),$$

where

$$Q(s) = a(t(s)) r(t(s)), \quad A(s) = a(t(s))$$

if (8.2.2) holds, and

$$Q(s) = a(t(s)) r(t(s)) \left(\int_{t(s)}^\infty \frac{d\sigma}{a(\sigma)} \right)^{\lambda+3}, \ A(s) = a(t(s)) \left(\int_{t(s)}^\infty \frac{d\sigma}{a(\sigma)} \right)^{\lambda+3}$$

if (8.2.3) holds. Notice that in both cases,

$$\frac{Q(s)}{A(s)} = r(t(s)).$$

The following lemmas will be used in the proofs of our main theorems. We will make use of the following condition. Note the similarity to conditions used in Chapter 3 and other places. For any function $Q \in AC^1_{\text{loc}}[s_0, \infty)$ with $Q > 0$ on $[s_0, \infty)$, let

$$g(s) = -\frac{Q'(s)}{Q^{(\lambda+5)/(\lambda+3)}(s)}.$$

We then will often ask that

$$\int_{s_0}^\infty |g'(s)| ds < \infty. \quad (8.2.7)$$

For a solution z of (8.2.6), we define

$$F(s) = Q^{-2/(\lambda+3)}(s)(z'(s))^2 + \frac{2}{\lambda+1}Q^{(\lambda+1)/(\lambda+3)}(s)|z(s)|^{\lambda+1}, \quad s \geq s_0, \tag{8.2.8}$$

and

$$G(s) = F(s)\left(\int_{s_0}^{s} |g'(s)|ds\right)^{-2(\lambda+1)/(\lambda-1)}, \quad s \geq s_0. \tag{8.2.9}$$

Our next lemma is really Lemma 2.3.1 restated using the notation for this setting.

Lemma 8.2.1. *Let z be a solution of (8.2.6). Then:*

(i) *z is defined on $[s_0, \infty)$;*
(ii) *we have the estimates*

$$|z(s)| \leq CQ^{-1/(\lambda+3)}(s)F^{1/(\lambda+1)}(s) \tag{8.2.10}$$

and

$$|z'(s)| \leq Q^{1/(\lambda+3)}(s)F^{1/2}(s), \quad s \geq s_0, \tag{8.2.11}$$

where $C = \left(\frac{\lambda+1}{2}\right)^{1/(\lambda+1)}$;
(iii) *for $s_0 \leq \bar{s} < s$, we have*

$$F(s) = F(\bar{s}) - \frac{2}{\lambda+3}g(\bar{s})z(\bar{s})z'(\bar{s}) + \frac{2}{\lambda+3}g(s)z(s)z'(s)$$
$$-\frac{2}{\lambda+3}\int_{\bar{s}}^{s} z(\sigma)z'(\sigma)g'(\sigma)d\sigma. \tag{8.2.12}$$

The proof of the following lemma is quite similar to that of Lemma 2.3.2 and will be omitted.

Lemma 8.2.2. *Let (8.2.7) hold. Then there exist a solution z of (8.2.6) and a constant C_0 such that*

$$0 < 3C_0/4 = C_1 \leq F(s) \leq C_2 = 3C_0/2 \quad \textit{for } s \in [s_0, \infty).$$

If $\lambda < 1$, then C_0 can be chosen to be arbitrarily small.

Lemma 8.2.3. *Suppose that (8.2.7) holds,*

$$\int_{s_0}^{\infty} \frac{A(s)ds}{Q^{(\lambda+1)/(\lambda+3)}(s)} = \infty, \tag{8.2.13}$$

and either

(a)

$$\int_0^\infty \frac{|r'(t)|dt}{r^2(t)} < \infty \tag{8.2.14}$$

holds,

(b) $\lambda < 1$ *and*

$$\int_{s_0}^\infty \frac{|A'(s)|ds}{Q(s)} < \infty \tag{8.2.15}$$

holds, or

(c) *condition* (8.2.15) *holds and* $\lim_{s\to\infty} g(s) = 0$.

Then we have the following:

(i) *If* (8.2.2) *holds, then there exists a solution* z *of* (8.2.6) *such that*

$$\int_{s_0}^\infty A(s)|z(s)|^{\lambda+1}ds = \infty. \tag{8.2.16}$$

(ii) *If* (8.2.3) *holds and*

$$r(t) \geq r_0 > 0, \tag{8.2.17}$$

then the conclusion in part (i) *holds.*

Proof. Lemma 8.2.2 yields the existence of a solution z of (8.2.6) and a constant C_0 such that

$$0 < 3C_0/4 = C_1 \leq F(s) \leq C_2 = 3C_0/2 \quad \text{for } s \geq s_0. \tag{8.2.18}$$

We will show that (8.2.16) holds. Suppose, to the contrary, that

$$\int_{s_0}^\infty A(s)|z(s)|^{\lambda+1}ds < \infty. \tag{8.2.19}$$

If we multiply $F(s)$ by $A(s)/Q^{(\lambda+1)/(\lambda+3)}(s)$, then, in view of (8.2.13) and (8.2.18), we see that

$$\int_{s_0}^\infty \frac{(z'(s))^2A(s)}{Q(s)} = \infty. \tag{8.2.20}$$

By Lemma 8.2.1(ii) and (8.2.18), we have

$$|z(s)z'(s)| \leq \left(\frac{\lambda+1}{2}\right)^{1/(\lambda+1)} (C_2)^{(\lambda+3)/(2(\lambda+1))} = M_1 \tag{8.2.21}$$

for $s \geq s_0$. Define

$$J(s) = \int_{s_0}^s \frac{(z'(s))^2A(s)}{Q(s)}ds - \int_{s_0}^s \frac{(d/ds)r(t(s))}{r^2(t(s))}z(s)z'(s)ds.$$

If (8.2.14) holds, then (8.2.20) and (8.2.21) imply

$$J(s) \geq \int_{s_0}^{s} \frac{(z'(s))^2 A(s)}{Q(s)} ds - M_1 \int_{s_0}^{s} \frac{|(d/ds)r(t(s))|ds}{r^2(t(s))}$$
$$\geq \int_{s_0}^{s} \frac{(z'(s))^2 A(s)}{Q(s)} ds - M_1 \int_{0}^{\infty} \frac{|(d/dt)r(t)|}{r^2(t)} dt \to \infty$$

as $s \to \infty$.

Let either (b) or (c) hold. Then (8.2.8), (8.2.19), and (8.2.20) imply the existence of $s_1 \geq s_0$ such that

$$\int_{s_0}^{s} \frac{(z'(s))^2 A(s)}{Q(s)} ds \geq \frac{C_1}{2} \int_{s_0}^{s} \frac{A(s)}{Q(s)^{(\lambda+1)/(\lambda+3)}} ds, \ s \geq s_1.$$

From this, (8.2.13), and (8.2.21), we have

$$J(s) \geq \frac{C_1}{2} \int_{s_0}^{s} \frac{A(s)}{Q(s)^{(\lambda+1)/(\lambda+3)}} ds - M_1 \int_{s_0}^{s} \frac{|(Q(s)/A(s))'|}{(Q(s)/A(s))^2} ds$$
$$\geq \frac{C_1}{2} \int_{s_0}^{s} \frac{A(s)}{Q^{(\lambda+1)/(\lambda+3)}(s)} ds - M_1 \int_{s_0}^{s} \frac{|g(s)|A(s)}{Q^{(\lambda+1)/(\lambda+3)}(s)} ds$$
$$-M_1 \int_{s_0}^{\infty} \frac{|A'(s)|}{Q(s)} ds.$$

By Lemma 3.3.1, $\lim_{s\to\infty} g(s) = M \in [0, \infty)$. If $M = 0$, then there exists $s_1 \geq s_0$ such that $|g(s)| \leq C_1/4M_1$ for $s \geq s_1$. If $M > 0$, then $\lambda < 1$, and according to Lemma 8.2.2, C_0 can be taken to be arbitrarily small. Thus, in view of (8.2.18) and (8.2.21), C_1/M_1 can be chosen arbitrarily large so that $|g(s)| \leq C_1/4M_1$ and we can set $s_1 = s_0$.

Hence, in either case, (8.2.13) and (8.2.15) imply

$$\lim_{s\to\infty} J(s) \geq \lim_{s\to\infty} \left[\frac{C_1}{2} \int_{s_1}^{s} \frac{A(s)}{Q^{(\lambda+1)/(\lambda+3)}(s)} ds - \frac{C_1}{4} \int_{s_1}^{s} \frac{A(s)ds}{Q^{(\lambda+1)/(\lambda+3)}(s)} \right.$$
$$\left. - M_1 \int_{s_0}^{s_1} \frac{|g(s)|A(s)}{Q^{(\lambda+1)/(\lambda+3)}(s)} ds - M_1 \int_{s_0}^{\infty} \frac{|A'(s)|}{Q(s)} ds \right] = \infty.$$

Thus, if either (8.2.14) or (8.2.15) holds, we have

$$\lim_{s\to\infty} J(s) = \infty. \tag{8.2.22}$$

It follows from (8.2.6) that

$$\infty > \int_{s_0}^{\infty} |z(s)|^{\lambda+1} A(s)ds \geq -\int_{s_0}^{s} \frac{z(s)z''(s)A(s)}{Q(s)} ds$$

$$= -\left.\frac{A(s)z(s)z'(s)}{Q(s)}\right|_{s_0}^{s} + \int_{s_0}^{s} z'(s)\left(\frac{z(s)A(s)}{Q(s)}\right)' ds$$

$$= -\frac{A(s)z(s)z'(s)}{Q(s)} + \frac{A(s_0)z(s_0)z'(s_0)}{Q(s_0)} + J(s), \quad s \geq s_0. \quad (8.2.23)$$

If z is oscillatory, let $\{s_k\}_1^\infty \to \infty$ be a sequence of zeros of z. Then, letting $s = s_k$ in (8.2.23) contradicts (8.2.22) for large k. If z is nonoscillatory, then $z(s)z'(s) > 0$ for large s. If (8.2.2) holds, then in view of (T_4), we see that $|y(t)|$ is increasing as well, so the conclusion follows from (8.2.4) since the integral on the left in (8.2.4) diverges. If (8.2.3) and (8.2.17) hold, then $|A(s)z(s)z'(s)/Q(s)|$ is bounded, and again (8.2.23) contradicts (8.2.22). □

The following result is due to Kiguradze and Chanturia [122, Theorem 20.4].

Lemma 8.2.4. *Let Q be nondecreasing,*

$$\int_{s_0}^{\infty} \left|\left(\frac{Q'(s)}{Q^{3/2}(s)}\right)'\right| ds < \infty \quad \textit{if } 0 < \lambda < 1$$

and

$$\int_{s_0}^{\infty} \left|\left(\frac{Q'(s)}{Q^{(\lambda+2)/(\lambda+1)}(s)}\right)'\right| ds < \infty \quad \textit{if } \lambda > 1.$$

Then (8.2.7) holds, and for every nontrivial solution z of (8.2.6), we have $\lim_{s\to\infty} F(s) = C \in (0, \infty)$.

Lemma 8.2.5. *Let one of the following conditions hold:*

(i) *$\lambda > 1$, g not be identically a constant on $\mathbb{R}_+$, and*

$$\int_{s_0}^{\infty} \frac{A(\sigma)}{Q^{(\lambda+1)/(\lambda+3)}(\sigma)} \times \left(\int_{s_0}^{\sigma} \left|\left(\frac{Q'(\alpha)}{Q^{(\lambda+5)/(\lambda+3)}(\alpha)}\right)'\right| d\alpha\right)^{2(\lambda+1)/(\lambda-1)} d\sigma < \infty;$$

(ii) $\lambda > 1$, (8.2.7) *holds, and*

$$\int_{s_0}^{\infty} \frac{A(s)ds}{Q^{(\lambda+1)/(\lambda+3)}(s)} < \infty; \tag{8.2.24}$$

(iii) $0 < \lambda < 1$, (8.2.24) *holds, Q is nondecreasing on* $[s_0, \infty)$, *and*

$$\int_{s_0}^{\infty} \left| \left(\frac{Q'(s)}{Q^{3/2}(s)} \right)' \right| ds < \infty.$$

Then every solution z of (8.2.6) *satisfies*

$$\int_{s_0}^{\infty} A(s)|z(s)|^{\lambda+1} ds < \infty. \tag{8.2.25}$$

Proof. (i) Let z be an arbitrary solution of (8.2.6). By Lemma 2.3.3, G is bounded, so (8.2.25) follows from (8.2.8) and the hypotheses of the lemma. Part (ii) is a special case of (i). To prove (iii), we see that by Lemma 8.2.4, F is bounded for every solution z of (8.2.6). The proof then follows from (8.2.8) and (8.2.24). □

8.3 Limit-Point and Limit-Circle Results

We are now ready to prove our nonlinear limit-point and nonlinear limit-circle results based on using the transformations discussed above.

Theorem 8.3.1. *Let $\lambda > 1$ and g not be identically a constant on $\mathbb{R}_+$.*

(i) *If* (8.2.2) *holds and*

$$\int_1^{\infty} (a(t)r(t))^{-(\lambda+1)/(\lambda+3)} \times \left(\int_0^t \left| \left[\frac{(a(\sigma)r(\sigma))'}{a^{2/(\lambda+3)}(\sigma)r^{(\lambda+5)/(\lambda+3)}(\sigma)} \right]' \right| d\sigma \right)^{2(\lambda+1)/(\lambda-1)} dt < \infty,$$

then (8.2.1) *is of the nonlinear limit-circle type.*
If

$$\int_0^{\infty} \left| \left[\frac{(a(t)r(t))'}{a^{2/(\lambda+3)}(t)r^{(\lambda+5)/(\lambda+3)}(t)} \right]' \right| dt < \infty, \tag{8.3.1}$$

$$\int_0^\infty \frac{dt}{(a(t)r(t))^{(\lambda+1)/(\lambda+3)}} = \infty, \tag{8.3.2}$$

and either

$$\int_0^\infty \frac{|r'(t)|}{r^2(t)} dt < \infty \tag{8.3.3}$$

or

$$\int_0^\infty \frac{|a'(t)|}{a(t)r(t)} dt < \infty \quad \text{and} \quad \lim_{s\to\infty} \frac{(a(s)r(s))'}{a(s)^{2/(\lambda+3)}r^{(\lambda+5)/(\lambda+3)}(s)} = 0, \tag{8.3.4}$$

then (8.2.1) *is of the nonlinear limit-point type.*

(ii) *If* (8.2.3) *and* (8.2.17) *hold and*

$$\int_1^\infty (a(t)r(t))^{-\frac{\lambda+1}{\lambda+3}} \times \left(\int_0^t \left| \left[\frac{[a(\sigma)r(\sigma)(\int_\sigma^\infty \frac{d\alpha}{a(\alpha)})^{\lambda+3}]'}{a^{\frac{2}{\lambda+3}}(\sigma)r^{\frac{\lambda+5}{\lambda+3}}(\sigma)(\int_\sigma^\infty \frac{d\alpha}{a(\alpha)})^{\lambda+3}} \right]' \right| d\sigma \right)^{\frac{2(\lambda+1)}{(\lambda-1)}} dt < \infty, \tag{8.3.5}$$

then (8.2.1) *is of nonlinear limit-circle type.*
If, in addition to (8.3.2),

$$\int_0^\infty \left| \left[\frac{\left[a(\sigma)r(\sigma)\left(\int_\sigma^\infty d\alpha/a(\alpha)\right)^{\lambda+3}\right]'}{a^{2/(\lambda+3)}(\sigma)r^{(\lambda+5)/(\lambda+3)}(\sigma)\left(\int_\sigma^\infty d\alpha/a(\alpha)\right)^{\lambda+3}} \right]' \right| d\sigma < \infty \tag{8.3.6}$$

holds, and either (8.3.3) *holds or*

$$\int_0^\infty \frac{1}{a(t)r(t)} \left| a'(t) - \frac{\lambda+3}{\int_t^\infty d\sigma/a(\sigma)} \right| dt < \infty \tag{8.3.7}$$

and

$$\lim_{s\to\infty} \frac{\left[a(s)r(s)\left(\int_s^\infty d\alpha/a(\alpha)\right)^{\lambda+3}\right]'}{a^{2/(\lambda+3)}(s)r^{(\lambda+5)/(\lambda+3)}(s)\left(\int_s^\infty d\alpha/a(\alpha)\right)^{\lambda+3}} = 0$$

hold, then (8.2.1) *is of the nonlinear limit-point type.*

Proof. Recall that $' = d/dt$ and $\cdot = d/ds$ and observe that

$$\int_{s_0}^{\infty} \frac{A(s)ds}{Q^{(\lambda+1)/(\lambda+3)}(s)} = \int_{s_0}^{\infty} \frac{a(t)ds}{(a(t)r(t))^{(\lambda+1)/(\lambda+3)}} = \int_{0}^{\infty} \frac{dt}{(a(t)r(t))^{(\lambda+1)/(\lambda+3)}},$$

$$\int_{s_0}^{s} \left| \left(\frac{\dot{Q}(s)}{Q^{(\lambda+5)/(\lambda+3)}(s)} \right)^{\cdot} \right| ds = \int_{s_0}^{s} \left| \left[\frac{(a(t)r(t))'a(t)}{(a(t)r(t))^{(\lambda+5)/(\lambda+3)}} \right]' \right| a(t)ds = \int_{0}^{t} \left| \left(\frac{(a(t)r(t))'}{a^{2/(\lambda+3)}(t)r^{(\lambda+5)/(\lambda+3)}(t)} \right)' \right| dt,$$

and

$$\int_{s_0}^{\infty} \frac{|\dot{A}(s)|ds}{Q(s)} = \int_{s_0}^{\infty} \frac{|a'(t)|a(t)}{a(t)r(t)} ds = \int_{0}^{\infty} \frac{|a'(t)|}{a(t)r(t)} dt.$$

The conclusions of part (i) then follow from (8.2.4) and Lemmas 8.2.3 and 8.2.5.

Similarly, in case (ii), we see that

$$\int_{s_0}^{\infty} \frac{A(s)ds}{Q^{(\lambda+1)/(\lambda+3)}(s)}$$

$$= \int_{s_0}^{\infty} \frac{a(t)\left(\int_t^{\infty} d\sigma/a(\sigma)\right)^{\lambda+3}}{\left[a(t)r(t)\left(\int_t^{\infty} d\sigma/a(\sigma)\right)^{\lambda+3}\right]^{(\lambda+1)/(\lambda+3)}} ds$$

$$= \int_{0}^{\infty} \frac{dt}{(a(t)r(t))^{(\lambda+1)/(\lambda+3)}},$$

$$\int_{s_0}^{s} \left| \left(\frac{\dot{Q}(s)}{Q^{(\lambda+5)/(\lambda+3)}(s)} \right)^{\cdot} \right| ds$$

$$= \int_{s_0}^{s} \left| \left[\frac{\left[a(t)r(t)\left(\int_t^{\infty} d\sigma/a(\sigma)\right)^{\lambda+3}\right]' a(t)\left(\int_t^{\infty} d\sigma/a(\sigma)\right)^2}{\left[a(t)r(t)\left(\int_t^{\infty} d\sigma/a(\sigma)\right)^{\lambda+3}\right]^{(\lambda+5)/(\lambda+3)}} \right]^{\cdot} \right| ds$$

$$= \int_{0}^{t} \left| \left[\frac{\left[a(t)r(t)\left(\int_t^{\infty} d\sigma/a(\sigma)\right)^{\lambda+3}\right]'}{a^{2/(\lambda+3)}(t)r^{(\lambda+5)/(\lambda+3)}(t)\left(\int_t^{\infty} d\sigma/a(\sigma)\right)^{\lambda+3}} \right]' \right| dt,$$

and

$$\int_{s_0}^{\infty} \frac{|\dot{A}(s)|ds}{Q(s)} = \int_0^{\infty} \left| \frac{a'(t) - (\lambda+3)\left(\int_t^{\infty} d\sigma/a(\sigma)\right)^{-1}}{a(t)r(t)} \right| dt.$$

We then use (8.2.5) and Lemmas 8.2.3 and 8.2.5. □

Theorem 8.3.2. *Let $\lambda > 1$, r be nondecreasing, and let one of the following assumptions hold:*

(i) *Condition* (8.2.2) *holds, ar is nondecreasing, and*

$$\int_0^{\infty} \left| \left[\frac{(a(t)r(t))'}{a^{\frac{1}{\lambda+1}}(t) r^{\frac{\lambda+2}{\lambda+1}}(t)} \right]' \right| dt < \infty. \tag{8.3.8}$$

(ii) *Condition* (8.2.3) *holds, $a(t)r(t)(\int_t^{\infty} d\sigma/a(\sigma))^{\lambda+3}$ is nondecreasing, and*

$$\int_0^{\infty} \left| \left[\frac{\left(a(t)r(t)\left(\int_t^{\infty} \frac{d\sigma}{a(\sigma)}\right)^{\lambda+3}\right)'}{a^{\frac{1}{\lambda+3}}(t) r^{\frac{\lambda+2}{\lambda+1}}(t) \left(\int_t^{\infty} \frac{d\sigma}{a(\sigma)}\right)^{\frac{\lambda^2+3\lambda+4}{\lambda+1}}} \right]' \right| dt < \infty. \tag{8.3.9}$$

Then (8.2.1) *is of the nonlinear limit-circle type if and only if*

$$\int_0^{\infty} \frac{dt}{(a(t)r(t))^{(\lambda+1)/(\lambda+3)}} < \infty. \tag{8.3.10}$$

Proof. Conditions (8.3.8) and (8.3.9) ensure that

$$\int_0^{\infty} \left| \left[\frac{Q'(s)}{Q^{(\lambda+2)/(\lambda+1)}(s)} \right]' \right| ds < \infty,$$

and hence Lemma 8.2.4 implies that (8.2.7) holds. The remainder of the proof is similar to the proof of Theorem 8.3.1. Notice that (8.2.14) holds since r is nondecreasing. □

Theorem 8.3.3. *Let $0 < \lambda < 1$ and one of the following assumptions hold:*

(i) (8.2.2), (8.3.1), (8.3.2), *and either* (8.3.3) *or* (8.3.4) *hold;*
(ii) (8.2.3), (8.2.17), (8.3.2), (8.3.6), *and either* (8.3.3) *or* (8.3.7) *hold.*

Then (8.2.1) *is of the nonlinear limit-point type.*

Proof. The proof is similar to the proof of Theorem 8.3.1 since conditions (8.3.1) and (8.3.6) guarantee that (8.2.7) holds. □

The proof of the following result is similar to the proof of Theorem 8.3.2.

Theorem 8.3.4. *Let $0 < \lambda < 1$, r be nondecreasing, and let one of the following assumptions hold:*

(i) *Condition* (8.2.2) *holds, ar is nondecreasing, and*

$$\int_0^\infty \left| \left(\frac{(a(t)r(t))'}{a^{1/2}(t)r^{3/2}(t)} \right)' \right| dt < \infty. \tag{8.3.11}$$

(ii) *Condition* (8.2.3) *holds, $a(t)r(t)(\int_t^\infty d\sigma/a(\sigma))^{\lambda+3}$ is nondecreasing, and*

$$\int_0^\infty \left| \left(\frac{\left(a(t)r(t)\left(\int_t^\infty d\sigma/a(\sigma)\right)^{\lambda+3}\right)'}{a^{1/2}(t)r^{3/2}(t)\left(\int_t^\infty d\sigma/a(\sigma)\right)^{(3\lambda+5)/2}} \right)' \right| dt < \infty. \tag{8.3.12}$$

Then (8.2.1) *is of the nonlinear limit-circle type if and only if* (8.3.10) *holds.*

Remark 8.3.1. It is interesting to compare Theorems 8.3.1(i) to Theorem 2.4.1 and Theorem 8.3.3(i) to Theorem 2.4.2. If (8.2.2) holds, we obtain the same results as we do by not using the transformation. If (8.2.3) holds, the results are different.

Next, we apply our results to the case $a \equiv 1$, i.e., to the equation

$$y'' + r(t)|y|^\lambda \operatorname{sgn} y = 0, \quad \lambda > 0, \quad \lambda \neq 1. \tag{8.3.13}$$

Corollary 8.3.1. *Let $\lambda > 1$ and*

$$\int_1^\infty r^{-(\lambda+1)/(\lambda+3)}(t) \times \left(\int_0^t \left| \frac{r''(\sigma)}{r^{(\lambda+5)/(\lambda+3)}(\sigma)} - \frac{\lambda+5}{\lambda+3} \frac{(r'(\sigma))^2}{r^{2(\lambda+4)/(\lambda+3)}(\sigma)} \right| d\sigma \right)^{2(\lambda+1)/(\lambda-1)} dt < \infty. \tag{8.3.14}$$

Then (8.3.13) *is of the nonlinear limit-circle type.*

Corollary 8.3.2. *Let one of the following assumptions hold:*

(i) $\lambda > 1$,

$$\int_0^\infty \left| \frac{r''(\sigma)}{r^{(\lambda+5)/(\lambda+3)}(\sigma)} - \frac{\lambda+5}{\lambda+3} \frac{(r'(\sigma))^2}{r^{2(\lambda+4)/(\lambda+3)}(\sigma)} \right| d\sigma < \infty, \tag{8.3.15}$$

and either

$$\int_0^\infty \frac{|r'(t)|}{r^2(t)} dt < \infty \quad \textit{or} \quad \lim_{t\to\infty} \frac{r'(t)}{r^{(\lambda+5)/(\lambda+3)}(t)} = 0; \tag{8.3.16}$$

(ii) $\lambda > 1$, *r is nondecreasing, and*

$$\int_0^\infty \left| \frac{r''(\sigma)}{r^{(\lambda+2)/(\lambda+1)}(\sigma)} - \frac{\lambda+2}{\lambda+1} \frac{(r'(\sigma))^2}{r^{(2\lambda+3)/(\lambda+1)}(\sigma)} \right| d\sigma < \infty; \tag{8.3.17}$$

(iii) $0 < \lambda < 1$, *r is nondecreasing, and*

$$\int_0^\infty \left| \left(\frac{r'(t)}{r^{3/2}(t)} \right)' \right| dt < \infty. \tag{8.3.18}$$

Then (8.3.13) *is of the nonlinear limit-circle type if and only if*

$$\int_0^\infty \frac{dt}{r^{(\lambda+1)/(\lambda+3)}(t)} < \infty. \tag{8.3.19}$$

Proof. The corollary follows from Theorems 8.3.1, 8.3.2, and 8.3.4. Note that (8.3.3) holds in cases (ii) and (iii). □

8.4 Examples and Discussion

We begin this section with some examples to illustrate our results and to compare them to what is already known.

Example 8.4.1. Consider the equation

$$y'' + t^\delta |y|^\lambda \operatorname{sgn} y = 0, \quad \lambda > 0, \quad \lambda \neq 1, \quad \delta \geq 0. \tag{8.4.1}$$

By 8.3.2 (all three parts hold), equation (8.4.1) is of the nonlinear limit-circle type if and only if

$$\delta > \frac{\lambda+3}{\lambda+1}.$$

Example 8.4.2. Consider the equation

$$y'' + t^\delta |y|^\lambda \operatorname{sgn} y = 0, \quad \lambda > 1, \quad \delta \geq 0. \tag{8.4.2}$$

Assume that

$$\delta < \frac{2(\lambda+3)}{\lambda-1}. \tag{8.4.3}$$

By [25, Corollary, 3.2](also see the discussion on p. 325 in [98]), equation (8.4.2) is of the nonlinear limit-circle type if and only if

$$\delta > \frac{\lambda + 3}{\lambda + 1}.$$

However, Corollary 8.3.1 above applies without requiring condition (8.4.3).

Example 8.4.3. Consider the equation

$$y'' + e^t |y|^\lambda \operatorname{sgn} y = 0, \quad \lambda \geq 3. \tag{8.4.4}$$

By Corollary 2 (any part), equation (8.4.4) is of the nonlinear limit-circle type, but Corollary 3.2 in [25] does not apply (condition (3.42) does not hold).

The results in this section do not hold for linear equations, i.e., the case $\lambda = 1$. However, it is interesting to blindly let $\lambda = 1$ in some of our conditions such as (8.3.1), (8.3.2), (8.3.4), (8.3.8), (8.3.10), (8.3.11), (8.3.15), (8.3.17), and (8.3.19), and compare what results to the following well-known theorems of Dunford and Schwartz [72] for the equation

$$(a(t)y')' + r(t)y = 0. \tag{8.4.5}$$

Theorem 8.4.1. *Assume that*

$$\int_0^\infty \left| \left[\frac{(a(u)r(u))'}{a^{1/2}(u)r^{3/2}(u)} \right]' + \frac{\{[a(u)r(u)]'\}^2}{4a^{3/2}(u)r^{5/2}(u)} \right| du < \infty. \tag{8.4.6}$$

If

$$\int_0^\infty [1/(a(u)r(u))^{1/2}] du < \infty, \tag{8.4.7}$$

then equation (8.4.5) *is of the* (*linear*) *limit-circle type, i.e., every solution* $y(t)$ *of* (8.4.5) *satisfies*

$$\int_0^\infty y^2(u) du < \infty.$$

Their corresponding limit-point result is the following.

Theorem 8.4.2. *Assume that* (8.4.6) *holds. If*

$$\int_0^\infty [1/(a(u)r(u))^{1/2}] du = \infty, \tag{8.4.8}$$

then equation (8.4.5) *is of the* (*linear*) *limit-point type, i.e., there is a solution* $y(t)$ *of* (8.4.5) *such that*

$$\int_0^\infty y^2(u)du = \infty.$$

For example, with $\lambda = 1$, conditions (8.3.1) and (8.3.8) both become

$$\int_0^\infty \left| \left[\frac{(a(t)r(t))'}{a^{1/2}(t)r^{3/2}(t)} \right]' \right| dt < \infty, \tag{8.4.9}$$

which is the same as (8.3.11). If $a(t) \equiv 1$, then conditions (8.3.15) and (8.3.17) are precisely the expanded form of condition (8.4.6) of Dunford and Schwartz above. Also notice that with $\lambda = 1$, conditions (8.3.2) and (8.3.10) reduce to (8.4.8) and (8.4.7), respectively.

Chapter 9

Notes, Open Problems, and Future Directions

9.1 Introduction

In this chapter, we present some open problems and directions for future research involving the nonlinear limit-point and limit-circle properties of solutions. We also indicate some of the sources of the results presented in the earlier chapters in this work.

9.2 Notes and Open Problems

Chapter 2. The source of the results in this chapter are mainly the papers of Bartušek and Graef [30, 31].

Problem 2.1. Under the hypotheses of Theorem 2.4.1, are all solutions of equation (2.2.1) bounded?

Problem 2.2. Under the conditions of Theorem 2.4.4, all solutions of equation (2.2.1) are oscillatory? Are all solutions bounded as well? If not, find a counterexample, and then see what conditions can be added to the theorem to ensure that they are.

Chapter 3. The results in this chapter are based primarily on the papers of Bartušek and Graef [32–35, 37, 38, 46].

Problem 3.1. Consider equation (3.5.1) with conditions (3.5.3) and (3.5.4) holding. By Lemmas 3.6.1 and 3.6.2, $\lim_{t\to\infty} F(t) = c$ and either $c \in \mathbb{R}_+$ or $c = +\infty$. Moreover, Theorem 3.10.1(i) implies that either $c = 0$ or $c = \infty$ if the solution is nonoscillatory. Prove or give a counterexample to the conjecture that c is either 0 or ∞ if and only if y is nonoscillatory. If the conjecture is true, then it would be possible to weaken the assumptions of the type (3.6.6) in Theorem 3.7.1 or (3.6.5) in Theorem 3.8.1.

Chapter 4. The results in this chapter are based primarily on the papers of Bartušek and Graef [44, 45].

Problem 4.1. If $b(t) < 0$ on $\mathbb{R}_+$ and $p \neq q$, show that equation (4.3.1) has noncontinuable solutions. (If $p = q$, it is known that all solutions of (4.3.1) are continuable (see [48, Theorem 1]).)

Problem 4.2. Show that the damped super-half-linear equation (4.5.1) has eventually trivial solutions if $b(t) > 0$.

Problem 4.3. Compare conditions (4.3.6) and (4.5.6) and use this information to formulate a result to hold for both the super-linear and sub-linear cases by combining Theorems 4.4.2 and 4.6.1. Construct a similar result for Theorems 4.4.3 and 4.6.2.

Chapter 5. The results in this chapter are taken from the papers of M. Bartušek and J. R. Graef [36, 39].

Problem 5.1. Condition (5.3.1) in Theorem 5.3.1 implies $\alpha = 0$ in equation (5.2.1) Prove an analogous result for the case $\alpha = 1$, i.e., for $r(t) \leq -r_0 < 0$.

Problem 5.2. Prove a result analogous to Theorem 5.3.3 in case $\alpha = 1$.

Problem 5.3. Prove a result analogous to Theorem 5.3.5 in case $\alpha = 1$.

Problem 5.4. Extend Theorem 5.3.6 to the more general case of $\lambda > p$ and n even in equation (5.2.1).

Problem 5.5. Extend the results in Section 5.5 to the case of equation (5.2.1) with $n > 2$.

Problem 5.6. Extend Theorem 5.5.6 to equation (5.4.1).

Problem 5.7. Any results for equation (5.4.1) with $r(t) < 0$ would be of interest.

Problem 5.8. Add a damping term to equations (5.2.1) and (5.4.1) and obtain results analogous to those in this chapter.

Problem 5.9. Prove any of the results in this chapter for equations with a delay (see Chapters 6 and 7).

Chapter 6. The results in this chapter are taken from the papers of Bartušek and Graef [40, 41].

Problem 6.1. Consider the equation

$$\left(a(t)|y''|^{p-1}y''\right)'' = r(t)|y(\varphi(t))|^{\lambda-1}y(\varphi(t)) \tag{9.2.1}$$

and formulate definitions of strong nonlinear limit-circle and strong nonlinear limit-point solutions for equation (9.2.1). Extend any of the results in Chapter 6 to this equation. Such results would be interesting to see even in the case $p = 1$.

Problem 6.2. Consider the equation

$$\left(a(t)|y^{(n)}|^{p-1}y^{(n)}\right)^{(n)} = r(t)|y(\varphi(t))|^{\lambda-1}y(\varphi(t)) \tag{9.2.2}$$

where $n \geq 2$. Formulate definitions of strong nonlinear limit-circle and strong nonlinear limit-point solutions for equation (9.2.2). Extend any of the results in Chapter 6 to this equation. Such results would be interesting to see even in the case $p = 1$.

Problem 6.3. Add a damping term to equation (6.2.1), (9.2.1), or (9.2.2) and obtain results corresponding to those in this chapter.

Chapter 7. The papers of Bartušek and Graef [42, 43] formed the basis for much of the content in this chapter.

Problem 7.1. Consider the equation

$$\left(a(t)|y''|^{p-1}y''\right)'' + r(t)|y(\varphi(t))|^{\lambda-1}y(\varphi(t)) = 0 \tag{9.2.3}$$

and formulate definitions of strong nonlinear limit-circle and strong nonlinear limit-point solutions for equation (9.2.3). Extend any of the results in Chapter 7 to this equation. Such results would be interesting to see even in the case $p = 1$.

Problem 7.2. Consider the equation

$$\left(a(t)|y^{(n)}|^{p-1}y^{(n)}\right)^{(n)} + r(t)|y(\varphi(t))|^{\lambda-1}y(\varphi(t)) = 0 \tag{9.2.4}$$

and formulate definitions of strong nonlinear limit-circle and strong nonlinear limit-point solutions for this equation. Extend any of the results in Chapter 7 to this equation. Such results would be interesting to see even in the case $p = 1$.

Problem 7.3. Add a damping term to equation (7.2.1), (9.2.3), or (9.2.4) and obtain results corresponding to those in this chapter.

Chapter 8. The basis for this chapter is the paper of Bartušek and Graef [29].

Problem 8.1. Consider the equation

$$(a(t)y')' + r(t)|y|^\lambda \operatorname{sgn} y = e(t). \tag{9.2.5}$$

Using the transformations (T_4) and (T_5), find the equation equivalent to (8.2.6). Determine the conditions on $e(t)$ so that results corresponding to Theorems 8.3.1–8.3.4 hold.

Problem 8.2. Consider the equation

$$(a(t)y^{(n)})^{(n)} + r(t)|y|^\lambda \operatorname{sgn} y = 0. \tag{9.2.6}$$

Find a transformation similar to (T_4) or (T_5) that can be applied to equation (9.2.6). Any results that could be obtained in this way would be new and very interesting.

9.3 Directions for Future Research

Similar to the Sturm and Sturm–Picone comparison theorems for oscillation, the formulation and proof of comparison type results for the nonlinear limit-point, nonlinear limit-circle, strong nonlinear limit-point, and strong nonlinear limit-circle properties of solution would be of interest. That is, if one equation has such a property, and the functions a and r satisfy certain inequalities, then the other equation has the same property. One attempt at doing this for linear equations can be found in [154].

Little is known about the relationships between the strong nonlinear limit-point and strong nonlinear limit-circle properties and the boundedness, oscillation, and convergence to zero of solutions. For the second-order equation

$$(a(t)y')' + r(t)|y|^\lambda \operatorname{sgn} y = 0 \tag{9.3.1}$$

with $r(t) > 0$, it is known that if (9.3.1) is of the nonlinear limit-circle type and

$$\int_0^\infty \frac{ds}{a(s)} = \infty, \tag{9.3.2}$$

then all solutions are oscillatory (see Theorem 2.4.4). Patula and Waltman [147] proved that if the linear equation

$$(a(t)y')' + r(t)y = 0 \tag{9.3.3}$$

is limit-circle and

$$\int_0^\infty \frac{ds}{a^{\frac{1}{2}}(s)} = \infty, \tag{9.3.4}$$

then all solutions of (9.3.3) are oscillatory. It is not known if Theorem 2.4.4 is true if (9.3.2) is replaced by (9.3.4).

If

$$\int_0^\infty \frac{(a(s)r(s))'_-}{a(s)r(s)} ds < \infty,$$

and y is a nonoscillatory nonlinear limit-circle type solution of equation (9.3.1), then $y(t) \to 0$ as $t \to \infty$ (see [25, Theorem 5.11]). It is not known if this is true for any of the kinds of equations studied in Chapters 2–8 in this book.

Clearly, since a strong nonlinear limit-circle solution is also a nonlinear limit-circle type solution, some of these relationships will hold in the case of strong solutions as well.

Essentially nothing is known about the relationships between the nonlinear limit-point and nonlinear limit-circle properties and the boundedness, oscillation, and convergence to zero of solutions for damped equations, higher order equations, and delay equations; this clearly needs investigating. Any results of this type would be of interest. Consideration of equations with forcing terms or perturbation terms would also be of interest.

Over the years some authors have considered the problem of perturbing a linear limit-circle type equation with a nonlinear perturbation term. Wong [173], for example, compared an unperturbed nth-rder linear equation to the same equation with a perturbation term of the form $p(t, y) = p_1(t)|y|^\rho$ with $\rho \geq 1$. Extensions of these types of results would be of interest especially for higher order and delay equations.

The extension of the limit-point/limit-circle problem to difference equations, or more generally to dynamic equations on time scales, has been an elusive one. One paper on this problem is that of Weiss [167] who considered the linear equation

$$(a(t)y^\Delta)^\nabla + r(t)y = 0, \quad t \in [t_0, \infty)_\mathbb{T} = [t_0, \infty) \cap \mathbb{T}, \tag{9.3.5}$$

where $\mathbb{T}$ is a time scale with $\inf \mathbb{T} = t_0$ and $\sup \mathbb{T} = \infty$, Δ is the delta derivative, and ∇ is the nabla derivative. (If the time scale $\mathbb{T}$ is the integers, then these are the usual forward and backwards difference operators.) Limit-circle and limit-point solutions of equation (9.3.5) are defined as follows.

Definition 9.3.1. A solution y of equation (9.3.5) is of the limit-circle type if

$$\int_{t_0}^{\infty} y^2(t)\Delta t < \infty, \tag{9.3.6}$$

and it is said to be of the limit-point type otherwise, i.e., if

$$\int_{t_0}^{\infty} y^2(t)\Delta t = \infty. \tag{9.3.7}$$

Equation (9.3.5) is said to be of the limit-circle type if every solution y of (9.3.5) defined on $\mathbb{T}$ satisfies (9.3.6) and to be of the limit-point type if there is at least one solution y defined on $\mathbb{T}$ for which (9.3.7) holds.

One result in [167] is the following; it is an extension of a result of Wong and Zettl [174, Theorem 2].

Theorem 9.3.1. *Assume that* $a(t) > 0$, $r(t) > 0$, *and* $\int_{t_0}^{\infty} \frac{1}{a(t)}\Delta t = \infty$. *If equation* (9.3.5) *is nonoscillatory, then equation* (9.3.5) *is of the limit-point type, i.e., it has a limit-point type solution.*

A form of an extension of a limit-point result of Levinson (see [134, Theorem IV] or [25, Theorem 1.10]) is also presented in [167, Theorem 4.2].

Huseynov [116] also considered equation (9.3.5) and proved a time-scale version of the Weyl Alternative (Theorem 1.2.2 above) in [116, Theorems 3.1 and 5.2]. In this regard, also see [62, 63, 94, 117, 177].

To date, there do not appear to be any results on the limit-point/limit-circle problem for equations with impulses. Nor are there any results for fractional differential equations. These are areas that are clearly worth exploring even in the linear case.

Bibliography

[1] R. P. Agarwal, S. L. Shieh and C.-C. Yeh, Oscillation criteria for second-order retarded differential equations, *Math. Comput. Modelling* **26** (1977), 1–11.

[2] W. O. Amrein, A. M. Hinz and D. B. Pearson, *Sturm–Liouville Theory: Past and Present*, Birkhäuser Verlag, Basel, 2005.

[3] F. V. Atkinson, Nonlinear extensions of limit-point criteria, *Math. Z.* **130** (1973), 297–312.

[4] M. Bartušek, On asymptotic properties of oscillatory solutions of the system of differential equations of fourth order, *Arch. Math.* (*Brno*) **17** (1981), 125–136.

[5] M. Bartušek, On oscillatory solutions of the system of differential equations with deviating argument, *Czechoslovak Math. J.* **33**(110) (1986), 529–532.

[6] M. Bartušek, On oscillatory solutions of differential inequalities, *Czechoslovak Math. J.* **42**(117) (1992), 45–52.

[7] M. Bartušek, *Asymptotic Properties of Oscillatory Solutions of Differential Equations of the n-th Order*, Folia Facultatis Scientiarum Naturalium Universitatis Masarykianae Brunensis: Mathematica, Masarykova Univerzita Brno Přírodovědecká Fakulta, Vol. 3, Masaryk University, 1992.

[8] M. Bartušek, On the structure of solutions of a system of three differential inequalities, *Arch. Math.* (*Brno*) **30** (1994), 117–130.

[9] M. Bartušek, On structure of solutions of a system of four differential inequalities, *Georgian Math. J.* **2** (1995), 225–236.

[10] M. Bartušek, Oscillatory criteria for nonlinear n-th order differential equations with quasiderivatives, *Georgian Math. J.* **3** (1996), 301–314.

[11] M. Bartušek, Asymptotic behavior of oscillatory solutions of n-th order differential equations with quasiderivatives, *Czechoslovak Math. J.* **47**(122) (1997), 245–259.

[12] M. Bartušek, On existence of singular solutions on n-th order differential equations, *Arch. Math.* (*Brno*) **36** (2000), 395–404.

[13] M. Bartušek, Singular solutions for the differential equation with p-Laplacian, *Arch. Math.* (*Brno*) **41** (2005), 123–128.

[14] M. Bartušek, M. Cecchi, Z. Došlá and M. Marini, On nonoscillatory solutions of third order nonlinear differential equations, *Dynam. Systems Appl.* **9** (2000), 483–500.
[15] M. Bartušek, M. Cecchi, Z. Došlá and M. Marini, On oscillatory solutions of quasilinear differential equations, *J. Math. Anal. Appl.* **320** (2006), 108–120.
[16] M. Bartušek and Z. Došlá, On solutions of a third order nonlinear differential equation, *Nonlinear Anal.* **23** (1994), 1331–1343.
[17] M. Bartušek and Z. Došlá, Oscillatory criteria for nonlinear third order differential equations with quasiderivatives, *Differential Equations Dynam. Systems* **3** (1995), 251–268.
[18] M. Bartušek and Z. Došlá, Remark on Kneser problem, *Appl. Anal.* **56** (1995), 327–333.
[19] M. Bartušek and Z. Došlá, On the limit-point/limit-circle problem for nonlinear third order differential equations, *Math. Nachr.* **187** (1997), 5–18.
[20] M. Bartušek, Z. Došlá and J. R. Graef, On L^2 and limit-point type solutions of fourth order differential equations, *Appl. Anal.* **60** (1996), 175–187.
[21] M. Bartušek, Z. Došlá and J. R. Graef, Limit-point type results for nonlinear fourth order differential equations, *Nonlinear Anal.* **28** (1997), 779–792.
[22] M. Bartušek, Z. Došlá and J. R. Graef, Nonlinear limit-point type solutions of n-th order differential equations, *J. Math. Anal. Appl.* **209** (1997), 122–139.
[23] M. Bartušek, Z. Došlá and J. R. Graef, The nonlinear limit-point/limit-circle problem for higher order equations, *Arch. Math. (Brno)* **34** (1998), 13–22.
[24] M. Bartušek, Z. Došlá and J. R. Graef, On the definitions of the nonlinear limit-point/limit-circle properties, *Differential Equations Dynam. Systems* **9** (2001), 49–61.
[25] M. Bartušek, Z. Došlá and J. R. Graef, *The Nonlinear Limit-Point/Limit-Circle Problem*, Birkhäuser, Boston, 2004.
[26] M. Bartušek and J. R. Graef, On L^2 solutions of third order nonlinear differential equations, *Dynam. Systems Appl.* **9** (2000), 469–482.
[27] M. Bartušek and J. R. Graef, Some limit-point/limit-circle results for third order differential equations, in *Proceedings of the Third International Conference on Dynamical Systems and Differential Equations*, Discrete and Continuous Dynamical Systems, pp. 31–38 (2001).
[28] M. Bartušek and J. R. Graef, On the limit-point/limit-circle problem for second order nonlinear equations, *Nonlinear Stud.* **9** (2002), 361–369.
[29] M. Bartušek and J. R. Graef, Some limit-point and limit-circle results for second order Emden-Fowler equations, *Appl. Anal.* **83** (2004), 461–476.
[30] M. Bartušek and J. R. Graef, The nonlinear limit-point/limit-circle problem for second order equations with p-Laplacian, *Dynam. Systems Appl.* **14** (2005), 431–446.

[31] M. Bartušek and J. R. Graef, Asymptotic properties of solutions of a forced second order differential equation with p-Laplacian, *Panamer. Math. J.* **16** (2006), 41–59.

[32] M. Bartušek and J. R. Graef, The strong limit-point property for Emden-Fowler equations, *Differential Equations Dynam. Systems* **14** (2006), 383–405.

[33] M. Bartušek and J. R. Graef, The strong nonlinear limit-point/limit-circle properties for sub-half-linear equations, *Dynam. Systems Appl.* **15** (2006), 585–602.

[34] M. Bartušek and J. R. Graef, The strong nonlinear limit-point/limit-circle properties for super-half-linear equations, *Panamer. Math. J.* **17** (2007), 25–38.

[35] M. Bartušek and J. R. Graef, Asymptotic behavior of solutions of a differential equation with p-Laplacian and a forcing term, *Differential Equations Dynam. Systems* **15** (2007), 61–87.

[36] M. Bartušek and J. R. Graef, Strong nonlinear limit-point/limit-circle properties for a class of fourth order equations, *Commun. Appl. Anal.* **11** (2007), 469–484.

[37] M. Bartušek and J. R. Graef, Strong nonlinear limit-point/limit-circle properties for forced Thomas–Fermi equations with p-Laplacian, *Panamer. Math. J.* **18** (2008), 73–88.

[38] M. Bartušek and J. R. Graef, Nonlinear limit-point/limit-circle properties of solutions of second order differential equations with P-Laplacian, *Internat. J. Pure Appl. Math.* **45** (2008), 501–518.

[39] M. Bartušek and J. R. Graef, The strong nonlinear limit-point/limit-circle properties for a class of even order equations, *Comm. Appl. Nonlinear Anal.* **15** (2008), 29–45.

[40] M. Bartušek and J. R. Graef, Strong nonlinear limit point/limit-circle properties for second order differential equations with delay, *Panamer. Math. J.* **20** (2010), 31–49.

[41] M. Bartušek and J. R. Graef, Asymptotic properties of second order differential equations with delay, *Math. Eng. Sci. Aero.* **1** (2010), 337–349.

[42] M. Bartušek and J. R. Graef, Limit-point/limit-circle properties for delay equations of the second order, *Panamer. Math. J.* **21** (2011), 1–17.

[43] M. Bartušek and J. R. Graef, Limit-point/limit-circle problem for sub-half-linear second order delay differential equations, *Dynam. Systems Appl.* **20** (2011), 261–278.

[44] M. Bartušek and J. R. Graef, Limit point/limit-circle results for equations with damping, *Abstr. Appl. Anal.* **2012** (2012), Article ID 979138, 19 pp.

[45] M. Bartušek and J. R. Graef, Limit-point/limit-circle results for superlinear damped equations, *Abstr. Appl. Anal.* **2013** (2013), Article ID 784761, 10 pp.

[46] M. Bartušek and J. R. Graef, Limit-point/limit-circle results for forced second order differential equations, *Adv. Dyn. Syst. Appl.* **10** (2015), 11–26.

[47] M. Bartušek and M. Medveď, Existence of global solutions for systems of second-order functional-differential equations with p-Laplacian, *Electron. J. Differential Equations* **2008**(40) (2008), 1–8.
[48] M. Bartušek and E. Pekárková, On existence of proper solutions of quasilinear second order differential equations, *Electron. J. Qual. Theory Differ. Equ.*, **2007**(5) (2007), 1–14.
[49] E. F. Beckenbach and R. Bellman, *Inequalities*, Springer-Verlag, Berlin, 1961.
[50] R. Bellman, *Stability Theory of Differential Equations*, McGraw-Hill, New York, 1953.
[51] O. Borůvka, *Linear Differential transformationen 2. Ordung*, VEB Deutscher Verlag, Berlin, 1967.
[52] J. Burlak, On the non-existence of L_2-solutions of nonlinear differential equations, *Proc. Edinburgh Math. Soc.* **14** (1965), 257–268.
[53] T. A. Burton and W. T. Patula, Limit circle results for second order equations, *Monatsh. Math.* **81** (1976), 185–194.
[54] M. Cecchi, Z. Došlá and M. Marini, On third order differential equations with property A and B, *J. Math. Anal. Appl.* **231** (1999), 509–525.
[55] M. Cecchi, Z. Došlá and M. Marini, On decaying solutions for functional differential equations with p-Laplacian, *Nonlinear Anal.* **47** (2001), 4387–4398.
[56] M. Cecchi, Z. Došlá and M. Marini, On nonoscillatory solutions of differential equations with p-Laplacian, *Adv. Math. Sci. Appl.* **11** (2001), 419–436.
[57] M. Cecchi, Z. Došlá and M. Marini, Monotone solutions of two-dimensional nonlinear functional differential systems, *Dynam. Systems Appl.* **17** (2008), 595–608.
[58] M. Cecchi, M. Marini and G. Villari, Integral criteria for a classification of solutions of linear differential equations, *J. Differential Equations* **99** (1992), 381–397.
[59] T. A. Chanturia, On Kneser problem for systems of ordinary differential equations, *Mat. Zametki* **15** (1974), 897–906 (in Russian).
[60] T. A. Chanturia, On singular solutions of nonlinear systems of ordinary differential equations, (Colloq., Keszthely, 1974), *Colloq. Math. Soc. János Bolyai* **15** (1976), 107–119.
[61] T. A. Chanturia, On existence of singular and unbounded oscillatory solutions of differential equations of the Emden–Fowler's type, *Diff. Urav.* **28** (1992), 1009–1022 (in Russian). English translation in *Differ. Equ.* **28** (1992), 811–824.
[62] J. N. Chen, Limit-circle and limit-point tests for second order difference equations, *Math. Appl.* (*Wuhan*) **15** (2002), suppl., 1–4.
[63] J. N. Chen and Y. Shi, The limit circle and limit point criteria for second order linear difference equations, *Comput. Math. Appl.* **47** (2004), 967–976.
[64] E. A. Coddington and N. Levinson, *Theory of Ordinary Differential Equations*, McGraw–Hill, New York, 1955.
[65] W. A. Coppel, *Stability and Asymptotic Behavior of Differential Equations*, Heath, Boston, 1965.

[66] J. Detki, The solvability of a certain second order nonlinear ordinary differential equation in $L^p(0,\infty)$, *Math. Balkanica (N.S.)* **4** (1974), 115–119 (in Russian).
[67] A. Devinatz, The deficiency index of certain fourth-order ordinary self adjoint differential operators, *Quart. J. Math. Oxford Ser.* (2) **23** (1972), 267–286.
[68] A. Devinatz, The deficiency index problem for ordinary self adjoint differential operators, *Bull. Amer. Math. Soc.* (N.S.) **79** (1973), 1109–1127.
[69] Z. Došlá, On oscillatory solutions of third-order linear differential equations, *Časopis Pěst. Mat.* **114** (1989), 28–34.
[70] Z. Došlá, On square integrable solutions of third order linear differential equations, *International Scientific Conference on Mathematics* (Herl'any 1999), University of Technology, Košice, pp. 68–72, (2000).
[71] O. Došlý and P. Řehák, *Half-Linear Differential Equations*, North-Holland Mathematics Studies, Vol. 202, Elsevier, Amsterdam, 2005.
[72] N. Dunford and J. T. Schwartz, *Linear Operators; Part II: Spectral Theory*, Wiley, New York, 1963.
[73] M. S. P. Eastham, On the L^2 classification of fourth-order differential equations, *J. London Math. Soc.* **3**(2) (1971), 297–300.
[74] M. S. P. Eastham, The limit-3 case of self-adjoint differential expressions of the fourth order with oscillating coefficients, *J. London Math. Soc.* **8**(2) (1974), 427–437.
[75] M. S. P. Eastham, Self-adjoint differential equations with all solutions $L^2(0,\infty)$, in: *Differential Equations, Proceedings Uppsala* 1977, Almqvist & Wiksell, Stockholm, pp. 52–61, 1977.
[76] M. S. P. Eastham and C. G. M. Grudniewicz, Asymptotic theory and deficiency indices for fourth and higher order self-adjoint equations: a simplified approach, in: *Ordinary and Partial Differential Equations*, eds. W. N. Everitt and B. D. Sleeman, Lecture Notes in Mathematics Vol. 846, Springer-Verlag, Berlin, pp. 88–99, 1981.
[77] M. S. P. Eastham and M. L. Thompson, On the limit-point, limit-circle classification of second ordinary differential equations, *Quart. J. Math. Oxford Ser.* **24**(2) (1973), 531–535.
[78] J. Elias, On the solutions of n-th order differential equation in $L^2(0,\infty)$, in: *Qualitative Theory of Differential Equations Szeged (Hungary)*, 1979, ed. M. Farkas, Colloquia Mathematica Societatis János Bolyai, Vol. 30, North-Holland, Amsterdam, pp. 181–191, 1981.
[79] W. D. Evans and A. Zettl, Interval limit-point criteria for differential expressions and their powers, *J. London Math. Soc.* **15**(2) (1977), 119–133.
[80] W. N. Everitt, On the limit-point classification of second order differential operators, *J. London Math. Soc.* **41** (1966), 531–534.
[81] W. N. Everitt, On the limit-point classification of fourth–order differential equations, *J. London Math. Soc.* **44** (1969), 273–281.
[82] W. N. Everitt, On the limit-circle classification of second–order differential expressions, *Quart. J. Math. Oxford Ser.* **23**(2) (1972), 193–196.

[83] W. N. Everitt, On the strong limit-point condition of second–order differential expressions, in: *Proceedings of the International Conference on Differential Equations, Los Angeles*, Academic Press, New York, pp. 287–307, 1974.
[84] W. N. Everitt, On the deficiency index problem for ordinary differential operators 1910–1976, in: *Differential Equations, Proceedings Uppsala 1977*, Almqvist & Wiksell, Stockholm, pp. 62–81, 1977.
[85] W. N. Everitt, A personal history of the m-coefficient, *J. Comput. Appl. Math.* **171** (2004), 185–197.
[86] W. N. Everitt, M. Giertz and J. Weidemann, Some remarks on a separation and limit-point criterion of second-order, ordinary differential expressions, *Math. Ann.* **200** (1973), 335–346.
[87] W. N. Everitt and V. K. Kumar, On the Titchmarsh–Weyl theory of ordinary symmetric differential expressions I: The general theory, *New Archief voor Wiskunde* **34**(3) (1976), 1–48.
[88] W. N. Everitt and V. K. Kumar, On the Titchmarsh–Weyl theory of ordinary symmetric differential expressions II: the odd-order case, *New Archief voor Wiskunde* **34**(3) (1976), 109–145.
[89] X. Fan, W. -T. Li and C. Zhong, A classification scheme for positive solutions of second order nonlinear iterative differential equations, *Electron. J. Differential Equations* **2000**(25) (2000), 1–14.
[90] M. V. Fedorjuk, Asymptotics of solutions of ordinary linear differential equations of n-th order, *Dokl. Akad. Nauk SSSR* **165** (1965), 777–779 (in Russian).
[91] M. V. Fedorjuk, Asymptotic method in the theory of one–dimensional singular differential operators, *Trudi Mosk. Mat. Obsch.* **15** (1966), 296–345. (English translation: *Trans. Moscow Math. Soc.* **15** (1966), 333–386.
[92] M. V. Fedorjuk, *Asymptotic Analysis*, Springer-Verlag, New York, 1993.
[93] K. O. Friedrichs, Über die ausgezeichnete Randbedingung in der Spektraltheorie der halbbeschränkten gewöhnlichen Differential operatoren zweiter Ordnung, *Math. Ann.* **112** (1935), 1–23.
[94] Z. Q. Fu, R. W. Lu and X. F. Han, Limit circle criteria for fourth-order linear difference equations. *J. Qingdao Univ. Nat. Sci.* **19** (2006), 12–17.
[95] V. Garbušin, Inequalities of the norms of a function and its derivatives in L_p metric, *Mat. Zametki* **1** (1967), 291–298 (in Russian).
[96] I. M. Glazman, *Direct Methods of Qualitative Spectral Analysis of Singular Differential Operators*, Davey, Jerusalem, 1965.
[97] J. R. Graef, Limit circle type criteria for nonlinear differential equations, *Proc. Japan Acad. Ser A Math. Sci.* **55** (1979), 49–52.
[98] J. R. Graef, Limit circle criteria and related properties for nonlinear equations, *J. Differential Equations* **35** (1980), 319–338.
[99] J. R. Graef, Limit circle type results for sublinear equations, *Pacific J. Math.* **104** (1983), 85–94.
[100] J. R. Graef, L. Hatvani, J. Karsai and P. W. Spikes, Boundedness and asymptotic behavior of solutions of second order nonlinear differential equations, *Publ. Math. Debrecen* **36** (1989), 85–99.

[101] J. R. Graef and P. W. Spikes, Asymptotic behavior of solutions of a second order nonlinear differential equation, *J. Differential Equations* **17** (1975), 461–476.
[102] J. R. Graef and P. W. Spikes, Asymptotic properties of solutions of a second order nonlinear differential equation, *Publ. Math. Debrecen* **24** (1977), 39–51.
[103] J. R. Graef and P. W. Spikes, The limit point-limit circle problem for nonlinear equations, in: *Spectral Theory of Differential Operators*, North-Holland Mathematics Studies, Vol. 55, North-Holland, Amsterdam, pp. 207–210, 1981.
[104] J. R. Graef and P. W. Spikes, On the nonlinear limit-point/limit-circle problem, *Nonlinear Anal.* **7** (1983), 851–871.
[105] J. R. Graef and P. W. Spikes, Some asymptotic properties of solutions of $(a(t)x')' - q(t)f(x) = r(t)$, in: *Differential Equations: Qualitative Theory (Szeged*, 1984), Colloquia Mathematica Societatis János Bolyai, Vol. 47, North-Holland, Amsterdam, pp. 347–359, 1987.
[106] M. K. Grammatikopoulos and M. R. Kulenović, On the nonexistence of L^2–solutions of n-th order differential equations, *Proc. Edinburgh Math. Soc.* **24** (1981), 131–136.
[107] I. S. Grandshteyn, I. M. Ryzhik, *Table of Integrals, Series and Products*, Academic Press, Boston, 1994.
[108] M. Greguš, *Third Order Linear Differential Equations*, D. Reidel Publishiny Company, Dordrecht, 1987.
[109] R. C. Grimmer and W. T. Patula, Nonoscillatory solutions of forced second-order linear equations, *J. Math. Anal. Appl.* **56** (1976), 452–459.
[110] T. G. Hallam, On the nonexistence of L^p solutions of certain nonlinear differential equations, *Glasgow Math. J.* **8** (1967), 133–138.
[111] B. J. Harris, Limit-circle criteria for second-order differential expressions, *Quart. J. Math. Oxford* **35**(2) (1984), 415–427.
[112] P. Hartman, *Ordinary Differential Equations*, 2nd edn. Birkhäuser, Boston, 1982.
[113] P. Hartman and A. Wintner, Criteria of non-degeneracy for the wave equation, *Amer. J. Math.* **70** (1948), 295–308.
[114] P. Hartman and A. Wintner, A criteria for the non degeneracy of the wave equation, *Amer. J. Math.* **71** (1949), 206–213.
[115] D. B. Hinton, Limit point–limit circle criteria for $(py')' + qy = \lambda ky$, in: *Ordinary and Partial Differential Equations*, eds. B. D. Sleeman and I. M. Michael, Lecture Notes in Mathematics, Vol. 415, Springer-Verlag, New York, pp. 173–183, 1974.
[116] A. Huseynov, Limit point and limit circle cases for dynamic equations on time scales, *Hacet. J. Math. Stat.* **39** (2010), 379–392.
[117] A. Huseynov, Weyl's limit point and limit circle for a dynamic systems, in: *Dynamical Systems and Methods*, eds. A. C. J. Luo, J. A. T. Machado and D. Baleanu, Springer, New York, pp. 215–225, 2012.
[118] R. M. Kauffman, On the limit-n classification of ordinary differential operators with positive coefficients, *Proc. London Math. Soc.* **35** (1977), 496–526.

[119] R. M. Kauffman, T. T. Read and A. Zettl, *The Deficiency Index Problem for Powers of Ordinary Differential Expressions*, Lecture Notes in Mathematics, Vol. 621, Springer-Verlag, New York, 1977.

[120] I. T. Kiguradze, On the oscillation of solution of the equation $d^m u/dt^m + a(t)|u|^n \operatorname{sgn} u = 0$, *Mat. Sb.* **65** (1964), 172–187 (in Russian).

[121] I. T. Kiguradze, *Some Singular Boundary-Value Problems for Ordinary Differential Equations*, University of Tbilisi, 1975 (in Russian).

[122] I. T. Kiguradze and T. A. Chanturia, *Asymptotic Properties of Solutions of Nonautonomous Ordinary Differential Equations*, Kluwer Academic Publishers, Dordrecht, 1993.

[123] I. Knowles, *The Limit-Point and Limit-Circle Classification of the Sturm–Liouville Operator* $(py') + qy$, Ph.D. thesis, Flinders University of South Australia, 1972.

[124] I. Knowles, On a limit-circle criterion for second-order differential operators, *Quart. J. Math. Oxford Ser.* **24**(2) (1973), 451–455.

[125] I. Knowles, On second-order differential operators of limit circle type, in: *Ordinary and Partial Differential Equations*, eds. B. D. Sleeman and I. M. Michael, Lecture Notes in Mathematics, Vol. 415, Springer-Verlag, New York, pp. 184–187, 1974.

[126] I. Knowles, Note on a limit-circle criterion, preprint.

[127] A. M. Krall, On the solutions of $(ry')' + qy = f$, *Monatsh. Math.* **80** (1975), 115–118.

[128] A. Kroopnick, L^2-solutions to $y'' + c(t)y' + a(t)b(y) = 0$, *Proc. Amer. Math. Soc.* **39** (1973), 217–218.

[129] A. J. Kroopnick, Note on bounded L^p-solutions of a generalized Liénard equation, *Pacific J. Math.* **94** (1981), 171–175.

[130] A. J. Kroopnick, Note on a bounded L^p-solution to $x'' - a(t)x^c = 0$, *J. Math. Anal. Appl.* **113** (1986), 451–453.

[131] M. K. Kwong, On the boundedness of solutions of second order differential equations in the limit circle case, *Proc. Amer. Math. Soc.* **52** (1975), 242–246.

[132] M. K. Kwong and A. Zettl, *Norm Inequalities for Derivatives and Differences*, Lecture Notes in Mathematics, Vol. 1536, Springer-Verlag, New York, 1992.

[133] G. Ladas, Connection between oscillation and spectrum for selfadjoint differential operators of order $2n$, *Comm. Pure Appl. Math.* **22** (1969), 561–585.

[134] N. Levinson, Criteria for the limit-point cases for second order linear differential operators, *Časopis Pěst. Mat.* **74** (1949), 17–20.

[135] M. Marini and P. L. Zezza, On the asymptotic behavior of the solutions of a class of second-order linear differential equations, *J. Differential Equations* **28** (1978), 1–17.

[136] S. Matucci, On asymptotic decaying solutions for a class of second order differential equations, *Arch. Math. (Brno)* **35** (1999), 275–284.

[137] M. Medveď and E. Pekárková, Existence of global solutions for systems of second-order differential equations with p-Laplacian, *Electron. J. Differential Equations* **2007**(136) (2007), 1–9.

[138] J. D. Mirzov, *Asymptotic Properties of Solutions of Systems of Nonlinear Nonautonomous Ordinary Differential Equations*, Folia Facultatis Scientiarum Naturalium Universitatis Masarykianae Brunesis: Mathematica, Masarykova Univerzita Brno Přirodovědecká Fakulta, Vol. 14, Masaryk University, 2004.

[139] O. Mustafa and Y. V. Rogovchenko, Limit-point type results for linear differential equations, *Arch. Inequal. Appl.* **1** (2003), 377–386.

[140] O. Mustafa and Y. V. Rogovchenko, Existence of square integrable solutions of perturbed nonlinear differential equations, in: *Dynamical Systems and Differential Equations* (*Wilmington, NC*, 2002), Discrete and Continuous Dynamical Systems 2003, suppl., American Institute of Mathematical Sciences Press, Springfield, MO, USA, pp. 647–655.

[141] O. Mustafa and Y. V. Rogovchenko, Limit-point criteria for superlinear differential equations, *Bull. Belg. Math. Soc. Simon Stevin* **11** (2004), 431–440.

[142] O. Mustafa and Y. V. Rogovchenko, Limit-point type solutions of nonlinear differential equations, *J. Math. Anal. Appl.* **294** (2004), 548–559.

[143] M. A. Naimark, *Linear Differential Operators, Part II*, George Harrap & Co., Ltd, London, 1968.

[144] F. Neuman, On a problem of transformations between limit-circle and limit-point differential equations, *Proc. Roy. Soc. Edinburgh Sect. A* **72** (1973), 187–193.

[145] F. Neuman, *Global Properties of Linear Ordinary Differential Equations*, Academia, Prague, 1991.

[146] R. B. Paris and A. D. Wood, On the L^2 nature of solutions of n-th order symmetric differential equations and McLeod's conjecture, *Proc. Roy. Soc. Edinburgh Sect. A* **90** (1981), 209–236.

[147] W. T. Patula and P. Waltman, Limit point classification of second order linear differential equations, *J. London Math. Soc.* **8**(2) (1974), 209–216.

[148] W. T. Patula and J. S. W. Wong, An L^p-analogue of the Weyl alternative, *Math. Ann.* **197** (1972), 9–28.

[149] I. A. Pavljuk, Necessary and sufficient conditions for boundedness in the space $L^2(0,\infty)$ for solutions of a class of linear differential equations of second order, *Dopovidi Akad. Nauk. Ukrain. RSR* **1960** (1960), 156–158 (in Ukrainian).

[150] M. Ráb, Asymptotic formulas for the solution of linear differential equations of the second order, in: *Differential Equations and Their Applications* (*Proc. Conf., Prague*, 1962), Academia, Prague, pp. 131–135, 1963.

[151] B. Schultze, On singular differential operators with positive coefficients, *Proc. Roy. Soc. Edinburgh Sect. A* **120** (1992), 361–365.

[152] D. B. Sears, On the solutions of a second order differential equation which are square integrable, *J. London Math. Soc.* **24** (1949), 207–215.

[153] J. Shao and W. Song, Limit circle/limit point criteria for second order sublinear differential equations with damping term, *Abstr. Appl. Anal.* **2011** (2011), Article ID 803137, 12 pp.

[154] K. J. Shinde and S. M. Padhye, Comparison theorem for limit point case and limit circle case of singular Sturm–Liouville differential operators, *J. Indian Math. Soc. (N.S.)* **80** (2013), 349–355.

[155] P. W. Spikes, On the integrability of solutions of perturbed nonlinear differential equations, *Proc. Roy. Soc. Edinburgh Sect. A* **77** (1977), 309–318.

[156] P. W. Spikes, Criteria of limit circle type for nonlinear differential equations, *SIAM J. Math. Anal.* **10** (1979), 456–462.

[157] S. Staněk, Bounds for solutions of nonlinear differential equation of the third order, *Acta Univ. Palac. Olomuc. Fac. Renum. Natur. Math.* **26 88** (1987), 47–55.

[158] L. Suyemoto and P. Waltman, Extension of a theorem of A. Winter, *Proc. Amer. Math. Soc.* **14** (1963), 970–971.

[159] M. Švec, On various properties of the solutions of third and fourth-order linear differential equations, in: *Differential Equations and Their Applications (Proc. Conf., Prague,* 1962), Academia, Prague, pp. 187–198, 1963.

[160] E. C. Titchmarsh, *Eigenfunction Expansions Associated with Second-Order Differential Equations, Part I,* Oxford University Press, Oxford, 1962.

[161] E. C. Titchmarsh, On the uniqueness of the Green's function associated with a second-order differential equation, *Canad. J. Math.* **1** (1949), 191–198.

[162] P. W. Walker, Deficiency indices of fourth-order singular differential operators, *J. Differential Equations* **9** (1971), 133–140.

[163] P. W. Walker, Asymptotics for a class of fourth order differential equations, *J. Differential Equations* **11** (1972), 321–334.

[164] J. Z. Wang, S. Q. Li and Y. Q. Geng, Limit circle criteria for nonlinear second-order differential equations, *Qufu Shifan Daxue Xuebao Ziran Kexue Ban* **23** (1997), 29–34.

[165] J. Weidmann, *Spectral Theory of Ordinary Differential Operators,* Lecture Notes in Mathematics, Vol. 1258, Springer-Verlag, New York, 1987.

[166] H. Weyl, Über gewöhnliche Differentialgleichungen mit Singularitäten und die zugehörige Entwicklung willkürlicher Funktionen, *Math. Ann.* **68** (1910), 220–269.

[167] J. Weiss, Limit-point criteria for a second order dynamic equation on time scales, *Nonlinear Dyn. Syst. Theory* **9** (2009), 99–108.

[168] A. Wintner, A criterion for the nonexistence of (L^2)-solutions of a nonoscillatory differential equation, *J. London Math. Soc.* **25** (1950), 347–351.

[169] J. S. W. Wong, Remark on a theorem of A. Wintner, *Enseignement Math.* **13**(2) (1967), 103–106.

[170] J. S. W. Wong, On a limit point criterion of Weyl, *J. London Math. Soc.* **1**(2) (1969), 35–36.

[171] J. S. W. Wong, Remarks on the limit-circle classification of second order differential operators, *Quart. J. Math. Oxford* **24**(2) (1973), 423–425.

[172] J. S. W. Wong, Square integrable solutions of L^p perturbations of second order linear differential equations, in: *Ordinary and Partial Differential Equations*, eds. B. D. Sleeman and I. M. Michael, Lecture Notes in Mathematics, Vol. 415, Springer-Verlag, New York, 282–292, 1974.

[173] J. S. W. Wong, Square integrable solutions of perturbed linear differential equations, *Proc. Roy. Soc. Edinburgh Sect. A* **73** (1974–1975), 251–254.

[174] J. S. W. Wong and A. Zettl, On the limit point classification of second order differential equations, *Math. Z.* **132** (1973), 297–304.

[175] S. C. Wu and F. W. Meng, The limit circle case of second-order nonlinear integro-differential equations, *Qufu Shifan Daxue Xuebao Ziran Kexue Ban* **27** (2001), 33–36.

[176] L. Xing, W. Song, Z. Zhang and Q. Xu, Limit circle/limit point criteria for second order superlinear differential equations with a damping term, *J. Appl. Math.* **2012** (2012), Article ID 361961, 11 pp.

[177] Z. W. Zheng, J. Shao and B. Zhang, Limit circle criteria for a class of dynamic equations on time scales, *Acta Math. Appl. Sinica* **33** (2010), 814–823.

Index

T

W

Printed in the United States
By Bookmasters